SOIL FORMATION

SECOND EDITION

Soil Formation
Second Edition

by

Nico van Breemen

and

Peter Buurman

Laboratory of Soil Science and Geology

KLUWER ACADEMIC PUBLISHERS

DORDRECHT / BOSTON / LONDON

A C.I.P. Catalogue record for this book is available from the Library of Congress.

ISBN 1-4020-0718-3 (HB)
ISBN 1-4020-0767-1 (PB)

Published by Kluwer Academic Publishers,
P.O. Box 17, 3300 AA Dordrecht, The Netherlands.

Sold and distributed in North, Central and South America
by Kluwer Academic Publishers,
101 Philip Drive, Norwell, MA 02061, U.S.A.

In all other countries, sold and distributed
by Kluwer Academic Publishers,
P.O. Box 322, 3300 AH Dordrecht, The Netherlands.

Printed on acid-free paper

Printed in the Netherlands

CONTENTS

PLATES

The cover

This is a typical example of an intrazonal podzol in poor quartz sands. The parent material is Miocene (upper Tertiary) marine sand from the southern Netherlands. The present vegetation is a sparse oak-birch forest. The sand is very poor in iron, which favoured formation of a deep E horizon. The top of the B-horizon has 2.5% C and less than 0.2% free Al+Fe. Its pH_{KCl} is 2.9. The sesquioxide maximum (0.35%) is in the lower B-horizon. The tonguing character of the E-horizon is probably due to a combination of (Late Glacial) frost polygons and roots. Thin humus bands in the subsoil accentuate pore discontinuities caused by tectonics. The tape measure is 1 m long. Note also humus bands (with DOC chemical signature) in the E horizon.

Photograph P. Buurman.

PART A

INTRODUCTION

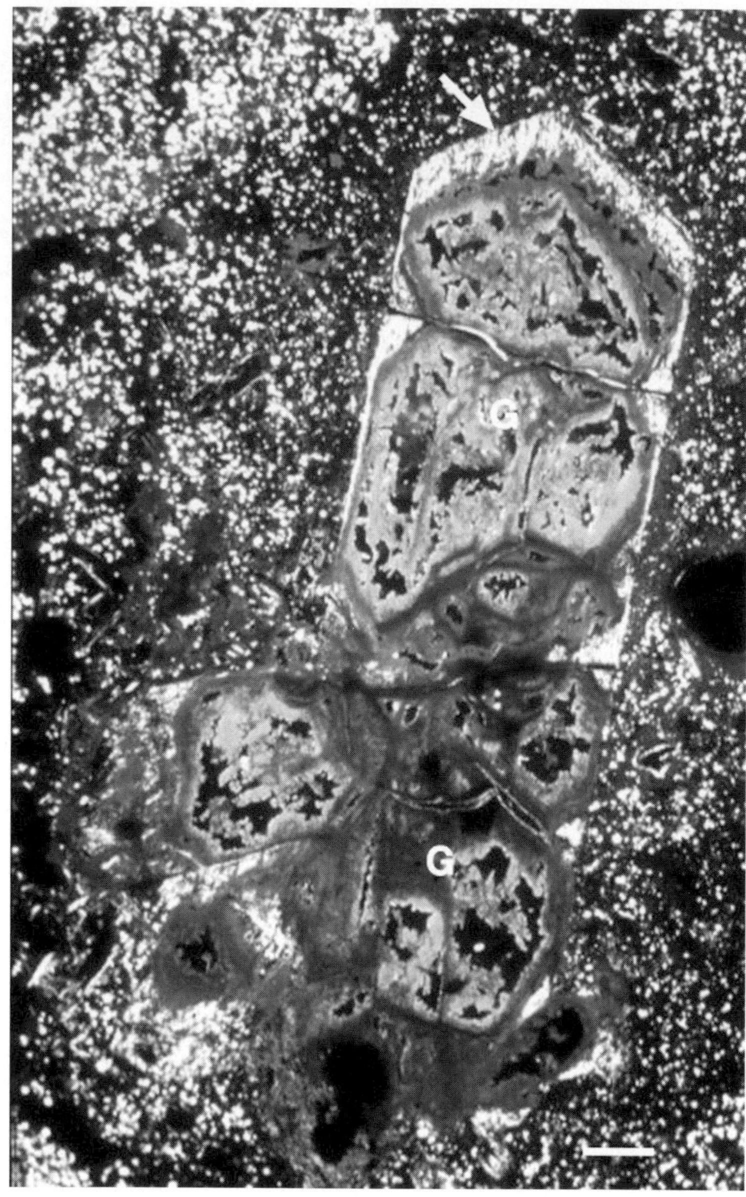

Plate A. Weathered pyroxene (pseudomorph) in volcanic material from Guadeloupe. The fringe (arrow) consists of 2:1 clay minerals due to hydrothermal alteration. The inside consists of pedogenic goethite (G). White spots in matrix are gibbsite. Crossed polarisers. Scale is 225 μm. Photograph A.G. Jongmans

PREFACE

Soils form a unique and irreplaceable essential resource for all terrestrial organisms, including man. Soils form not only the very thin outer skin of the earth's crust that is exploited by plant roots for anchorage and supply of water and nutrients. Soils are complex natural bodies formed under the influence of plants, microorganisms and soil animals, water and air from their parent material, i.e. solid rock or unconsolidated sediments. Physically, chemically and mineralogically they usually differ strongly from the parent material, and normally are far more suitable as a rooting medium for plants. In addition to serving as a substrate for plant growth, including crops and pasture, soils play a dominant role in the biogeochemical cycling of water, carbon, nitrogen and other elements, influencing the chemical composition and turnover rates of substances in the atmosphere and the hydrosphere.

Soils take decades to millennia to form. We tread on them and do not usually see their interior, so we tend to take them for granted. But improper and abusive agricultural management, careless land clearing and reclamation, man-induced erosion, salinization and acidification, desertification, air- and water pollution, and appropriation of land for housing, industry and transportation now destroy soils more rapidly than they can be formed.

To appreciate the value of soils and their vulnerability to destruction we should know what soil formation *IS*, how it proceeds, and at what rate. This book deals with the extremely complex sets of linked physical, chemical and biological processes involved in soil formation. The physical nature of soils is determined by the spatial arrangement of innumerable particles and interstitial spaces. These form a continuous structure that stores and transports gas, water and solutes, and that spans nine orders of scale, form nanometre to metre.

Chemically, soils are made up of a number of crystalline and amorphous mineral phases, plus organic matter. The soil organic matter ranges from recently formed, largely intact, plant litter and their increasingly transformed decomposition products to the amorphous organic, variably organised stuff called humus. The physical chemistry of soil is determined by a large, variably charged interface between the many solid phases and the soil solution. This interface continuously exchanges ions supplied or withdrawn via flowing water and soil organisms. Soil biology is structured as intricate meshes of plant roots and numerous decomposers. Plant roots and associated mycorrhizal fungi are active sinks of water and nutrients and sources of energy-rich organic substances, driven by solar energy. The decomposer food web that cleans up all plant and animal waste, is made up of innumerable, largely unknown species of microorganisms and a myriad of soil animals, forming one of the most biologically diverse but poorly known sub-ecosystems on earth.

These complex entities we call soils slowly evolve from a largely inert geological substrate. Initially, system complexity and pools of plant available nutrients tend to increase during primary vegetation succession over centuries to millennia. More slowly, soils tend to loose all primary weatherable minerals and most of their plant-available nutrients.

Most texts dealing with soil processes disregard such slow effects of soil genesis, and focus on the recurrent soil physical, soil chemical or soil biological processes: the movement of water and solutes, ion exchange, and the fates of plant nutrients that are passed along the autotrophic-heterotrophic cycle. By considering subsystems of the whole soil at a given stage of soil development, such processes can be studied in sufficient detail for quantitative simulation modelling. The study of soil genesis is based on a good understanding of these subsystems, and quantitative models of subsystems are useful to gain insight in complex soil forming processes. Typical examples are the many models that describe soil hydrology, ion exchange chromatography and the dynamics of soil organic matter. The utility of such models to test hypotheses about the combination of chemical, physical and biological processes that encompass soil formation, however, is very limited.

We have chosen to treat the subject of soil genesis by a semi-quantitative description of the processes responsible for the formation of major genetic soil horizons, that are recognised by the world-wide soil classification systems of FAO-UNESCO and USDA (Soil Taxonomy). In this way we could deal with all major combinations of processes that underlie soil formation. These "combination sets" of processes are described in nine chapters of part C, which forms the bulk of this book. Part C starts with a methodological chapter on how to quantify chemical and physical changes in soil profiles, and concludes with a chapter that shows examples of complex, often polycyclic soils that have been subjected to different climates. The most important individual physical, chemical and biological processes that, in various combinations, are involved in soil genesis are treated in Part B. The introductory chapter sets the stage by discussing the *what's, why's* and *how's* of soil genesis and its study.

This book was written as an advanced course for students majoring in soil science or related fields, who have followed introductory soil courses and are familiar with the fundamentals of pedology, soil chemistry and soil physics. Because it is an advanced course, the text also presents new, and sometimes controversial, ideas on certain soil forming processes. The text is suitable for self-study. Each chapter contains questions and problems. Questions are meant to help students to better understand the subject matter. Problems at the end of each chapter illustrate and integrate the material, largely by presenting data from real soils, taken from the literature. Answers to all questions and problems can be found at the end of each chapter. A glossary of specialist terms and an index complete the book. In addition to serving as a textbook, we hope that colleagues and readers from other disciplines interested in soils will find the text valuable as a review and a reference source of the subject.

Acknowledgements
The present text grew out of a teaching manual written by N van Breemen and R Brinkman in the 1980's. A.G (Toine) Jongmans chose and prepared most of the micromorphological illustrations. We thank them for their contributions, but feel responsible for any remaining errors and ambiguities.
In this revised edition, we have tried to remove the printing errors that marred the first printing. Especially Chapter 11 has undergone considerable revision to accommodate new ideas.
We have chosen to start each chapter on an odd page. This created space to add a number of field and micromorphological pictures that illustrate the various processes. We thank Dr A.G. Jongmans for the high-quality micromorphological pictures.

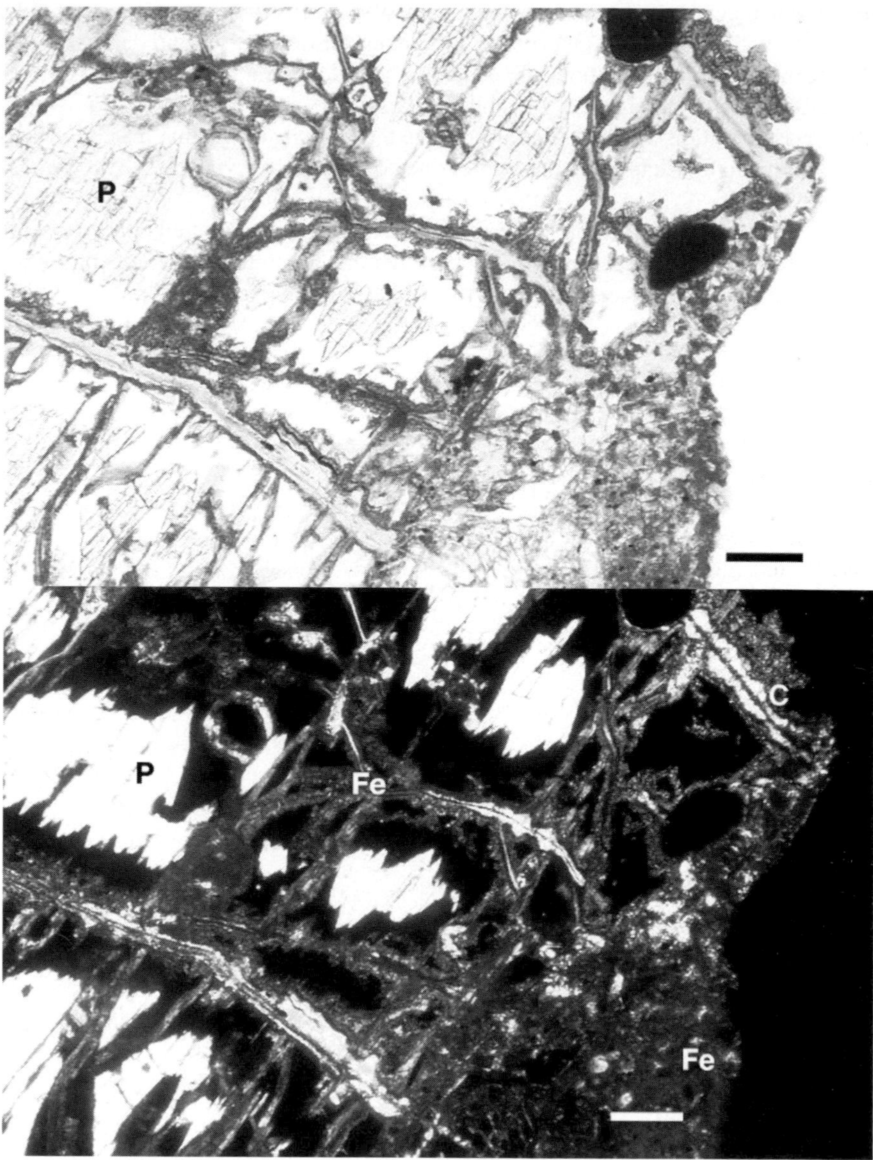

Plate B. Weathered pyroxene with 2:1 clay minerals and iron. Top: in normal light; bottom: crossed polarisers. Remnants of pyroxene (P) are surrounded by empty zones. Former cracks are now filled with highly birefringent 2:1 clay minerals (C; white in normal light; white under crossed polarisers) and iron (Fe; dark in both photographs). Scale is 84 μm. Origin: Guadeloupe. Photographs A.G. Jongmans.

CHAPTER 1

WHY SOIL GENESIS?

1.1. What is soil genesis?

Soil formation or soil genesis refers to changes of soil properties with time in one direction: the content of one component or mineral in a certain horizon decreases or increases, sedimentary layering disappears, etc. Mostly, such changes are slow and can be seen only after decades to millennia. So, most soil properties that change during soil formation are relatively stable. Sometimes, however, effects of soil formation can be seen within weeks or months. Examples are the quick drop in pH when sulphides oxidise to sulfuric acid upon exposure to air and the formation of gley mottles when a soil becomes very wet. Most rapid processes are cyclic, however, and are not considered part of soil formation

Soils may be moist or dry, and warm or cold, depending on weather and season. Seasonal variations in weather also drive biological processes, which in turn change soil properties. Examples of such biological processes are plant growth and uptake of water and nutrients, supply of fresh plant litter, and decomposition of plant litter by micro-organisms and soil fauna. These cause temporal variations in soil pH, in contents of certain fractions of soil organic matter (e.g., microbial biomass), and of soluble and adsorbed nutrients. Most of such soil properties change in a cyclic way: they are reversible on an annual or seasonal basis, and do not constitute a unidirectional change in soil properties. Therefore they are not considered part of soil genesis.

Question: 1.1. *Which of the following soil properties may vary strongly within a year, and which can change strongly only over much longer times (decades to millennia)? (a) soil temperature in °C, (b) cation exchange capacity (CEC), (c) dissolved salts, (d) clay mineralogy, (e) soil water retention characteristics, (f) soil organic matter (SOM) content.*

The rapid, cyclic processes are part of the complex set of processes that cause the unidirectional changes typical of soil formation. E.g., seasonal snow melt causes strong percolation by water. Over centuries to millennia, this causes a marked decrease of weatherable minerals by leaching. Part B treats many of such short-term physical, chemical and biological processes that are important in soil genesis.

SOIL FORMING PROCESSES

The properties of any soil are, at least theoretically, determined by five SOIL FORMING FACTORS (V.V.Dokuchaev, 1898 cited by H. Jenny, 1980):

parent material, topography, climate, biota, and time.

Any particular combination of these factors will give rise to a certain SOIL FORMING PROCESS, a set of physical, chemical and biological processes that create a particular soil. The factors *hydrology* and *human influence* have been added later.

If we could fully characterise and quantify each factor and describe all relevant processes in a simulation model, we could exactly predict the resulting soil profile. As you will come to realise in this course, reality is too complex for that!

It will be clear that a soil is not a static object that can be described once and for all, but a natural entity with a time dimension. A soil comprises living and non-living components. It can be considered as part of an ecosystem. Therefore, a soil should not normally be studied in isolation: the interactions with the rest of the ecosystem to which it belongs should be taken into account. Many publications about soil formation deal with one or a few aspects only, for example, soil chemistry or soil mineralogy. These refer to a subsystem of the soil, which itself is a subsystem of the ecosystem.

1.2. Why study soil genesis?

The study of soil genesis brings order to the overwhelming variety of soils that are observed in the world, and links the field of soil science to other scientific disciplines. A basic understanding of the main soil forming factors and soil forming processes (see box above), helps to order soil information. This can be very useful during soil survey or when setting up a system of soil classification. Also when you study plant-soil interactions or investigate consequences of large-scale human perturbations (climate change, acid rain, salinization or alkalinisation due to improper irrigation and drainage) a good general understanding of soil genesis is indispensable. Last but not least, the study of soil formation is the way to satisfy your curiosity about the many different and wondrous phenomena that can be observed in soil profiles all over the world.

Question 1.2. *The Soil Forming Factor (state factor) approach is often used to study the effect of one factor, by seeking sequences of soils where one factor varies, and the others remain constant. a) Give an example of a chronosequence (soil age varies), a climosequence (climate varies) and a toposequence (elevation varies). b) The state factor approach assumes that (i) the factors are independent, and (ii) state factors influence the soil, but not vice versa. Criticise these assumptions.*

1.3. How to study soil genesis?

WHAT HAPPENED?

When you try to explain the morphology and underlying physical, chemical and mineralogical properties of a certain soil profile, you have to distinguish two kinds of questions.

First: "WHAT physical, chemical/mineralogical and biological properties of a soil profile are due to soil formation? ", or: "In what respect does the soil differ from its parent material?"

HOW DID IT HAPPEN?

Second: "HOW did the soil form?", or: "Which physical, chemical and biological processes have formed the soil?"

A problem with the *"WHAT?"* question is that we first have to distinguish between properties caused by variations in parent material (geogenesis), and the effects of soil formation upon a given parent material by the action of soil forming factors (pedogenesis). A related problem is that we can rarely be sure of the nature of the parent material of a soil profile, and of variations of parent material with depth. Is a clayey surface soil over a sandy subsoil the result of weathering of more sandy parent material, or did it result from sedimentation of finer material? Furthermore, geogenesis and pedogenesis may alternate, which sometimes blurs the distinction between "parent material" and "soil". Ways to test assumptions about parent material will be discussed in Chapter 5.

Question 1.3. *Give examples of parent materials or landscapes where the distinction between geogenesis and pedogenesis is difficult, and where it is relatively easy.*

Often, *"HOW?"* questions are difficult too, because most processes cannot be observed directly: they take place under the soil surface, and most processes are so slow that their effects are not noticeable within the few years normally available for research. A large part of this book is devoted to tricks to get around these problems. But as with all sciences dealing with the past, it is fundamentally impossible to be really certain about an answer, because we have not been there to observe what happened.

WHAT has happened can often be inferred in the field from morphological differences between the C-horizon and the overlying soil horizons. Samples from different soil horizons can be studied further in the laboratory, e.g. microscopically, or by chemical, physical and mineralogical methods. Specific analyses to identify certain features of soil formation will be presented in later chapters. With knowledge about the reference situation (= the unchanged parent material), one can identify and sometimes quantify the changes that have taken place in the soil.

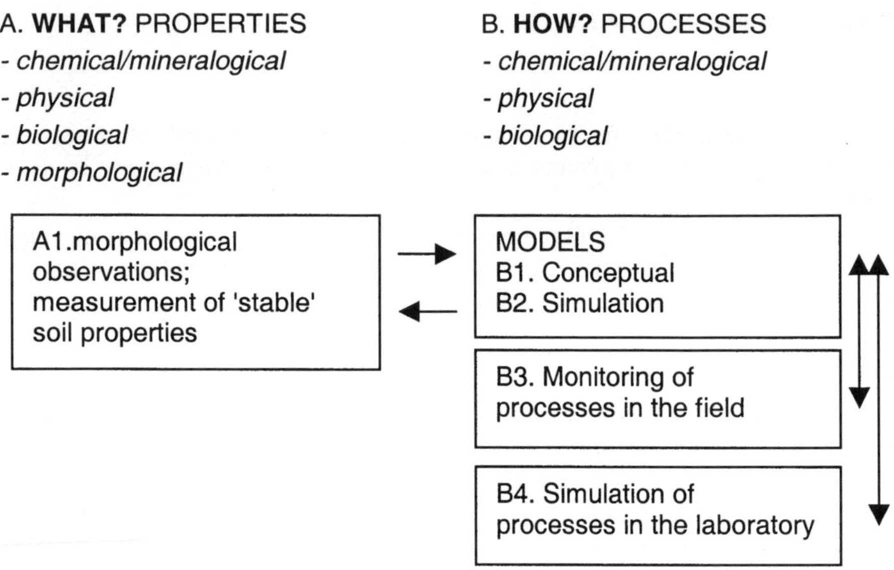

Figure 1.1. Soil forming processes and their effects on soil properties (*italics*), and approaches to the study of soil forming processes (boxes). Arrows show how the research activities are linked. The initial hypothesis (conceptual model B1) is based on morphological properties and other data on the soil profile (A1). The hypothesis can be tested by a combination of B3, B4, and simulation modelling (B2), and further research (A1) on the soil profile itself.

HOW soil formation takes place, can be extrapolated from processes observed over shorter time scales in the field or in the laboratory: mineral weathering, ion exchange, oxidation-reduction reactions, peptisation and coagulation of colloids, transport of solutes or suspended solids, nutrient uptake by plants, decomposition of organic matter, burrowing activities of soil organisms, etc. Soil formation can also be simulated in artificial soil columns in the laboratory, or by computer simulations of one or more processes acting on a hypothetical soil. To test such models one can use two kinds of data. First: data on the relatively stable properties, such as texture and mineralogy. Second: data from repeated measurements (monitoring) of seasonally variable, dynamic properties such as soil moisture content, the composition of the soil solution, or the soil gas phase, etc. The relationships between the different approaches to study soil processes are shown diagrammatically in Figure 1.1.

Question 1.4. *A silt-textured soil profile in a dry region has a water table within 1 m of the soil surface and has a white crust on the soil surface. You hypothesise that the white layer is a crust of easily soluble salt (your answer to:* WHAT?*), that has formed by evaporation of slightly saline ground water that has risen by capillary action to the soil* *surface (your answer to: HOW?). Discuss how you would test your hypothesis. Use Figure 1.1 and refer to A1, and B2-4.*

Tools to recognise WHAT happened, and explanations of HOW it happened are given in Part 3. Examples of reconstructions of HOW and WHAT in complex situations, in which we use the iterations indicated in Figure 1.1, are given in Chapter 15.

1.4. Answers

Question 1.1

Soil temperatures undergo daily and annual cycles. Clay mineralogy can change only very slowly. Soil organic matter (SOM) content varies very slightly (by a few % of total SOM) on a seasonal basis. Differences in SOM content between soils result from soil formation. The same is true for CEC and water retention, which depend mainly on SOM, clay content, and clay mineralogy. Dissolved salts vary seasonally in most soils. Very high salt contents as in saline soils, however, are the result of soil formation.

Question 1.2

a) Chronosequence: beach ridges or river terraces of different age; climosequence: continental-scale transects in loess landscapes; toposequence: any elevation gradient where parent material stays constant. **b**) (i) The state factors are at best only fairly independent: climate varies with elevation; biota vary strongly with climate. (ii) Soil properties influenced by biota may strongly feed back to the vegetation. Such feedbacks are even used by some plants (e.g. peat moss, *Sphagnum*) to outcompete other plants (Van Breemen, 1995).

Question 1.3

Geogenesis and pedogenesis can be distinguished easily in soils on Quaternary loess, and in residual soils derived from underlying igneous or metamorphic rock.

Difficult cases: soils in layered volcanic ash, in active floodplains, and on very old, extended land surfaces that have undergone repeated cycles of erosion and sedimentation.

Question 1.4

A1. Analyse the white crust and the groundwater (does the crust material dissolve in water? Has a water extract of the crust a composition similar to that of the ground water?).

B2. Calculate the capillary rise permitted by the texture of the soil, and compare that to the depth of the ground water below the surface.

B3. Does the thickness of the crust or the concentration of water-extractable salts in the surface layer increase with time?

B4. Fill a tube (say, 1 m long) in the lab with the soil from the field, create a water table at the bottom and supply slightly saline water to make up for evaporation losses. Does a salt crust develop?

1.5. References

Jenny, H., 1980. The Soil Resource. Springer Verlag, 377 pp.

Van Breemen, N., 1995. How *Sphagnum* bogs down other plants. Trends in Ecology and Evolution, 10:270-275.

PART B

BASIC PROCESSES

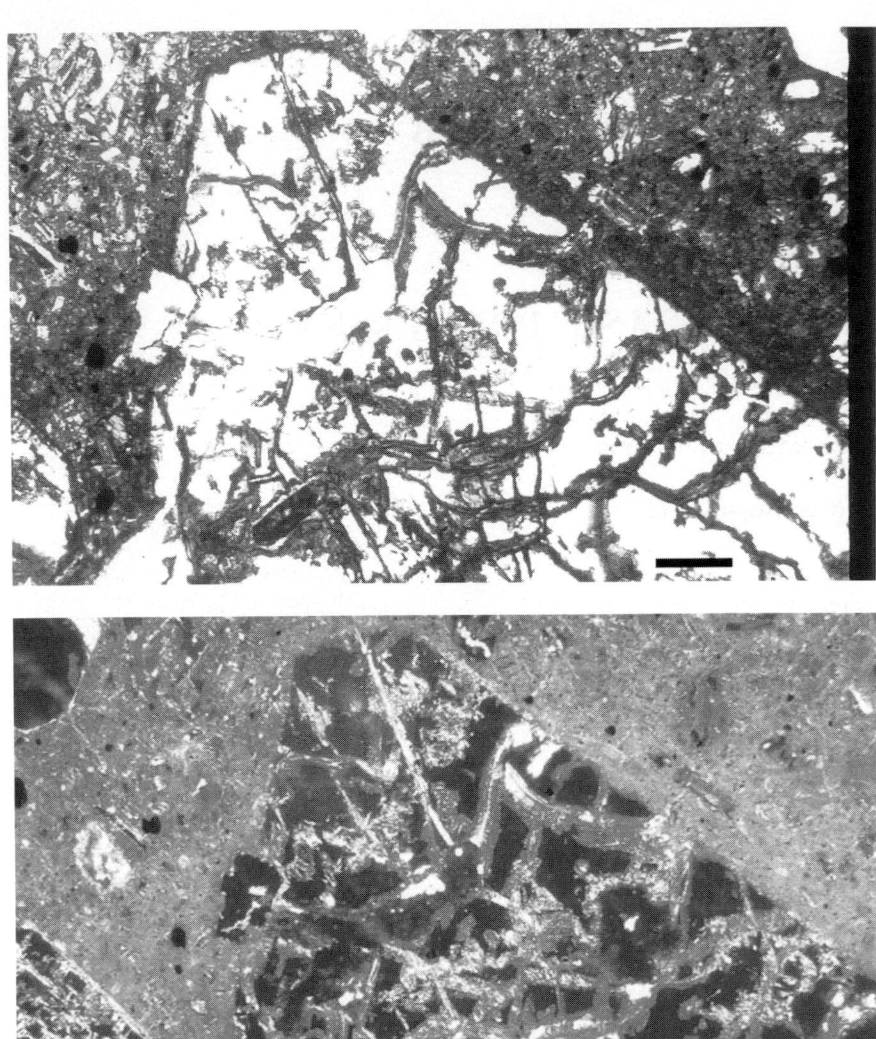

Plate C. Weathered pyroxene in andesitic saprolite from Guadeloupe. Top: normal light, bottom: with crossed polarisers. The original mineral has disappeared, leaving a network of iron oxide accumulations in former cracks. Scale bar is 135 μm. Photographs A.G. Jongmans.

CHAPTER 2

SOIL PHYSICAL PROCESSES

The main soil physical processes influencing soil formation are movement of water plus dissolved substances (solutes) and suspended particles, temperature gradients and fluctuations, and shrinkage and swelling. Here, a short outline of such processes will be given. For more details on the principles involved in the movement of water through soils, reference is made to introductory soil physics texts such as given in 2.7.

2.1. Movement of water

To understand the behaviour of water in a porous medium such as a soil, two characteristics are of paramount importance: **(1)** the relationship between the pressure potential and the volumetric moisture content, and **(2)** the relationship between the pressure potential and the hydraulic conductivity. These are shown in Figs 2.1 and 2.2 for soil materials of different textures.

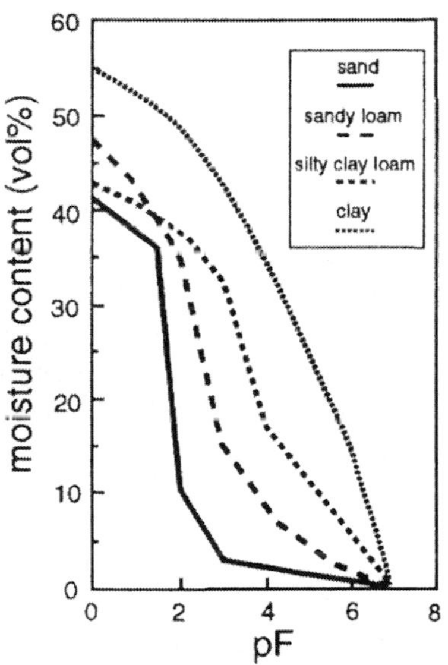

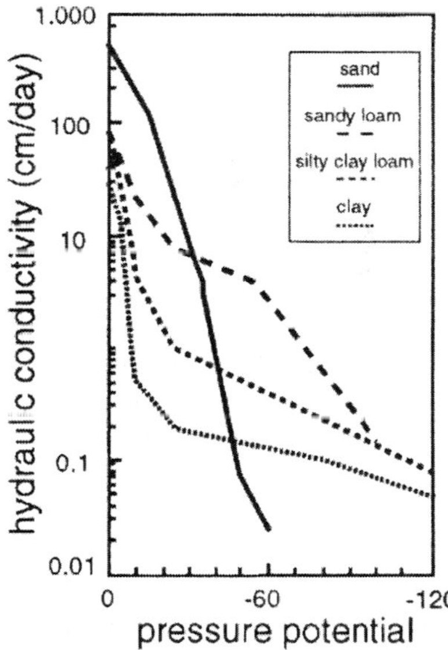

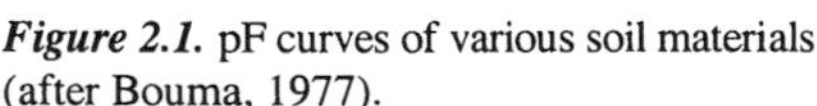
Figure 2.1. pF curves of various soil materials (after Bouma, 1977).

Figure 2.2. Hydraulic conductivity in relation to pressure potential (in cm) for various soil materials (after Bouma, 1977).

The force with which water is retained in the soil is expressed either by the pressure potential **h**, which is most conveniently expressed in the height of a water column (cm) relative to the groundwater level. In unsaturated soils above the groundwater level, the pressure potential **h** is negative, as a result of the capillary action caused by simultaneous adhesion between water and soil particles and cohesion between water molecules. In the unsaturated zone, the pressure potential is often indicated by the logarithm of its positive value, the pF value (pF = -log h; h in cm). So a pressure potential of -100 cm is equivalent to pF = 2.

Question 2.1. *Change the values on the X-axis of Figure 2.2 into pF values. Below which moisture content is the hydraulic conductivity less than 1 cm/day for a) sand, and b) clay? (Compare Figures 2.1 and 2.2).*

DARCY'S LAW

The generally slow, vertical laminar flow of water in soils can be described by Darcy's law:

$$q = -K \,.\, \text{grad}\, H = -K \,.\, \delta(h + z) / \delta z \,, \qquad \text{in which:}$$

q is the flux density (volume of water conducted through a cross-sectional surface area of soil per unit time ($m^3.m^{-2}.s^{-1}$ or $m.s^{-1}$),
K is the hydraulic conductivity ($m.s^{-1}$),
grad H (m/m) is the gradient in hydraulic head, **H**, in the direction of the flow,
and **z** the height above a reference level.

The hydraulic head is composed of the gravitational potential (numerically equal to z) and the pressure potential h. K depends on number and sizes of pores that conduct water.

The volume of water that can be conducted per unit time by an individual tubular pore at a given hydraulic gradient increases with the 4th power of the pore radius. Therefore, soils with coarse interstitial pores, such as sands, have a much higher **saturated** (h=0) hydraulic conductivity than (structureless) finer-textured soils which have narrower interstitial pores. As shown in Figure 2.2, **K** decreases with decreasing water content (Θ) or pressure potential (h). The decrease in **K** with decreasing h is very steep in coarse-textured soils. Therefore, in coarse textured soils, water movement is particularly slow in non-saturated conditions.

Question 2.2. *Why does the hydraulic conductivity decrease with the water content of the soil?*

Question 2.3. *Why is unsaturated water movement in coarse-textured soils particularly slow? Consider pore-sizes and pore geometry.*

DIRECTION OF TRANSPORT

If the hydraulic head decreases with depth (e.g. if the soil surface is wetter than the subsoil), grad H is positive, and water will move downwards ($q<0$). Upward flow, or capillary rise ($q>0$), occurs if grad H is negative, e.g. if h has been lowered near the soil surface by evapo-(transpi)ration. In climates characterised by excess precipitation over evapo-transpiration on an annual basis, there will be a net downward movement of water through the soil profile. In such conditions, soil formation is characterised in part by a downward movement of solutes and suspended particles.

When water is supplied slowly to an unsaturated soil, it may become distributed homogeneously throughout the wetted part of the soil, and the wetting front may be horizontal and sharp. Usually, however, downward water movement is far more complex. The main reasons for irregular flow are (1) a complex pore system and (2) wetting front instability (Hillel 1980).

EFFECT OF PORES

Complex pore systems are typical for most medium- and fine-textured soils, which usually have a pronounced soil structure. Such so-called *pedal* soils contain well-developed structural elements with coarse planar voids between the aggregates (inter-aggregate pores) and much finer intra-aggregate pores. In addition, many fine-textured soils contain large (mm to cm wide) vertical or oblique cylindrical pores formed by roots or burrowing soil fauna.

Question 2.4. *Which of the following properties characterise a 'complex pore system' refer to 1) a large range in pore sizes, 2) variable pore architecture, 3) unequal pore distribution?*

At low rainfall intensity, water infiltrating into a relatively dry soil will enter the smaller pores and will move relatively slowly. At high rainfall intensity, or if the soil is already moist, water flow will bypass the smaller pores and move downwards very quickly through any large, vertical, previously air-filled pores (*bypass flow).* The uneven displacement of water present earlier in a soil by infiltrating water is called *hydrodynamic dispersion.*

Question 2.5. *Which flow type causes the strongest hydrodynamic dispersion: slow infiltration or bypass flow?*

WETTING FRONT INSTABILITY

Wetting front instability occurs most notably when infiltrating water moves from a fine-textured to a coarse-textured layer. Rather than advancing as a smooth front, the water has to build up pressure before it can pass the contact. It concentrates at certain locations and breaks through locally into the coarse-textured layer. Thereafter, flow through the coarse textured layer occurs through "chimneys" or "pipes" (Hillel,1980). Concentrated water flow in "pipes" surrounded by dry soil can also be caused by the presence of hydrophobic, water-repellent surface soil. This is demonstrated in Figures 2.3 and 2.4.

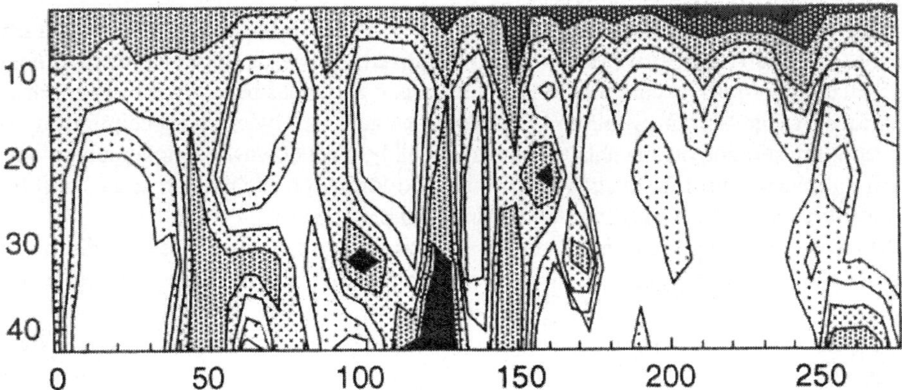

Figure 2.3. Water content in a sand dune from Terschelling after heavy rainfall. Dark hatching: 9-10% water; white <1%; steps of 1%. Horizontal and vertical scale in cm. From Ritsema and Dekker (1994), reproduced with permission of Elsevier Science, Amsterdam.

Question 2.6. *Do you expect wetting front instability to occur when a coarser layer overlies a finer one? Why?*

WATER REPELLENCE

Many peaty and sandy, but sometimes also clayey soils, may become strongly hydrophobic once they are dried out (Figure 2.4). Water repellence can be caused by a range of hydrophobic organic materials (e.g., long-chain aliphatic molecules), and tends to increase with organic matter content. Although water repellence is most common in surface soils, some subsoils low in organic matter can also become strongly water repellent. Podzol E horizons are almost always water-repellent when dry. Water repellence tends to decrease upon continuous wetting. Zones of preferential concentrated water flow, alternating with dry patches of soil, resulting from very short-range spatial variability in water repellence, tend to persist and cause highly irregular soil horizons. The Water Drop Penetration Test (WDPT) allows a gradation of water repellence. Materials are classified as non-repellent when water drops penetrate within 5 seconds; slightly repellent when the penetration time is 5-60 seconds; strongly to severely repellent when penetration times are between 50 and 3600 seconds, and extremely water repellent above this value (Dekker and Jungerius, 1990).

2.2. Movement of solutes

Soluble salts accumulate in the soil when evapo-transpiration exceeds rainfall on a yearly basis. Percolation of water is not determined simply by the total annual precipitation and potential evapo-transpiration. The amount of water percolating below the root zone depends on (1) the size of individual rainfall events, (2) the actual evapo-transpiration, and (3) the water holding capacity of the root zone. Problem 2.2 illustrates this point.

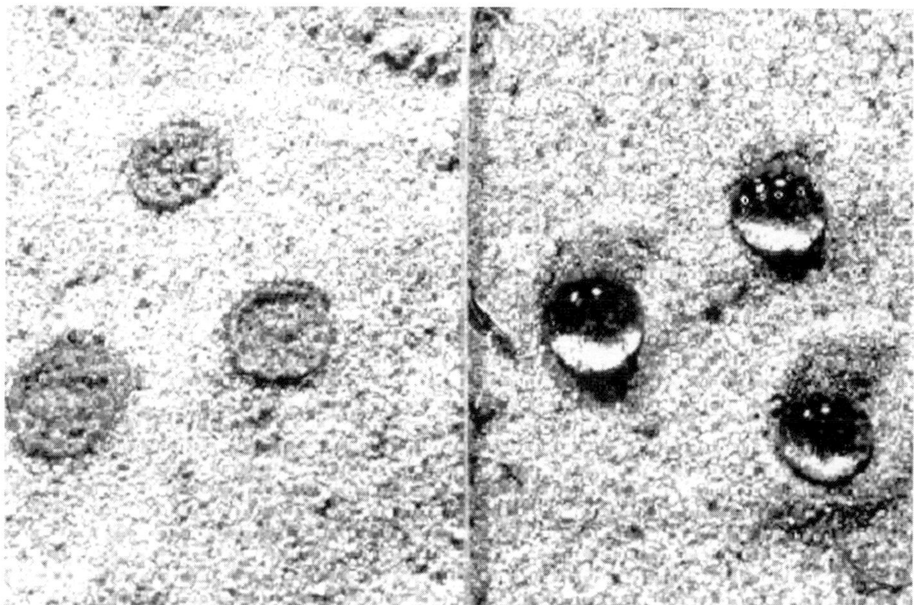

Figure 2.4. The effect of water repellence. Water drops have difficulty penetrating the water-repellent soil (right). From Dekker and Jungerius, 1990.

Infiltrating rainwater is low in solutes, and it is under-saturated with respect to many minerals. As a result, minerals dissolve (weather), and solutes percolate downward. Mineral weathering is discussed in more detail in Chapter 3.

Some solutes do not react with the solid phase of the soil and move unhindered through the soil with the percolating water. Many solutes, however, interact with solid soil materials and therefore move more slowly than the water does. The selective retention of solutes in the soil and the resulting change of soil water chemistry with depth is called a 'chromatographic' effect.

Part of the interaction between solid phase and solute is by ion exchange. This causes a change in the composition of the percolating water (and, at the same time, in the composition of the exchange complex) with depth. The principles involved in 'ion exchange chromatography' in soil columns have been well established. The models describing such processes are of particular relevance when exotic or 'foreign' solutes are added to the soil with percolating water, as when heavy doses of fertilisers are applied, or in case of pollution. Particularly in climates with a slight to moderate excess of rainfall over (actual) evapotranspiration, removal of water by roots of transpiring plants causes an increase in solute concentration with depth. As a result, super-saturation with respect to certain minerals may be reached at some depth, so that solutes may precipitate to form those minerals (see Chapter 3).

***Question** 2.7. What is meant with the 'ion exchange chromatography' of soil columns?*

2.3. Temperature effects

Because soils have a large heat capacity and low heat conductivity, temperature fluctuations are strongly buffered, which means that the amplitude of daily and seasonal temperature differences strongly decreases with depth. Temperature has four main effects on soil forming processes: 1) on the activity and diversity of biota; 2) on the speed of chemical reactions, 3) on physical weathering of rocks, and 4) on the distribution of fine and coarse fractions. The first two processes are discussed in Chapter 4 and 3, respectively.

Daily temperature variations are strongest in rocks that are exposed at the soil surface in extreme climates. The difference between day and night temperature may be more than 50°C in some arid climates. Temperature differences have a strong influence on physical weathering of rocks. Heating causes mineral grains to expand. Most rocks consist of more than one kind of mineral, and each mineral has its own coefficient of expansion. This will cause fractures at the contact between grains that expand at a different rate. Individual mineral grains are loosened from the rock. In addition, the temperature at the surface and in the interior of the rock is not equal, and the surface layer expands more strongly than the internal part. This causes chipping off of the surface layer (exfoliation). Both processes lead to physical diminution of the parent material, be it a solid rock or a sediment.

Frost action has a similar effect. The volume of ice at 0°C is larger than that of water at the same temperature, and because freezing starts at the surface of rocks, structure elements, or mineral grains, freezing water causes cracks to expand and the material to fall apart into smaller units, mainly of sand and silt size.
Frost action also results in a redistribution of fine and coarse material. Frost heaval causes stones to gradually move to the surface because ice lenses at their bottom lift them (Plate D, p. 40). Coarse pavements of polar deserts are formed this way. Frost heaval in combination with frost polygons causes coarse material to accumulate in frost (shrinkage) cracks and to form stone polygons (Plate E, p. 40). The finer central parts of such polygons are favoured by the vegetation, which causes further differentiation.

2.4. Shrinkage and swelling of soil aggregates and clays

Removal of water from, or its addition to a soil may result in strong changes in volume: shrinkage and swelling. We distinguish three phases of shrinkage processes. A first phase is restricted to irreversible removal of water from sediments that have been deposited under water. This shrinkage and the chemical processes that are associated with it, are usually called *'soil ripening'*. The second and third phases, which are related to cyclical drying and rewetting, occur in different intensity in all soils.

SOIL RIPENING

Clayey sediments deposited under water are soft and have a high water content. When drainage or evapo-transpiration removes water, such sediments increase in consistency and undergo various chemical changes. These processes together are called *soil ripening.* This term was coined in analogy to the traditional term 'ripening' used for cheese. In this process, a firm cheese substance is formed by pressing moisture from an originally wet, very soft mass of milk solids, followed by a period of further moisture loss by evaporation. The nearly liquid starting material is called 'unripe'; the much firmer end product is called 'ripe'.

Clayey underwater sediments have a high pore volume that is completely water-filled, of the order of 80 percent. This corresponds to about 1.5 g of moisture per g dry mass of sediment. The sediment normally has a very low hydraulic conductivity, less than 1 mm per day under a potential gradient of 10 kPa/m (1 m water head per m). Therefore, the material dries out very slowly by drainage alone.

Question 2.8. *Check the statement that water-filled pore volume of 80% corresponds to about 1.5 g of water per g of dry sediment. Assume that organic matter (with a particle density of 1 g/cm^3) makes up 10% of the volume of all solid matter and that the particle density of the mineral fraction is 2.7 g/cm^3. Remember that pore volume + volume organic matter + volume mineral matter = 100%.*
For such a sample, calculate the dry bulk density (= the mass of dry soil per unit volume of soil in the field) and compare your result with the bulk density of well ripened sediments, which is 1.2 - 1.4 g/cm^3. Briefly describe the cause of the difference.

Question 2.9. *Why do clayey under-water sediments have a very low hydraulic conductivity, in spite of a high total porosity?*

EFFECT OF FAUNA AND VEGETATION

Roots need oxygen to grow and to function. Water-saturated, unripened sediments lack oxygen, and most plants cannot grow on such sediments. Certain plants, such as reed (*Phragmites*) and alder (*Alnus)* in temperate climates, and wetland rice (*Oryza*), or mangrove trees (*Rhizophora* and others) in the tropics, can supply oxygen to the roots through air tissue (aerenchyma). They can grow on unripened, water saturated sediments and extract moisture from it. Such plants are sometimes used to reclaim freshly drained lands. Saline or brackish tidal sediments in temperate climates lack trees, but also grasses and herbs contribute strongly to water loss by uptake through roots and by transpiration. Burrowing by soil animals may greatly contribute to vertical permeability in unripened sediments. Tropical tidal flats usually contain many channels formed by crabs. Such sediments have a very high permeability, and even ebb-tide drainage can contribute significantly to the ripening process.

The first biopores may be formed while the sediment is still under shallow water, or lies between high and low tide. The mud lobster (*Thalassina anomalis*), for example, may produce large, mainly vertical tunnels in tidal areas in the tropics. Roots of swamp vegetation may leave, after decomposition, biopores of different diameters. Such wetland pore systems may become fossilised by precipitation of iron oxides along the pore walls, forming hard, reddish brown pipes (see Chapter 7). After drainage or empoldering, terrestrial fauna and flora take over the perforation activities. The biopores in clayey material may increase the saturated hydraulic conductivity by factors ranging up to 10^4 compared to that of unperforated sediment. Such biopores help the removal of surface water. Evaporation, too, is accelerated by the presence of biopores extending to the soil surface. Upon drainage, the wider pores are rapidly emptied, even though the surrounding material may remain water-saturated. Next, oxygen penetrates into the soil through the air-filled pores, and oxidation of the sediment can start (Chapter 7).

Shrinkage can be illustrated by plotting the void ratio of a soil (volume of pores/volume of solids) against the moisture ratio (volume of water/volume of solids) (Figure 2.5). Voids include air-filled and water-filled voids.

NORMAL SHRINKAGE

As long as clayey material is still plastic, *normal shrinkage* takes place, and the soil shrinks without cracking and air entry: the decrease in volume of aggregates equals the volume of water lost. Moisture ratio and void ratio decrease and remain equal. During soil ripening, this shrinkage phase is usually short, and cracks can form already after only a small part of the water has been removed. As soon as cracks form, the soil enters the stage of residual shrinkage: air may enter the pores of the aggregates, but the aggregate volume still decreases.

Vertical cracks form, first in a very coarse polygonal pattern, creating prismatic blocks that may be up to several dm across. The size of the prisms decreases with increasing clay content. Subsequently, horizontal cracks are formed as well, and the prisms are broken up into progressively smaller blocks with sharp edges. When drying is rapid and hydraulic conductivity is low, the blocks may become very small, down to about one cm in size.

Part of the normal shrinkage is irreversible: upon rewetting, the material swells again, but far less than the original volume of the unripened sediment. To understand this, we have too look at the organisation of clay plates in the aggregates. Microstructures of clay aggregates show a honeycomb-like structure (Figure 2.7). The walls of the honeycomb consist of stacks of clay platelets. As discussed later in this paragraph, drying increases the number of platelets in a stack. Because this increase is partly irreversible, the flexibility of the honeycomb structure decreases, and swelling upon rewetting is less than original shrinkage.

Table 2.1 shows how strongly the drying increases the number of layers in the clay particles, and decreases rehydration. Further drying causes residual shrinkage, which can be accompanied by crack formation. Part of the cracks that form are permanent, even in very wet conditions, and become part of the permanent, heterogeneous pore system. If the pores formed have a low stability because of strong swelling or because of easy dispersibility of the clay, the permeability remains low, and ripening proceeds very slowly. The stability of pores depends partly on the chemistry of the interstitial water, as will be discussed in Chapter 3.

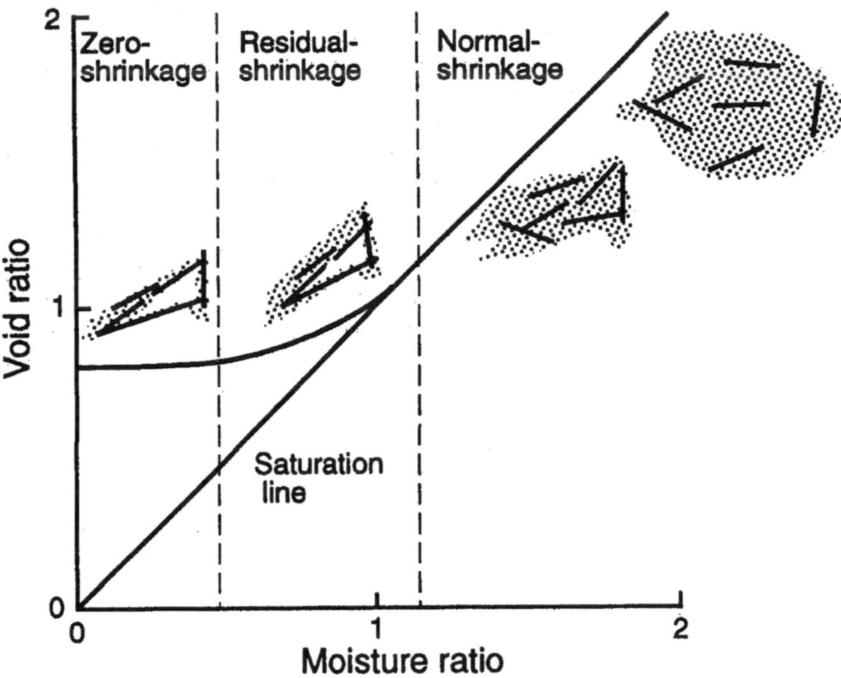

Figure 2.5. Three shrinkage phases of soil aggregates expressed by changes in void ratio and moisture ratio. From Bronswijk, 1991.

Table 2.1. Dehydration and rehydration of Wyoming Ca-smectite at 10^{-3} M $CaCl_2$ (after Ben Rahiem et al., 1987). *See also Figures 2.6 and 2.7.*

pF	layers in clay particle	d[1]-spacing (nm)	water content (g/g clay)
		dehydration (1st drying)	
1.5	55	1.86	4.90
3	55	1.86	1.28
4	225	1.86	0.60
6	400	1.86	0.23
		upon rehydration	
4 → 1.5	65	1.86	1/60
4 → 3	65	1.86	0.92
6 → 1.5	90	1.86	1.10
6 → 3	170	1.86	0.60

[1]. d-spacing is the repetition distance of clay plates; see also Figures 3.6 and 3.7.

RECOGNITION AND QUANTIFICATION OF RIPENING

In the first instance, the extent to which a soil horizon is ripened can be estimated manually in the field. Soil material that runs between the fingers without squeezing is completely unripened. Firm clay that cannot be squeezed out between the fingers is completely ripened. Several ripening classes, based on the ripening factor *n*, have been defined between these extremes, as shown in Table 2.2.

The ripening factor is the amount of water bound to a unit mass of clay fraction. This amount cannot be measured directly. It can be estimated from the moisture content of a soil sample if the contents of clay and organic matter are known as well, and if assumptions are made about the amounts of water bound to other soil components. The organic matter fraction in a soil binds about three times as much water per unit mass as the clay fraction under the same conditions. The (sand + silt) fraction of a clayey soil binds about 0.2 g water per g (sand + silt) under wet conditions. From these empirical relationships, the following formula for the n value was derived:

$$n = (A - 0.2R)/(L + 3H) \tag{2.1}$$

where **n** is the mass of water per unit mass of clay + organic matter,
A the water content of the soil on a mass basis (%, with respect to dry soil)
R the mass fraction of (silt + sand) (% of solid phase)
L the mass fraction of clay (% of solid phase)
H the mass fraction of organic matter (% of solid phase).

The n-value can be used to more precisely quantify ripening. The higher the n-value, the less ripened the soil. The n-value is used in the Soil Taxonomy to distinguish unripe soils (Hydraquents) from other mineral soils. Hydraquents should have an n-value over 1 in all subhorizons at a depth of 20 to 100 cm. The n-value decreases rapidly upon drying, so Hydraquents may quickly change in classification once they are drained.

RESIDUAL AND ZERO SHRINKAGE

Ripened fine-textured soils may exhibit appreciable shrinkage and swelling upon drying and wetting. This is illustrated in the left-hand part of Figure 2.5.

In such soils, we recognise two phases of shrinkage: a phase which includes further reduction of aggregate volume upon water loss *(residual shrinkage)*, and a second phase in which further water loss does not affect the aggregate volume *(zero shrinkage).*

Recent research has shown that normal and residual shrinkage can be appreciable in clay soils (15-60% particles <2 μm) in the Netherlands (Bronswijk, 1991). Some clay soils show very strong normal shrinkage, i.e. aggregates shrink considerably without any entrance of air into the aggregates. This implies that large inter-aggregate pores (cracks) may be formed while aggregates remain water-saturated, with water in very fine intra-aggregate pores. Under such conditions, rapid bypass flow through shrinkage cracks can take place. Furthermore, changing void ratio does imply that the architecture of the pore system is not constant, but may change with the water content.

Soils differ in swelling behaviour, depending on the nature of the solid phase and the ionic composition of the ion exchange complex and the soil solution. 2:1 clay minerals with a low charge deficit (see Chapter 3.2) have a high potential for swelling. Swelling increases with the relative amount of Na^+ ions on the adsorption complex and with decreasing electrolyte level of the interstitial water.

Table 2.2. Field characteristics of ripening classes (Pons and Zonneveld, 1965).

Field characteristics	Ripening class	limiting n-value
very soft[1], runs between the fingers without squeezing	completely unripened	
		2.0
soft, is easily squeezed out between fingers	practically unripened	
		1.4
moderately soft, runs between fingers when squeezed firmly	half ripened	
		1.0
moderately firm, can just be pushed out between fingers when squeezed firmly	almost ripened	
		0.7
firm, cannot be squeezed out between fingers	ripened	

[1] soft and firm in the sense of soft mud and firm, wet clay; not in the sense of the definitions for dry or moist material as in the Soil Survey Manual (Soil Survey Staff, 1951 or later editions.

As shown by Tessier (1984) and Wilding and Tessier (1988) (Figures 2.6 and 2.7), individual (TOT) plates of smectite are stacked together face to face to form clay particles, or "quasi crystals", which comprise between 5 and 10 (Na-saturated) to more than 50 (Ca-saturated) plates. Under wet conditions (h= -10 cm) with low electrolyte levels, Na-saturated smectites have a swollen diffuse double layer (the space between individual plates containing exchangeable cations), up to 10 nm thick. Under the same conditions, Ca-smectite plates have a spacing of 1.86 nm (see also Table 2.1). The clay particles are arranged in a honeycomb structure (Fig 2.6, 2.7), with water-filled pores of up to 1 μm wide. This structure is flexible as an accordion. As the wet clay dries out, water is lost from the bellows and the accordion closes. Upon rehydration, water re-enters the bellows. The ease with which the bellows open and close depends on the clay mineral and the physico-chemical conditions. In a Ca-smectite, the opening and closing of the bellows cause most of the shrinkage and swelling. In a Na-smectite the uptake and removal of interlayer water is important as well.

Flexibility of the clay particles is highest in Na-smectites, lower in Ca-smectites, and still lower in the brittle, rigid clay particles ("domains") of illite and in the coarse particles ("crystallites") of kaolinites. This explains the decreasing shrinkage and swelling potential in that order. (More information on smectites, illites, kaolinites, and other clay minerals is given in Chapter 3.2).

The data in Fig 2.6 refer to pure clay-water mixtures. In actual soils the swelling is much less because of the presence of physically inert coarser minerals, increased cohesion of mineral particles due to binding to various organic and inorganic substances and, at least at some depth, overburden pressure.

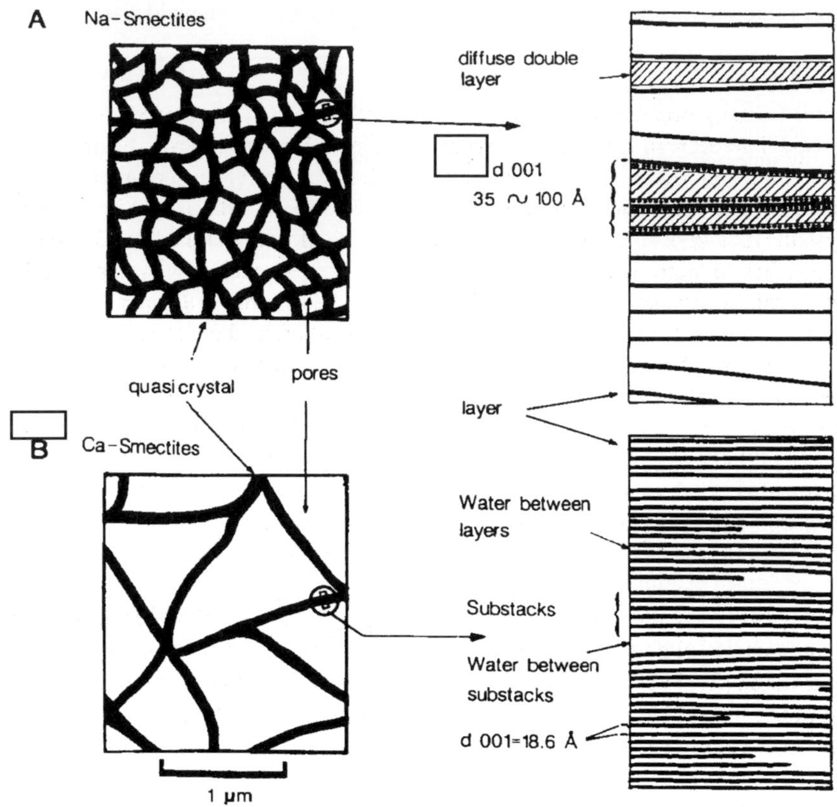

Figure 2.6. Schematic representation of the microstructure of smectite saturated with (A) NaCl, or (B) $CaCl_2$ at 10^{-3} M chloride concentration. From Wilding and Tessier, 1988.

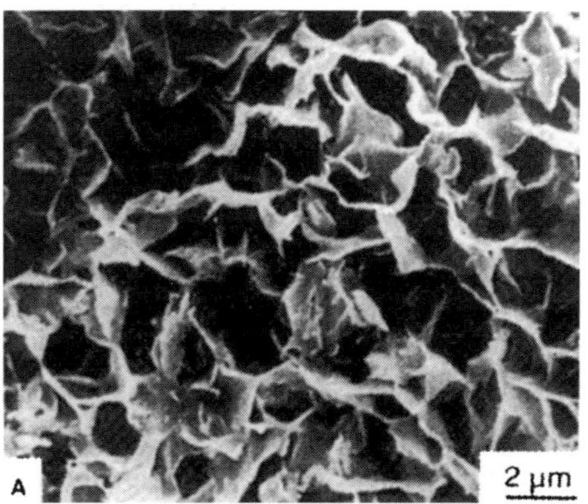

Fgure 2.7. Microstructure of a smectite at 1M NaCl; water saturated.
From Wilding and Tessier, 1988.

2.5. Problems

Problem 2.1

Many soil-forming processes involve movement of water and suspended or dissolved substances. Therefore a qualitative insight on effects of macro-porosity, initial soil moisture conditions and rate of application of water will be helpful to understand soil formation. Figure 2A depicts columns of a hypothetical soil with a very homogenous pore system (A) and of actual soil with a subangular blocky structure (B to E). The grey hatching refers to water in moist soil. From left to right in each row is depicted how infiltrating rain- or irrigation water (black) displaces the original soil water. The colour of the inflow and outflow arrows indicates the composition of the water. The spaghetti indicates a system of macropores that is depicted as air-filled (white) or filled with the infiltrating water (black).
The rows B to E refer to one of the following situations, in random order:
(1) moist soil that every day receives a short, highly intensive rain storm (1cm/day);
(2) moist soil ponded with irrigation water, resulting in saturated water flow;
(3) moist soil, constantly receiving low-intensity rain (drizzle) of 1 cm/day; and
(4) water-saturated soil ponded with irrigation water, resulting in saturated flow.

a. Which of the rows B-E belongs to which of the situations 1 to 4? The difference between cases with saturated and unsaturated flow follows from the presence of open, water-filled macropores. During intensive rainstorms, an appreciable part of infiltrating water will be drained by the macropores, because the infiltration capacity of the soil matrix is exceeded.

b. Briefly explain your answer

c. By adding an inert tracer (Cl^-, not present in the original soil water) to the infiltrating water, the moments of first breakthrough of infiltrating water (first trace of Cl^- in drainage water at bottom of column) and of complete displacement (equal Cl^- concentrations in input water and drainage water) were determined. The following results were obtained (again in random order):

	first trace	complete displacement
i	1 day	5 days
ii	1 hour	2 days
iii	9 days	21 days
iv	3 days	24 days

Which of the rows B-E belongs to which of the situations i to iv? The cases with a rapid breakthrough of Cl^- obviously relate to water-saturated flow. If you cannot figure out the cases with early appearance of the first Cl^- (after 1 hour) and somewhat slower drainage (Cl^- after 1 day) have a look at the arrows indicating bottom water composition.

d. Briefly explain your answer.

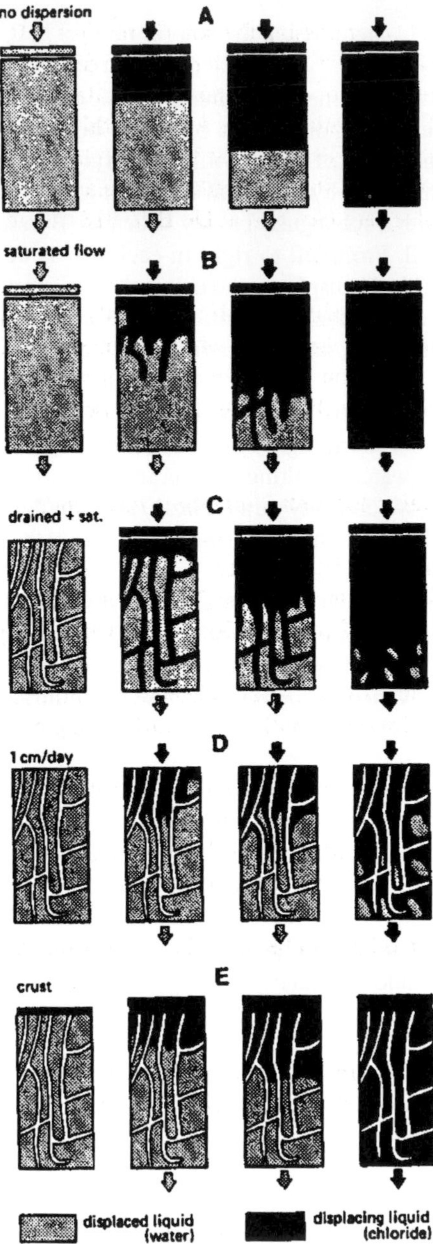

Figure 2A. Displacement of soil solution by infiltrating water in a structured soil. Explanations see tex
Bouma, 1977.

Problem 2.2
Table 2B gives calculated amounts of water passing the soil (in mm/yr) at 0, 10, 20 and 30 cm depth below the land surface for (1) three soil materials with different volumetric soil moisture content (%), (2) three levels of total annual precipitation, and (3) two regimes of rainfall intensity. Annual precipitation values with monthly distributions and monthly potential evapo-transpiration (PET) typical for De Bilt (Netherlands), Rabat (Marocco) and Kintapo (Uganda) were used in the calculations (taken from Feijtel and Meyer, 1990). Annual total PET values are 626 mm at De Bilt, 818 mm at Rabat and 1521 mm at Kintapo.

In the low rainfall intensity regime, the monthly rainfall was evenly distributed; in the high intensity regime, every month had only three days with rain, providing 20 % of the monthly rainfall on day 2, 30 % on day 14 and 50 % on day 26 of each month. Water uptake by roots is assumed to be evenly distributed over a 30-cm deep root zone.

Questions
a. What is the total annual precipitation for each of the three locations?

b. Draw graphs (depth on the y-axis, percolating water on the x-axis) for any three of the different climate-soil combinations to illustrate the effects of climate and soil water holding capacity on the amount of water percolating at various depths.

c. Briefly explain the reasons for the effects of rainfall intensity and soil moisture holding capacity for the amount of water percolating at various depths.

d. Provided percolating water would always completely displace previously existing soil solution, which rainfall regime would be most favourable for leaching of solutes, followed by chemical precipitation of translocated solutes at a certain depth? Which would be more conducive to leaching of the whole soil profile?

e. How does your answer to question **d** have to be modified for a strongly pedal soil?

Problem 2.3
Criticise the following statement: In dry climates, where potential evapo-transpiration exceeds annual precipitation, leaching of solutes to below the root zone does not take place.

Problem 2.4
Figures 2C and 2D (from Bronswijk, 1991) show shrinkage characteristics for two subsoils of heavy clay soils from the Netherlands. The vertical axis gives the void ratio. In which soil do aggregates shrink strongly during drying, giving rise to formation of cracks *between* the aggregates, and allowing bypass flow?

Table 2B. Amounts of water passing through the soil (in mm/yr) at 0, 10, 20 and 30 cm depth below the land surface for (1) three different soil materials (as indicated by differences in available soil moisture, volumetric %), (2) three levels of total annual precipitation (De Bilt, Rabat and Kintapo), and (3) two regimes of rainfall intensity.

location	rainfall intensity	available soil moisture (%)	water (mm/yr) passing at depth (cm)				actual evapo-tran-spiration (mm)
			0	10	20	30	
de Bilt	low	25.0	765	452	296	243	522
		10.0	765	452	309	261	504
		1.5	765	452	318	270	495
	high	25.0	765	698	658	630	135
		10.0	765	702	665	640	125
		1.5	765	702	665	640	125
Rabat	low	25.0	497	290	200	147	350
		10.0	497	302	218	176	320
		1.5	497	310	225	189	307
	high	25.0	497	410	389	370	127
		10.0	497	440	413	398	99
		1.5	497	448	424	409	87
Kintampo	low	25.0	1517	856	566	392	1125
		10.0	1517	868	581	407	1110
		1.5	1517	877	589	416	1101
	high	25.0	1517	1296	1183	1111	406
		10.0	1517	1317	1213	1147	370
		1.5	1517	1326	1218	1151	366

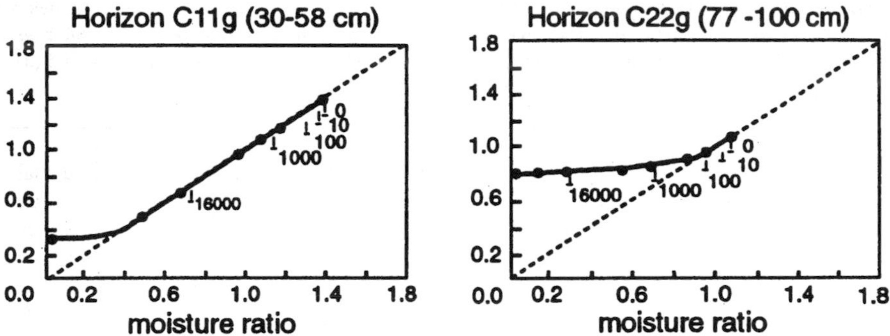

Figures 2C and 2D. Shrinkage characteristics of the C11g of profile Bruchem (left) and of the C22g horizon of profile Schermerhorn (right).

Problem 2.5

Figures 2E and 2F refer to a soil reclaimed from the IJsselmeerpolders (The Netherlands). Fig. 2E indicates the elevations of the soil surface and of reference plates, that had been installed in the soil at 40, 80, 120 and 200 cm depth just after reclamation, as a function of time. Fig. 2F shows changes with time of the volume of soil cracks in mm (= liter/m^2) in the same soil.

In both figures, the lines were calculated by computer simulation of the ripening process, while x indicates a measured value. Only use calculated values, to be measured from the figures, in answering the questions. Assume that no changes took place below an original depth of 150 cm.

a. Estimate the subsidence (in cm) that took place between 1968 and 1979 in the layers with the original depths 0-40 cm, 40-80 cm, 80-125 cm and 125-200 cm.

b. How much water has been removed from the soil (in mm) between 0 and 150 cm depth over the period 1968-1979? Take crack volume into account. Assume that no water loss took place below the depth indicated by the 50 cm dotted line (150 cm below original soil surface).

c. Explain (1) the stepwise character of the lines, and (2) the presence of "sills", small peaks, on the steps in Fig. 2F.

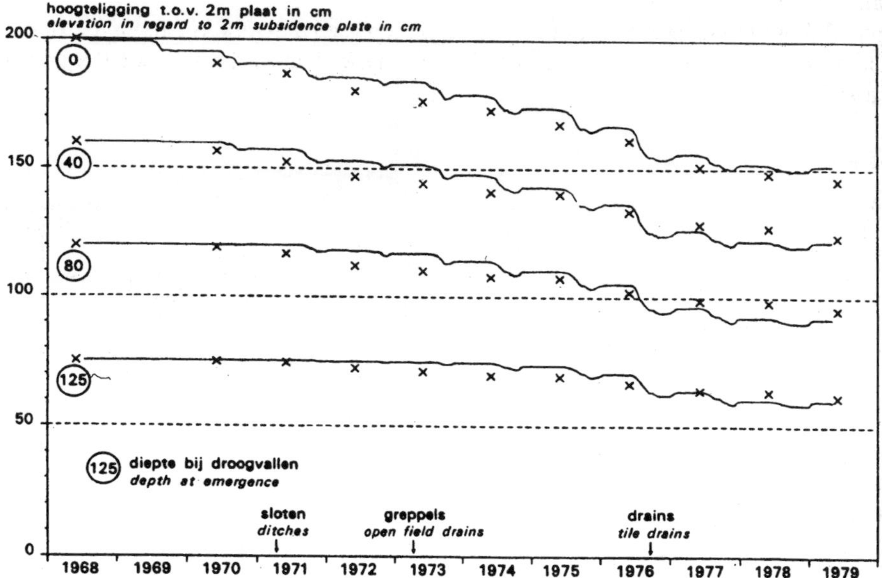

Figure 2E. Changes with time in elevation of the surface and of the depth of reference plates, for a polder soil undergoing ripening in the IJsselmeerpolders, The Netherlands. From Reinierce, 1983.

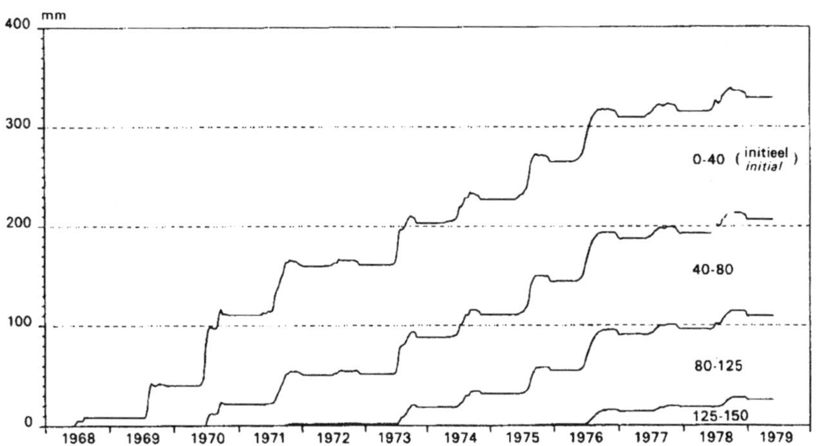

Figure 2F. Changes in the content of cracks in the soil of Figure 2E. The crack content of the four horizons is cumulative, so that the top line gives the total crack content of the soil, recalculated in mm subsidence. From Reinierce, 1983.

Problem 2.6

Figure 2G shows changes in water content upon repeated dehydration and hydration of a Ca-smectite (left) and a Na-smectite. Discuss these data in the light of the text of section 2.4.

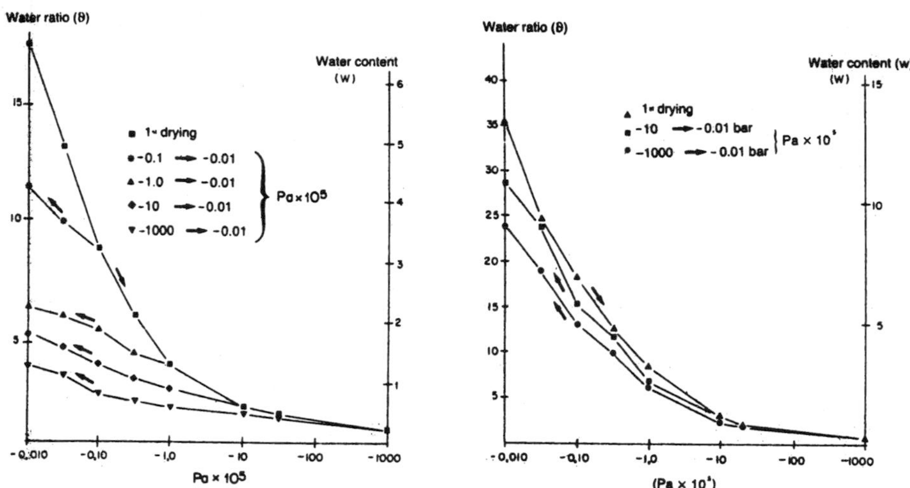

Figure 2G. Water ratio (volumetric ratio) and water content (mass ratio) as a function of dehydration of Wyoming Ca-smectite (left) and Greek Na-smectite (right) in a 10^{-3} M $CaCl_2$ solution from 0.01 to 1000 bars suction (**h** = -10 to -10^4 cm) and subsequent rehydration. From Wilding and Tessier, 1988.

Problem 2.7

Data for the following problem refer to samples taken from different depths in an IJsselmeerpolder soil (Dronten). The surface soil has undergone appreciable ripening, the subsoil is still very soft (Fig. 2H). Data on the solid soil material are in Table 2J.

a. Estimate the n-value of each of the samples in its water-saturated state.

b. Assume that the 0-22 cm surface soil (A11) ever had the same shrinkage curve as the 22-42 cm subsoil (ACg). Show, by means of a sketch in a diagram similar to those in Fig 2H, how successive swelling and shrinkage curves upon repeated wetting from **h** = -16000 to 0 cm and drying from **h** = 0 to -16000 cm, <u>could</u> look. Assume that during the third drying the curve for the 0-22 cm surface soil would be reached.

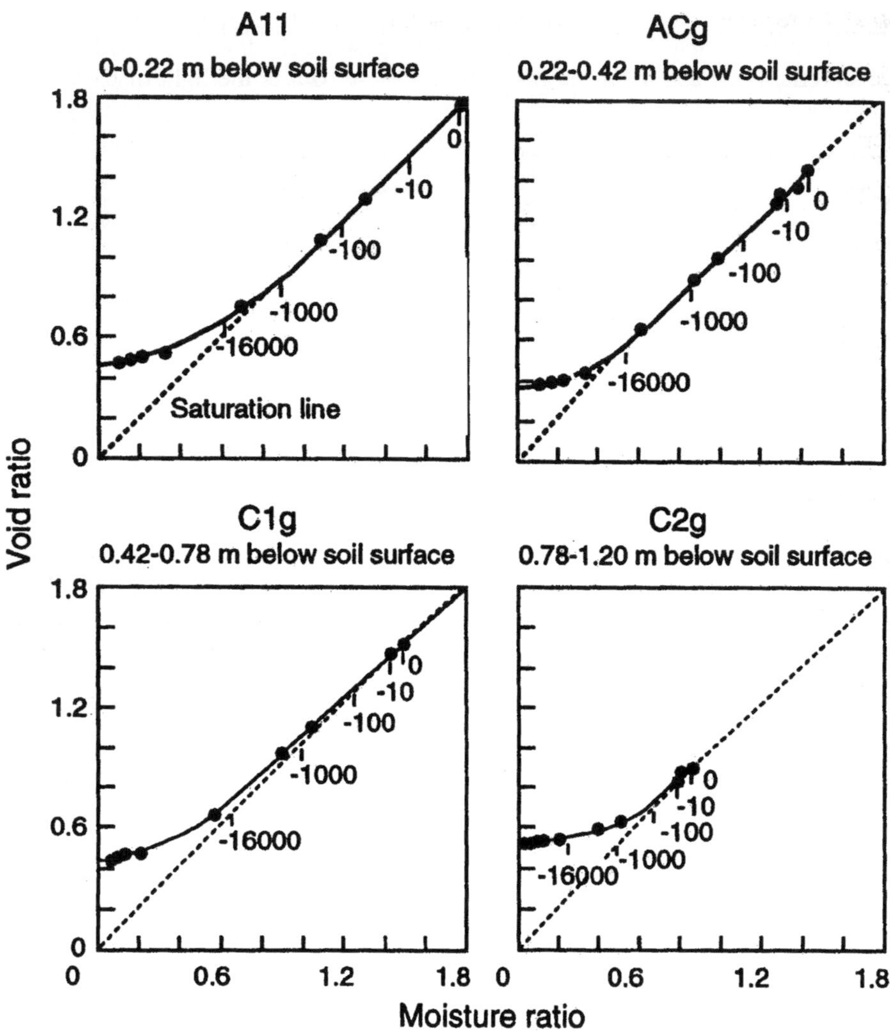

Figure 2H. Soil shrinkage characteristics at four depths in a recent IJsselmeerpolder. The values printed in the graphs refer to the pressure potential (indicate by short verticals at different values of moisture ratio), expressed in cm (equivalent height of water column).
From Bronswijk, 1991.

Table 2J. Characteristics of the solid soil material of the Dronten profile. From Bronswijk, 1991.

depth	clay	org. matter	particle density
(cm)	mass fraction (%)		(g/cm^3)
0-22	37	9.9	2.66
22-42	46	8.1	2.66
42-78	35	6.6	2.63
78-120	16	5.8	2.59

2.6. Answers

Question 2.1
A pressure potential of -10 cm equals a pF of 1; a potential of -100 cm a pF of 2. For sand, the hydraulic conductivity is less than 1 cm/day at a pressure potential of -50 (pF = 1.7) and for clay this value is -10 (pF = 1.0). In Figure 1.1, the water contents at these pressures are approximately 35% for sand, and 50% for clay.

Question 2.2
The hydraulic conductivity decreases with decreasing pore size. At lower water contents, the larger pores are filled with air.

Question 2.3
Coarse-textured soils have few fine pores and these are restricted to contacts between sand grains and therefore not continuous.

Question 2.4
A 'complex pore system' refers to all three characteristics.

Question 2.5
Bypass flow causes the strongest dispersion, because there is hardly any interaction with water in small pores.

Question 2.6
Wetting front instability does not occur when a coarse layer overlays a fine one, because the finer pores in the underlying layer cause rapid transport into that layer.

Question 2.7
'Ion exchange chromatography' refers to the fact that different cations in solution are retained to a different degree by the adsorption complex, causing differences in the speed

by which they move through the soil. The same principle is used in chemical analysis by chromatography.

Question 2.8

If 20 cm^3 of a volume of 100 cm^3 consist of solid matter, this solid matter consists of 2 cm^3 of organic matter and 18 cm^3 of mineral matter. Together, this weighs 18*2.7 + 2*1 g = 50.6 g. The weight of 80 cm^3 water is 80 g. The water/solid weight ratio is therefore about 1.58. The difference with the factor of 1.5 lies in the assumptions.
The dry bulk density of such a soil equals 0.506. This is much lower than that of a ripened soil, because there is very much water between the particles, and shrinkage would increase the bulk density. For (dry) bulk density, the dry weight is divided by the wet volume.

Question 2.9

Clayey under-water sediments have a very low hydraulic conductivity because all pores are inter-particle pores, which are very small.

Problem 2.1

a/b. Situations B and C depict saturated flow (also macropores filled with water). These must refer to situations with water standing on the soil. A moist soil has open voids, through which water can percolate rapidly, so C matches with 2. Irrigation water on wet soil causes slower replacement of water already present in large pores, so figure B matches with situation 4.
A short rainstorm causes some percolation along major channels, while a slow drizzle does not. Therefore, figure D belongs to situation 1 and figure E to situation 3.
c/d. Rapid breakthrough of a tracer indicates bypass flow. The most rapid flow occurs in figure C, which should match situation ii; the next is figure B, which should match with situation i. For situations D and E, breakthrough occurs earlier in D (between the third and the fourth row, while breakthrough in E is clearly in the fourth row. Complete displacement will be faster in E. Therefore, figure D belongs to situation iv and figure E to situation iii.

Problem 2.2

a. The total annual precipitation is the amount that passes through depth 0.

b/c. The graph on the next page is drawn for the data from De Bilt. They illustrate the speed with which water is withdrawn from the soil and the effect of rainfall intensity on wetting depth. The amount of water passing through a specific layer increases with decreasing moisture-holding capacity of that layer. Of intense rainfall, a smaller percentage disappears through evapotranspiration.

d. Leaching of the topsoil in combination with accumulation at some depth means that large amounts of water should be transported through the topsoil, and much smaller amounts through the deeper layers. This is the case at low rainfall intensity and high available soil moisture. Maximum removal of solutes occurs with high rainfall intensity and low available soil moisture.

e. In a strongly pedal soil, most water at high rainfall intensity would be removed through major channels, without removing solutes from the bulk soil (bypass flow). In such soils, moderate rainfall intensity and saturated flow would remove more solutes.

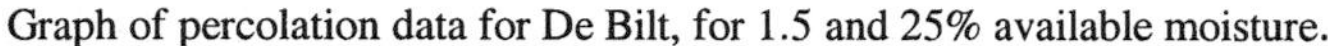

Graph of percolation data for De Bilt, for 1.5 and 25% available moisture.

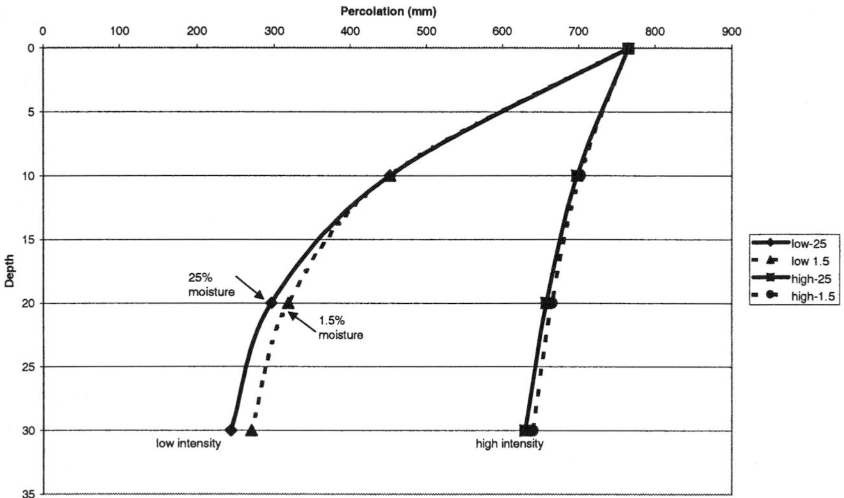

Problem 2.3

Removal of solutes does not depend on the total annual evapo-transpiration of the total rainfall, but on seasonal effects. If during the wet season precipitation is higher than evapo-transpiration, movement of solutes below the root zone may take place. Potential evapo-transpiration is not a good criterion anyway, because during the dry season water may not be available for evapo-transpiration.

Problem 2.4

If a soil sample shrinks upon drying, without crack formation, the void ratio decreases because the amount of pore space decreases while the amount of solids remains constant. If cracks are formed, the pore space increases dramatically and may not be affected by further drying. This means that the soil of Figure 2D has crack formation.

Problem 2.5

a. The subsidence per layer is estimated by measuring the thickness of that layer in 1968 and in 1979.

Layer	top in 1968 cm	top in 1979 cm	difference cm (cumulative subsidence)	subsidence per layer cm	cm crack volume 1979
0-40	200	155	45	5	12
40-80	160	120	40	15	10
80-125	120	95	25	10	8.5
125-200	75	60	15	15	2.5
			Total subs.	45	33

b. Water removal equals subsidence + crack volume: 78 cm.

c. The stepwise character of the lines and the presence of sills is explained by seasonal variation. During winter the soil is rewetted and swells slightly. During the next summer, drying causes further subsidence, but some of the subsidence is compensated by swelling during the next winter, etc. The pore volume is the mirror image of this process.

Problem 2.6

Sodium smectite has a higher water ratio than calcium smectite. In sodium smectite, rewetting causes almost complete rehydration, even if the soil has been dried strongly. In calcium smectite, the reversal is only partial. In addition, the reversal is strongly determined by the degree of drying. This means that in Ca-smectites, stable stacks of plates are formed, which cannot be reactivated through wetting.

Problem 2.7

a. The n-value is calculated according to formula (2.1).
For the 0-22 cm (A11) horizon: If the void ratio is 1.8, there is 18 cm^3or 18g water to 10 cm^3 solids. The solids weigh 10*2.66g = 26.6g. The water content (mass fraction) is therefore 18/26.6*100% = 67.7 %. The solids contain 37% clay, 9.9 % organic matter, and 53.1 % sand and silt. If these values are substituted into formula 2.1, the result is:

$$n = (67.7-0.2*53.1)/(37+3*9.9) = 0.86.$$

Values for the other horizons are calculated similarly.

b. Successive swelling and shrinkage should give a number of lines that iterate *between* the lines of the ACg and A11 curves, every time closer to that of A11.

2.7. References

Ben Rahiem, H., C.H. Pons and D. Tessier, 1987. Factors affecting the microstructure of smectites: role of cation and history of applied stresses. p 292-297 in L.G. Schultz, H. van Olphen and F.A. Mumpton (eds) Proc. Int. Clay Conf. Denver, The Clay Mineral Society, Bloomington, IN.

Bouma, J., 1977. Soil survey and the study of water in unsaturated soil. Soil Survey Papers No 13, Soil Survey Inst., Wageningen, The Netherlands, 107 pp.

Bronswijk, J.J.B., 1991. Magnitude, modeling and significance of swelling and shrinkage processes in clay soils. PhD Thesis, Wageningen Agricultural University, 145 pp.

Dekker, L.W., and P.D.Jungerius, 1990. Water repellency in the dunes with special reference to the Netherlands. Catena supplement 18, p173-183. Catena Verlag, Cremlingen, Germany

Driessen, P.M. and R. Dudal (eds.), 1991. The major soils of the World. Agricultural University, Wageningen, and Katholieke Universiteit Leuven, 310 pp.

FAO, 1989. Soil Map of the World at scale 1:5,000,000. Legend. FAO, Rome.

Feijtel, T.C.J. and E.L. Meyer, 1990. Simulation of soil forming processes 2nd ed., Dept of Soil Science and Geology, WAU, Wageningen, The Netherlands, 74 pp.

Hillel, D., 1980 Applications of soil physics. Academic Press, 385 pp.

Pons, L.J. and I.S. Zonneveld, 1965. Soil ripening and soil classification. ILRI publ. 13, Veenman, Wageningen, 128 pp.

Reinierce, K. 1983. Een model voor de simulatie van het fysische rijpingsproces van gronden in de IJsselmeerpolders. (in Dutch). Van Zee tot Land no. 52, 156 pp.

Ritsema, C.J., and L.W. Dekker, 1994. Soil moisture and dry bulk density patterns in bare dune sands. Journal of Hydrology 154:107-131.

Smits, H., A.J. Zuur, D.A. van Schreven & W.A. Bosma. 1962. De fysische, chemische en microbiologische rijping der gronden in de IJsselmeerpolders. (in Dutch). Van Zee tot Land no. 32. Tjeenk Willink, Zwolle, 110 p.

Tessier, D., 1984. Etude experimentale de l'organisation des materiaux argileux. Dr Science Thesis. Univ. de Paris INRA, Versailles Publ,. 360 pp.

United States Soil Conservation Service. 1975. Soil Taxonomy; a basic system of soil classification for making and interpreting soil surveys. Agric.Handbook no. 436. USA Govt. Print Off., Washington.

Wilding, L.P. and D. Tessier, 1988. Genesis of vertisols: shrink-swell phenomena. p.55-81 in: L.P. Wilding and R. Puentes (eds) Vertisols: their distribution, properties, classification and management. Technical Monograph no 18, Texas A&M University Printing Center, College Station TX USA.

General reading:

Hillel, D., 1980. Fundamentals of soil physics. Academic Press, New York, 413 pp.

Hillel, D., 1980. Applications of soil physics. Academic Press, New York, 385 pp.

Koorevaar, P., G. Menelik and C. Dirksen, 1983. Elements of soil physics. Elsevier, Amsterdam

Plate D. Stone cover due to frost heaving, Iceland. Photograph P. Buurman.

Plate E. Frost polygons with clear sorting of coarse material along the cracks, Iceland. Note also hummocky vegetation. Photograph P. Buurman

CHAPTER 3

SOIL CHEMICAL PROCESSES

This chapter deals with a number of chemical concepts relevant to soil forming processes. After studying this chapter you will be familiar with chemical aspects of weathering of primary minerals, the nature of important crystalline and amorphous weathering products, cation exchange processes in relation to colloidal properties of clay, complexation of metals by organic ligands, and redox processes in soils.

3.1. Chemical weathering and formation of secondary minerals

PRIMARY AND SECONDARY MINERALS

Chemical weathering is the transformation of minerals at the earth's surface to solutes (=dissolved substances) and solid residues. Minerals formed at higher temperature and pressure may be thermodynamically unstable at the lower temperatures and pressure of the earth's surface. Most igneous and metamorphic rocks are made up mainly of silicates and consist of a combination of silicate configurations (such as $Si_2O_5^{2-}$, $Si_2O_6^{4-}$ and SiO_4^{4-}) and metal ions, often with free silica (SiO_2) forms such as quartz. These-forming minerals are also called *primary* minerals. Primary minerals weather to iron and aluminium oxides, clay minerals and amorphous silicates (secondary minerals). These secondary minerals, that together with highly resistant primary minerals, such as quartz, form the typical minerals of highly weathered soils, are also called *soil minerals*. Thermodynamic stability diagrams show what minerals are stable under which conditions. According to Figure 3.1, the primary minerals analcime, albite, K-feldspar, and muscovite are unstable at low pH and K^+ concentration of the soil solution, and will eventually transform to smectite or kaolinite. At very low concentrations of dissolved silica (H_4SiO_4) , the clay minerals smectite and kaolinite will transform to gibbsite, $Al(OH)_3$.

CONGRUENT DISSOLUTION

Weathering takes place under the influence of weathering agents: water, various organic and inorganic acids and complexing agents, and oxygen. An example is the dissolution of olivine in an acid environment. In that case olivine dissolves completely to Mg^{2+} and H_4SiO_4 (H_4SiO_4 is the dominant form of dissolved silica in natural waters). Weathering without leaving a residue is called congruent dissolution.

$$Mg_2SiO_4(s)+4H^+(aq) \rightarrow 2Mg^{2+}(aq) + H_4SiO_4(aq) \qquad (3\text{-}1)$$

Question 3.1. *The table below shows solute concentrations observed in waters from different soils and weathering environments associated with granite. Plot the data points in Fig 3.1. What different secondary minerals would you expect as a function of climate and rate of water movement in fissures in granite? P-E = precipitation - total evapo-transpiration.*

sample	P-E mm/yr	water type	pH	(Na^+) mol/L	(H_4SiO_4) mol/L
1	50	stagnant water in bed rock	9.5	0.1	0.002
2	50	spring water	7.0	0.01	0.0015
3	500	spring water	5.5	0.001	0.0001
4	2500	spring water	4.5	0.0001	0.00001

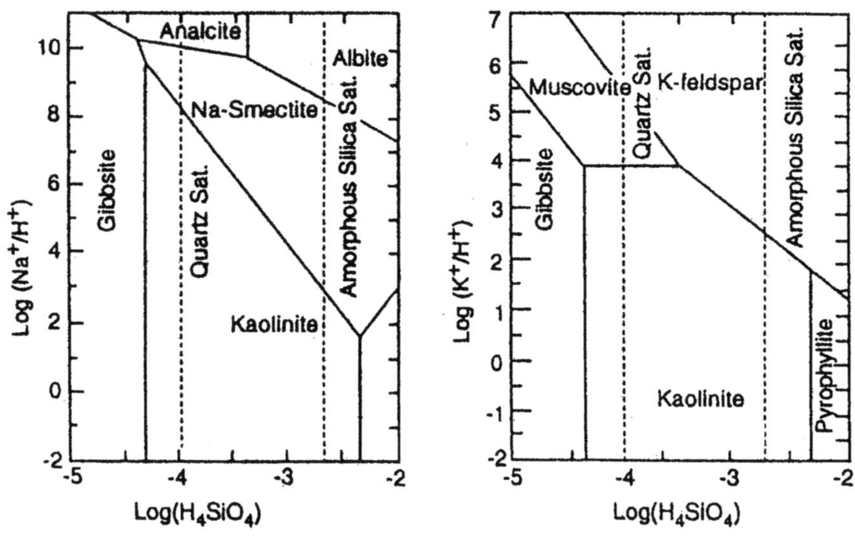

Figure 3.1. Stability diagram at 25°C and 1 bar total pressure for the system Na_2O- and K_2O-Al_2O_3-SiO_2-H_2O, with ratio of the molar concentrations in solution (actually thermodynamic activities) of Na^+, or K^+ to H^+ and of H_4SiO_4 as variables, plotted logarithmically (White, 1995; for chemical formulae of minerals, see Appendix 2).

Different acids can provide the H^+. In soils, carbon dioxide is usually abundant, and is often the main proton donor involved in weathering:

$$4CO_2(g,aq) + 4H_2O \rightarrow 4HCO_3^-(aq) + 4H^+(aq) \quad (3\text{-}2)$$

Question 3.2. *a. Combine reactions 3-1 and 3-2 to show how weathering of olivine leads to the production of dissolved bicarbonate (HCO_3^-), along with Mg^{2+}. b. Write a reaction equation for the dissolution of olivine by sulfuric acid, e.g. from acid rain.*

INCONGRUENT DISSOLUTION

In relatively dry climates, the concentrations of dissolved weathering products can become relatively high. In that case, part of the Mg released during dissolution of olivine is precipitated as $MgCO_3$. Weathering that leaves a solid weathering product is called *incongruent* dissolution. Another example of incongruent dissolution is the weathering of K feldspar under the influence of CO_2+H_2O to kaolinite plus dissolved H_4SiO_4, K^+ and HCO_3^-.

O_2 is an important weathering agent of minerals that contain elements in a lower oxidation state, such as Fe(II) or S(-II). The Fe(II) form of olivine, fayalite (Fe_2SiO_4), often leaves a residue of brown hydrated iron oxide, e.g. goethite, upon weathering. Iron oxides related to pyroxene weathering are illustrated in Plates A (p. 2) and C (p. 14).

Question 3.3 *Write the reaction equation for incongruent weathering of olivine to magnesite ($MgCO_3$), with CO_2 as the weathering agent*

Question 3.4 *Write the reaction equation of incongruent dissolution of K-feldspar to kaolinite (for mineral formulas, see Appendix 2).*

Question 3.5 *a. Write the reaction equation for weathering of fayalite to goethite. b. How could the weathering of fayalite proceed in the absence of O_2, so in an anoxic environment?*

Weathering by incongruent dissolution is common, because most (silicate) minerals contain iron and aluminium. In soils, Fe and Al form very poorly soluble oxyhydrates and Al can be incorporated in highly insoluble clay minerals such as kaolinite. Such so-called secondary minerals form the stable mineral assemblage in soils. Minerals differ in their 'weatherability', in the ease with which they weather. The weatherability of a mineral depends on i) its equilibrium solubility in an aqueous solution and ii) its rate of dissolution. Most minerals with a high solubility dissolve quickly (Table 3.1). However, the K concentration in a solution in equilibrium with mica is higher than the Ca concentration in a solution in equilibrium with calcite, but mica dissolves far more slowly than calcite, so its weatherability is much lower.

The rate of dissolution of primary minerals is so slow that equilibrium concentrations are rarely reached, and the bulk of the soil solution is usually far under-saturated with respect to the primary minerals. That is why kinetics of dissolution is important in describing mineral weathering.

Table 3.1. Two aspects of 'weatherability' (equilibrium solubility in mmol/l, and dissolution rate), and pH dependency of solubility, of primary and secondary minerals (From Van Breemen and Brinkman, 1978).

mineral	solubility at $pCO_2 = 10^{-1}$ kPa and $(H_4SiO_4) = 10^{-3}$ mol/L	factor of solubility increase per unit pH-decrease; pH<7	dissolution rate
primary minerals			
Mg^{2+} from forsterite	50	100	moderately slow
K^+ from mica[1]	7	10	extremely slow
Si from quartz,	0.1	1	very slow
secondary minerals in soils from wet to moderately dry regions			
Mg^{2+} from smectite[1]	0.05	10	very slow
Al^{3+} from gibbsite	$2 * 10^{-3}$	1000	slow
Al^{3+} from kaolinite	10^{-4}	1000	very slow
secondary minerals in soils from dry and arid regions			
Ca^{2+} from calcite	1	100	moderately fast
Ca^{2+} or SO_4^{2-} from gypsum	10	1	fast
Na^+ or Cl^- from halite	$6.1 * 10^3$	1	fast

[1] in equilibrium with kaolinite

Question 3.6 *a. What can you conclude about the differences in weatherability between the primary and secondary minerals listed in Table 3.1? b. Check the pH dependency of the solubility for gypsum, forsterite, calcite and gibbsite (hint: write the dissolution reaction with H^+ as reactant, if possible). c. If a suspension of gibbsite at $pCO_2 = 10^{-1}$ kPa has a pH of 5; what will be the solubility of gibbsite at pH 3?*

KINETICS OF WEATHERING

The speed of weathering kinetics depends on (1) the dissolution process itself, and (2) the removal of the dissolved material.

Question 3.7. *Why is removal of dissolved material important for the kinetics of weathering?*

During incongruent dissolution, secondary minerals may grow directly on top of the weathering mineral (and form a protective coating that slows down further weathering), or they can be formed elsewhere in the soil or weathering zone.

Question 3.8. *Describe how (a) the rate of removal of weathering products (water percolation) or (b) the degree of aeration could determine whether or not weathering of fayalite is slowed down by the formation of a protective coating.*

The slowest step determines the overall speed (rate) of chemical weathering. This so-called rate-limiting step can be the dissolution step of one of the elements in the weathering mineral, or it can be the removal of dissolved products. We will consider the dissolution step in somewhat more detail. Figure 3.2 illustrates factors that determine the dissolution rate of a silicate mineral, with reference to the energy level of silicon in the system.

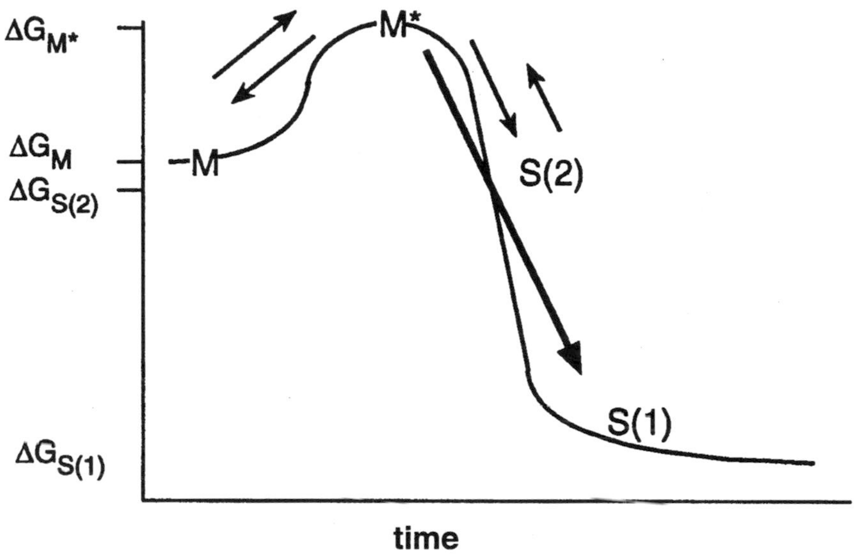

Figure 3.2. Change in the Gibbs free energy (ΔG) of a silica group during its dissolution from mineral-bound (M) to dissolved (S) form. The length of arrows is proportional to the rate of change.

The vertical axis refers to the Gibbs free energy (ΔG) of one mole of Si. A spontaneous reaction always proceeds in the direction of lower ΔG of the whole system. In the primary mineral, Si has a higher energy level (ΔG_M) than in solution (ΔG_S). The higher the concentration in solution, the higher the value of ΔG_S. A Si group at the mineral surface has to be activated into the higher energy state M* before it can dissolve. This so-called activated surface complex is an incompletely bound species (e.g. $Si(OH)_4$), often associated with an adsorbed species, e.g. H^+ or OH^-. At a given set of conditions (pH, temperature, etc.) the activated sites are usually in equilibrium with the mineral (suggested by equal rate arrows). The density of activated complexes at the mineral surface depends on e.g. pH, temperature etc. At a concentration S(1), where the free energy of the group equals $\Delta G_{S(1)}$, the solution is highly under-saturated. The activated surface complexes of a mineral placed in such a highly under-saturated solution will dissolve faster than they are formed. At a high degree of under-saturation, the precise value of S(1) has little effect: the rate of dissolution of the mineral is determined by the rate of formation of the activated surface complexes, and is independent of the concentration of the surrounding solution. At a higher concentration, close to equilibrium S(2), the activated surface groups can be rebuilt from the solution, and more frequently so as the concentration of dissolved weathering products is higher (as ΔG_S is higher). So, the rate of dissolution of a mineral in contact with a slightly under-saturated solution depends on the degree of undersaturation:

$$R = k\,(C_M - C_S)^n \qquad (3\text{-}3)$$

where R is the dissolution rate (expressed, e.g. as moles $m^{-2}\ s^{-1}$; m^{-2} refers to the surface area of reacting mineral), k is the rate constant, C_M is the equilibrium concentration, C_S is the concentration in the ambient solution. In equation 3-3, n determines the order of the reaction, which varies generally between 0 and 2 (Velbel, 1986).

***Question 3.9**. Draw a physical analogue of Figure 3.2 by means of a rectangular solid block of wood on a table. a) Draw the positions of the block in situations M, M* and S. b) How can you simulate the difference between the transitions $M \rightarrow M^* \rightarrow S(1)$ and $M \rightarrow M^* \rightarrow S(2)$ by changing the form of the block?*

***Question 3.10**. Criticise the statement: "The rate of mineral weathering is always proportional to the annual precipitation, because higher rainfall dilutes the soil solution, causing a stronger undersaturation of the soil solution with respect to weatherable minerals."*

If the solution is under-saturated with the mineral by more than one to two orders of magnitude, formation of M* may become rate limiting. In laboratory experiments with feldspars in highly under-saturated solutions, the dissolution rate is pH dependent (Fig 3.3). This is explained by adsorption of H^+ at low pH, and of OH^- at high pH, during formation of the activated complex. In nature, an increase in pH above 8 does not increase the weathering rate as suggested in Figure 3.3, because natural high-pH environments have high concentrations of dissolved Na and H_4SiO_4

ETCH PITS AND MICRO PORES

The surface of many weathering minerals changes by formation of micrometer-size etch pits (see Fig 3.4) or much smaller micropores (diameters of 10-20 nm). Etch pits develop if minerals are in contact with highly under-saturated solutions. Micropores normally develop at dislocations (crystal defects) when the surrounding solution is slightly under-saturated with respect to the mineral. The amount of mineral which dissolves by etch pitting may constitute a large part of the total volume. Thus, etch pitting tends to increase the weathering rate per unit mass of mineral ($mol.g^{-2}.s^{-1}$). Formation of micropores does not involve much mineral volume, but the mineral surface may increase by several orders of magnitude. Because most micropore surface

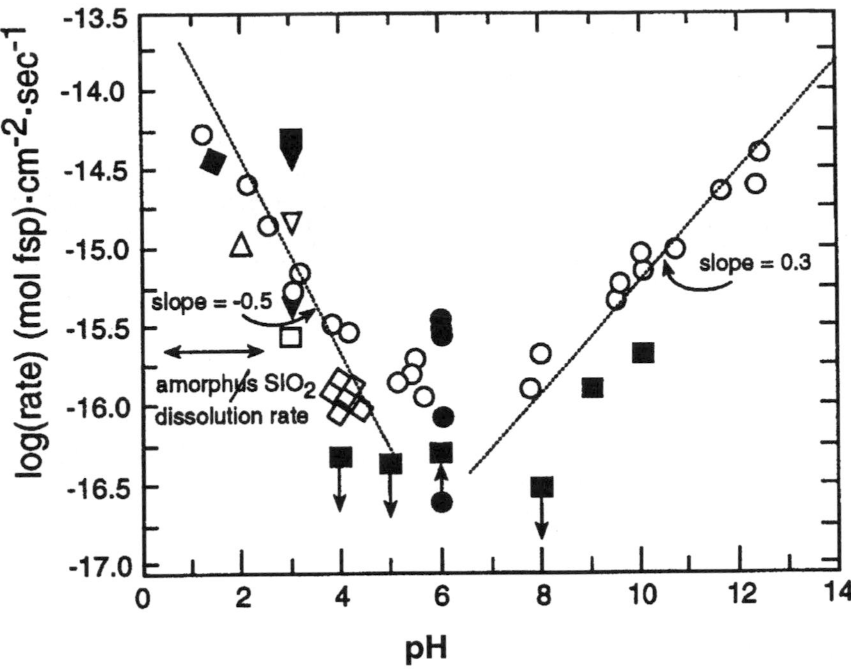

Figure 3.3. Albite dissolution rates as a function of pH (Blum and Stillings, 1995). Symbols refer to different laboratory experiments.

area is not in direct contact with the bulk soil solution, diffusion of weathering products out of the micropores may become rate-limiting. Thus, contrary to etch pitting, formation of micropores does not increase the weathering rate. In most laboratory experiments (see Fig 3.3) solutions are highly under-saturated, micropores do not form, and dissolution rates are high. In the field, where solutions are often less under-saturated, micropores form, and weathering rates (which are averaged over total surface area) may decrease by several orders of magnitude as reaction products pile up inside the micropores (eq. 3-3). This explains why weathering rates of feldspars in the field are generally 2 to 4 orders of magnitude lower than those observed in the laboratory (Anbeek, 1994). In laboratory experiments, dissolution rates of feldspar (expressed as Si released) are in the order of 10^{-11} to 10^{-14} $mol.m^{-2}.s^{-1}$.

Question 3.11. *Would you expect (i) etch-pitting or (ii) micro pores at the surface of feldspars in a) narrow cracks in granitic bedrock or, b) a soil under tropical rain forest?*

Low rates of removal of reaction products may also result from diffusion through protective coatings of secondary minerals, or through stagnant water layers or slow flow of thin water layers over the surface of grains in (hydrologically) under-saturated soils.

The rate of weathering of most silicate minerals also tends to increase with increasing percolation, but this is the result of an increase in R with increasing dilution (eq. 3-3), and of an increase in the proportion of the exposed mineral surface that comes into contact with under-saturated water.

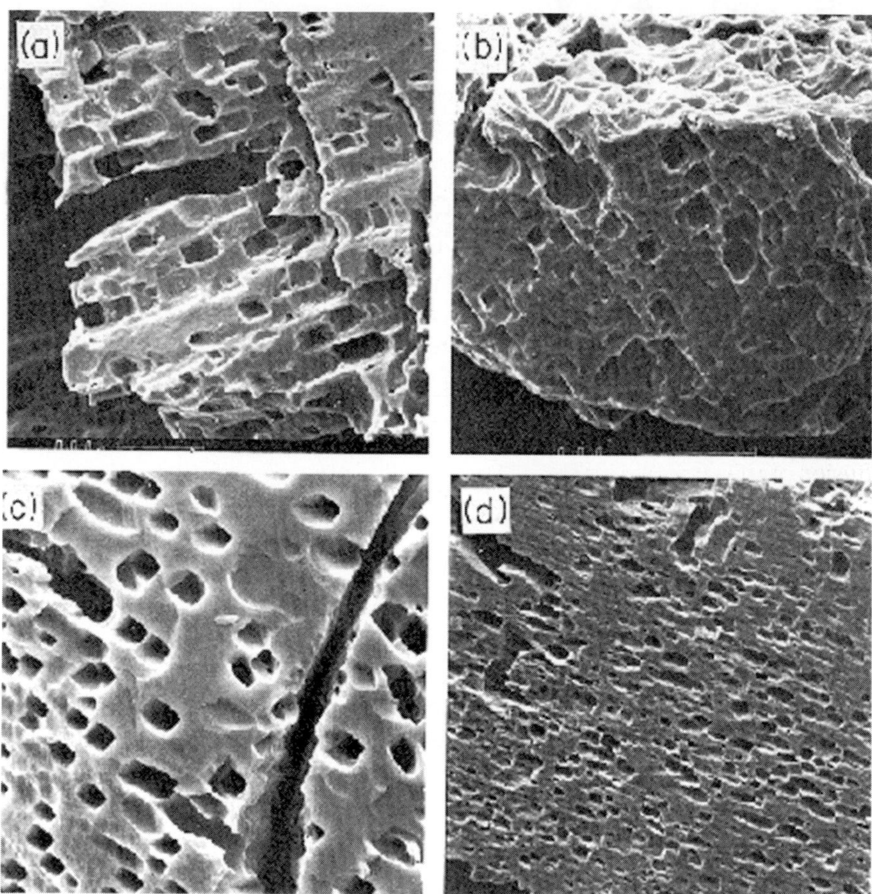

Figure 3.4. Scanning electron micrographs of prismatic etch pits in feldspars: a) oligoclase-andesine from a saprolite; b) oligoclase from a surface soil; c) albite; and d) microcline treated with HF-H_2SO_4 in the laboratory. Picture widths are 10 microns (a,b,d) and 1 micron (c). From Berner and Holdren, 1979; reprinted with permission from Elsevier Science, Amsterdam.

CALCITE WEATHERING
Silicate weathering is usually too slow to achieve a soil solution saturated with the primary mineral in question, even at very slow rates of percolation of water. The rate of dissolution of calcite, however,

$$CaCO_3\,(s) + CO_2\,(g) + H_2O\,(l) \rightarrow Ca^{2+} + 2HCO_3^- \qquad (3\text{-}4)$$

is in the order of 10^{-10} mol.m^{-2}.s^{-1} , quick enough for soil solutions and ground water in calcite-rich soils to remain practically saturated with calcite at common rates of percolation. In other words, the rate of supply of 'fresh' rainwater infiltrating the soil is usually slow relative to the rate of dissolution of calcite. Therefore, the rate of decalcification of a soil can be calculated from (1) the equilibrium concentration of calcite, and (2) the annual percolation rate.
In the absence of strong (e.g. organic) acids, the solubility of $CaCO_3$ depends on the CO_2 pressure in the soil atmosphere and on the temperature (Fig. 3.5). The solubility is highest at high CO_2 pressures and at low temperatures. The CO_2 pressure in the soil atmosphere is often between 0.1 and 1 kPa, i.e. several orders of magnitude higher than in the air (3 x 10^{-2} kPa). The relatively high CO_2 pressures are due to a combination of respiratory activity by microorganisms and plant roots, and a relatively low rate of diffusion of CO_2 from the soil.

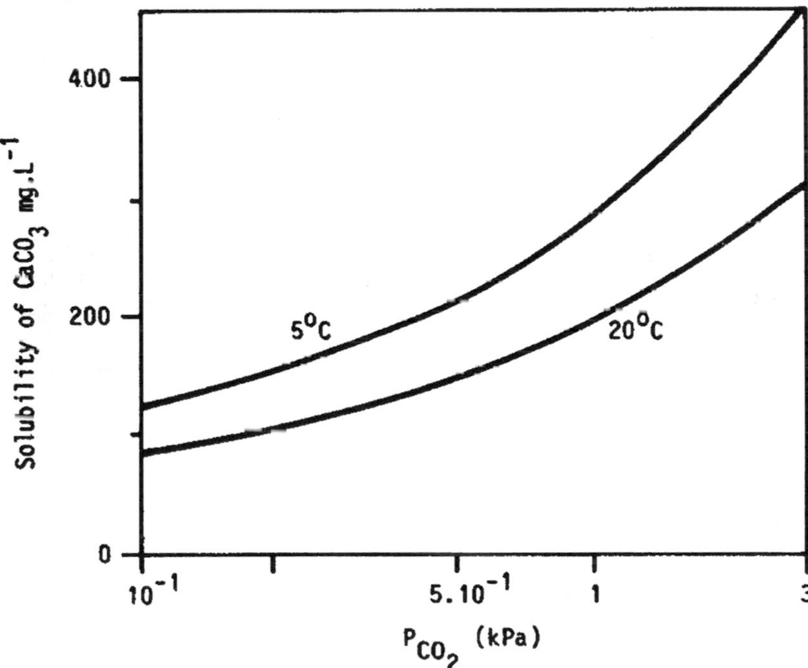

Figure 3.5. Solubility of calcium carbonate depending on temperature and CO_2 partial pressure. From Van Breemen and Protz, 1988.

***Question 3.12.** a. List at least three (largely independent) factors that determine the rate of removal of calcite from a soil. b. Explain why the $CaCO_3$ content of the soil is not one of those factors. Is that different in case of feldspar weathering?*

In the presence of calcium carbonate the pH will be so high (7 to 8) that solubility of silicate minerals is low and silicate weathering is negligible. After decalcification, the pH can fall to lower levels, where primary silicates may dissolve at higher rates.

***Question 3.13.** The boundary between decalcified loess and its much more calcareous parent material is often visible because the decalcified material is more strongly brown coloured than the parent material. What causes the stronger brown colour?*

Weathering of silicates by water plus CO_2 always produces dissolved silica (H_4SiO_4) and bicarbonate, which are removed in the percolating water. The process is also called *desilication*, because in addition to various cations (depending on the minerals in question, mainly K^+, Na^+, Ca^{2+} and Mg^{2+}), dissolved silica (H_4SiO_4) is removed during weathering of silicate minerals. Aluminium has a very low solubility at near-neutral pH. If the silica concentration in the soil solution remains very low, as in soils in areas with very high precipitation, $Al(OH)_3$ (gibbsite) can be formed as a secondary mineral.

METAL COMPLEXATION

Certain water-soluble organic acids (H_nL) can strongly complex metal ions:

$$M^{n+}(aq) + H_nL(aq) \leftrightarrow ML(aq) + nH^+(aq) \qquad (3\text{-}5)$$

Complexation decreases the concentration of free M^{n+}, which increases the solubility of minerals containing M, and causes higher total concentrations of M (= $[M^{n+}] + [ML]$) in solution at equilibrium.

Low-molecular weight organic acids (e.g. citric, oxalic, vanillic and p-hydroxy benzoic acids, see Figure 4.11) are commonly formed in soils. Their carboxylic and phenolic groups may form a kind of claw (Greek: chela) with very strong affinity for trivalent metal ions such as Al or Fe. Dissolved or solid compounds of such acids and aluminium or iron are called chelates. In the absence of chelators, concentrations of dissolved Al in equilibrium with Al-containing minerals strongly increase with decreasing pH (see also Question 3.6b):

$$Al(OH)_3 + 3H^+ \leftrightarrow Al^{3+} + 3H_2O \qquad (3\text{-}6)$$

$$K = [Al^{3+}] / [H^+]^3 \qquad (3\text{-}6a)$$

Only at very low pH (<4), the solubility of Al is high enough to cause an appreciable mobility of Al in the soil. At higher pH (4-5) chelators can increase the concentration of total dissolved Al by several orders of magnitude, and thus cause appreciable mobility of Al. Because small organic ligands will be decomposed eventually by microbes, Al and Fe will normally be precipitated as secondary minerals, also when organic acids act as weathering agents.

3.2. Soil minerals and their physico-chemical properties

As mentioned in the previous section, minerals that formed at great depth in the earth are not usually stable at the earth's surface. Roughly, *weathering* is the adaptation of a parent rock to circumstances that prevail at the surface: primary minerals of the earth's crust are transformed into secondary minerals typical of soils.

Question 3.14. *What essential differences exist in chemical and physical conditions between the earth's surface and the upper mantle?*

Weathering includes the dissolution of minerals, the oxidation of some mineral compounds, the leaching of dissolution products, and the recombination of weathering products to secondary minerals that are stable at the earth's surface. Thousands of different minerals make up the variety of igneous and metamorphic rocks of the crust. In soils, this number is reduced considerably. As in primary rocks, minerals in soils occur in assemblages: minerals belong to a typical environment, and possible combinations are restricted by thermodynamics. If we leave out the minerals that are not stable in the long run, we can distinguish the following assemblages in soils:

SOIL MINERAL ASSEMBLAGES

* *Assemblages of young volcanic soils and young soils on feldspar-rich and mafic rocks.*
  These contain mainly amorphous aluminium silicates and amorphous iron oxyhydrates. (See Chapter 12).

* *Assemblages of young, moderately weathered soils in temperate regions.*
  These contain a mixture of clay minerals, usually dominated by illite and vermiculite, and poorly crystalline iron oxyhydrates (mainly goethite). The main residual mineral is quartz.

* *Assemblages of old, strongly weathered (tropical) soils.*
  These contain only clays that are stable at low H_4SiO_4 concentrations, such as kaolinite and gibbsite, together with iron oxides (mainly hematite). (See Chapter 13).

* *Assemblages of steppe soils*
  In addition to the clay minerals illite and members of the smectite family, these soils usually contain free calcium carbonate (calcite). (See Chapter 9).

* *Assemblages of arid soils and saline coastal soils.*
  Calcite is usually present. Clay minerals are usually smectites, sometimes with palygorskite and sepiolite in arid soils. In addition, a large number of 'soluble salts' can be found: gypsum together with various chlorides, sulphates, bicarbonates and nitrates of sodium, potassium, magnesium, and calcium. (See Chapter 9).

* *Assemblages of hydromorphic soils*
  The clay minerals of such assemblages depend on the degree of water saturation and on the climate. Typical, however, are various iron sulphides, different iron and manganese (hydr)oxides (see later in this chapter). In addition, in acid sulphate soils, various sulphates may be found, e.g. jarosite, $KFe_3(SO_4)_2(OH)_6$. (See Chapter 7.4).

Question 3.15. *Calcite, gypsum, and soluble salts are not a common constituent of primary rocks. How could they have formed by weathering?*

Because soils contain a large variety in primary minerals, and each of these primary minerals has its own equilibrium conditions and weathering kinetics, primary and secondary minerals in soils are hardly ever fully in equilibrium. For the completion of many reactions, the time of soil formation is too short. In the tropics, where times of soil formation may be extremely long (up to millions of years on the shields of Africa, Australia, and South America), climate changes may prevent attainment of real equilibria. Present perhumid tropical climates with their rain forest vegetation are less than 10,000 years old in most places, and minerals of drier climates in the past, are now adapting to wetter circumstances, and vice-versa. Typical non-equilibrium situations exist in alluvial deposits, where the mineral assemblage, including that of 'soil minerals' is obtained by mixing of material from a whole watershed.
For environments with mild weathering, there is a relation between primary minerals and their weathering products. Common weathering products of a number of minerals are given in Table 3.2.

Table 3.2. Weathering products of some minerals

primary mineral	formula	secondary minerals
Olivine	$(Mg,Fe)_2SiO_4$	Fe (hydr)oxides
Pyroxenes	$Ca(Mg,Fe)_3(Al,Fe)_4(SiO_3)_{10}$	Fe (hydr)oxides, smectite
Amphiboles	$Ca_3Na(Mg,Fe)_6(Al,Fe)_3(Si_4O_{11})_4(OH)_2$	Fe (hydr)oxides, smectite
Feldspars	$(K,Na)Si_3O_8$ - $CaAl_2Si_2O_8$	Kaolinite, gibbsite, allophane
Serpentine	$(Mg,Fe)_3Si_2O_5(OH)_4$	(smectite), Fe-compounds
Muscovite	$KAl_2(AlSi_3O_{10})(OH)_2$	illite
Biotite	$K(Fe,Mg)_2(AlSi_3O_{10})(OH)_2$	vermiculite
Chlorite	$(Mg,Fe,Al)_6(Si,Al)_4O_{10}(OH)_8$	vermiculite, smectite
Garnet	$(Fe^{++},Al,Mn,Ca)_3Al_2Si_3O_{12}$	Fe, Al, Mn (hydrous) oxides
Apatite	$Ca_5(PO_4)_3(OH,F,Cl)$	Fe-phosphate, Al-phosphate
Volcanic glass		allophane

Table 3.2 illustrates that clay minerals and iron compounds are by far the most important secondary minerals. Because both iron compounds and clay minerals have a large surface area and have charged surfaces at pH values that normally occur in soils, these minerals strongly influence many soil characteristics. Silica, although a common weathering product, is usually removed from the soil as soluble H_4SiO_4 or recombined into secondary Al-silicates. Secondary accumulations of solid silica (e.g. opal, chalcedony) are scarce (See Chapter 14).

STRUCTURE AND CHARGE OF CLAY MINERALS

Most clay minerals are phyllosilicates, or sheet silicates. They consist of tetrahedral Si_2O_5 layers (T) and octahedral $Mg(OH)_2$ or $Al(OH)_3$ layers (O), combined in T-O (kaolinite, halloysite, serpentine), T-O-T (muscovite, biotite, smectite, illite, vermiculite), or T-O-T-O (chlorite) sequences (Figures 3.6 and 3.7).

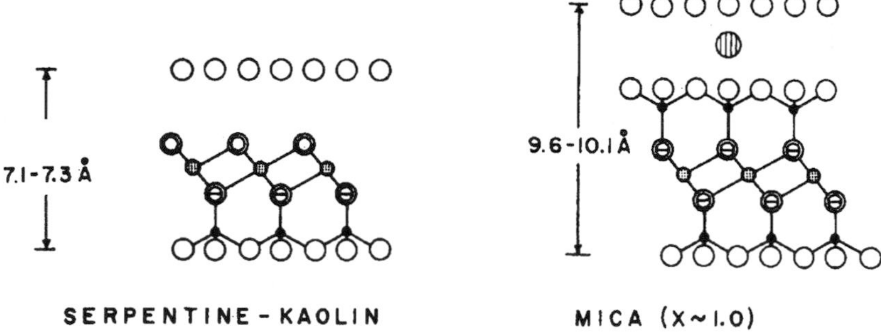

Figure 3.6. Structure of the 1:1 minerals kaolinite/serpentinite and the 2:1 mineral mica (illite). From Brindley and Brown, 1980. Legend see Figure 3.7.

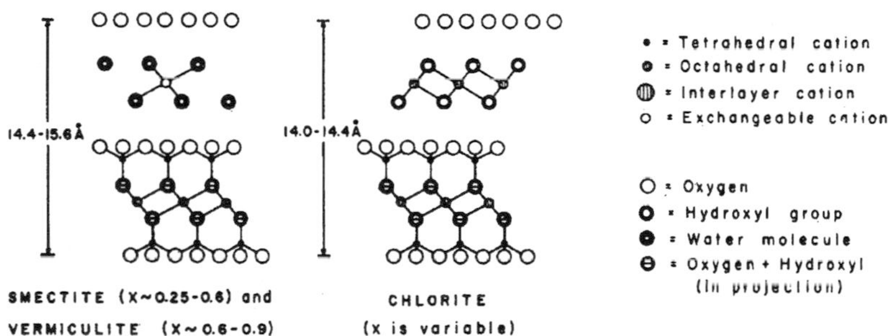

Figure 3.7. Structure of the 2:1 minerals smectite, vermiculite, and the 2:1:1 mineral chlorite. From Brindley and Brown, 1980.

Cation exchange is an important consequence of the charged nature of clay minerals. It is dealt with in general texts on the subject, e.g. Bolt and Bruggenwert, 1976. The aspects that are relevant to the processes of soil formation are summarised briefly here. Due to isomorphous substitution of higher-valence ions by lower-valence ions in the tetrahedral sheet (e.g., Al^{3+} for Si^{4+}) and in the octahedral sheet (e.g., Mg^{2+} or Fe^{2+} for Al^{3+}), clay minerals have negatively charged plate surfaces. This negative charge is compensated by cations between the clay plates. Such cations can either be structural,

or can be exchangeable for other cations in the soil solution. If, in 2:1 clay minerals, the adsorbed cation fits in the silica 6-rings of the plate surface in such a way that the plates are brought very close together, such cations effectively glue clay plates together and are not exchangeable anymore (e.g. K in illites, which becomes structural). Cation exchange characteristics and colloidal behaviour of clay minerals strongly depend on isomorphous substitution and compensating cation.
Table 3.3. presents the structural formulae of some common phyllosilicates. Exchangeable cations are indicated by 'X'. Figure 3.8 illustrates that the amount of exchangeable cations first increases with increasing layer charge, but decreases in clays with high layer charge, where interlayer cations are increasingly structural.

Table 3.3. Isomorphous substitution and exchangeable cations in some dioctahedral (dioct.) and trioctahedral (trioct.) phyllosilicates. Primary phyllosilicates are italicised. Montmorillonite belongs to the smectites.

mineral	**charge compensating cations**	**octahedral atoms**	**tetrahedr. atoms**	**structural O and OH**	**variable H_2O**
kaolinite 1:1, dioct.	none	Al_2	Si_2	$O_5(OH)_4$	none
halloysite 1:1, dioct.	none	Al_2	Si_2	$O_5(OH)_4$	$2H_2O$ or none
serpentine 1:1, trioct	none	Mg_3	Si_2	$O_5(OH)_4$	none
pyrophyllite 2:1, dioct.	none	Al_2	Si_4	$O_{10}OH)_4$	none
montmorillonite 2:1, dioct.	$X_{0.33}$	$Al_{1.5}Mg_{0.63}$	$Si_{3.91}Al_{0.09}$	$O_{10}(OH)_2$	yH_2O
muscovite 2:1, dioct.	$K_{1.0}$	Al_2	Si_3Al	$O_{10}(OH)_2$	none
biotite 2:1, trioct.	K $X_{1.17}$	Fe_3 $M^{3+}{}_{4.01}M^{4+}{}_{0.26}$ $M^{2+}{}_{1.67}$	Si_3Al $Si6_{6.6}Al_{1.4}$	$O_{10}(OH)_2$ $O_{10}(OH)_2$	none none
illite 2:1, dioct.	$K_{0.58}X_{0.17}$	$Al_{1.55}FeIII_{0.2}Mg_{0.25}$	$Si_{3.5}Al_{0.5}$	$O_{10}(OH)_2$	yH_2O
talc 2:1, trioct.	none	Mg_3	Si_4	$O_{10}(OH)_2$	none
vermiculite 2:1, trioct.	$X_{0.66}$	$Mg_{2.61}Al_{0.39}$	$Si_{2.95}Al_{1.05}$	$O_{10}(OH)_2$	yH_2O
chlorite 2:1:1, trioct.	none	$Mg_{4.65}FeII_{0.4}Al_{0.9}$	$Si_{3.2}Al_{0.8}$	$O_{10}(OH)_8$	none

Question 3.16. *Soils have a watery environment and Mg concentrations in the soil solution are usually low. What does this imply about the chance to encounter some of the minerals in Table 3.3 in soils?*

Question 3.17. *Can you deduct, by consulting Table 3.3, the meaning of 'trioctahedral' and 'dioctahedral'?*

Question 3.18. *Check the charge balance for three of the minerals in Table 3.3*

Question 3.19. *Compare Table 3.3 with Figures 3.6 and 3.7 to understand the difference between 1:1 and 2:1 phyllosilicates. Which of the minerals in Table 3.3 does not belong to either of these groups?*

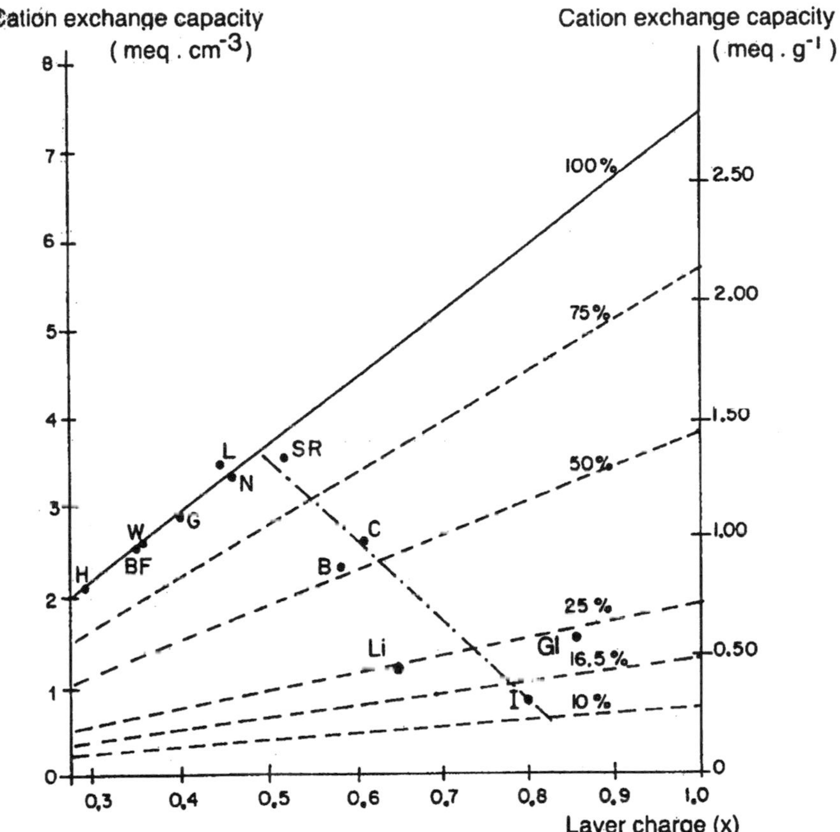

Figure 3.8. Relation between cation exchange capacity and layer charge of 2:1 clays. Li, I = illites; Gl = Glauconite; others are smectites. The diagonal lines indicate the proportion of the interlayer cations that is exchangeable. From Wilding and Tessier, 1988.

CLAY MINERAL DEFINITIONS

In Table 3.3, we have named various clay minerals by their isomorphous substitution and compensating cations. In X-ray diffraction, clay minerals are distinguished by the behaviour of their (001) spacing upon various treatments, and not by their chemistry (The 001 spacing is the repetition distance of the plate, indicated with vertical bars in Figures 3.6 and 3.7). This causes a discrepancy between the structural/chemical and the 'operational' definition (based on 001 behaviour) of 2:1 clays. It is important to keep this in mind when studying the clay weathering sequences of the following section. The sequences are based on operational definitions. This means that a transition from 'vermiculite' to 'smectite' during weathering does not necessarily imply a change in tetrahedral and octahedral substitution, as you would conclude from Table 3.3..

FORMATION OF CLAYS

Clay minerals form by two main pathways: a) precipitation from solution, and b) fragmentation and breakdown of primary phyllosilicates with retention of phyllosilicate crystal structure.

Formation of clay minerals by precipitation from the soil solution is a common phenomenon in many soils. It is the main source of kaolinite and gibbsite (not a phyllosilicate) in Oxisols. Also most smectites in soils are formed this way. Because minerals can only form from super-saturated solutions, clay minerals can only precipitate where solutions have relatively high concentrations of the relevant solutes.

Question 3.20. *a) What is the source of solutes that precipitate to form clay minerals in soils? b). Do you expect clay mineral formation in sandy soils with strong percolation?*

The primary phyllosilicate muscovite is common in igneous and metamorphic rocks. It can be transformed to illite by physical breakdown accompanied by progressive removal of interlayer K^+. Transformation of one phyllosilicate into another usually proceeds from the plate edges inwards, and the progress of the transformation is not equal at all sites. Intermediate products, which have characteristics of more than one typical phyllosilicate are called *interstratifications*. The transition, by removal of interlayer potassium, of muscovite through illite to vermiculite, is illustrated in Figure 3.9.

Biotite (Table 3.2, 3.3) transforms directly into vermiculite. At low pH, and in the presence of complexing organic acids, such as in podzol A and E horizons (see Chapter 11), illites and vermiculites can be transformed into aluminium smectites (beidellites). The full weathering sequence is then:

$$\text{mica} \rightarrow \text{MI} \rightarrow \text{illite} \rightarrow \text{IV} \rightarrow \text{vermiculite} \rightarrow \text{VS} \rightarrow \text{smectite} \qquad \text{(A)}$$

in which the letter combinations denote interstratified minerals (e.g., IV = illite-vermiculite interstratification).

If the environment is somewhat less acidic, and there is an addition of dissolved aluminium that can polymerise, an additional $Al(OH)_3$ (gibbsite) layer can be built between clay plates, which converts smectites or vermiculites into 'soil chlorites':

$$\text{smectite/vermiculite} + Al(OH)_3 \rightarrow \text{Al-interlayered 'soil chlorites'} \quad \text{(B)}$$

Question 3.21. *How does the CEC of the mineral phase change in the weathering sequence (A) and the upon chloritisation (B)? Indicate the site of the Al interlayer in Figure 3.7.*

Because chemical and physical conditions in soils change with depth, also the clay mineral assemblage in many soils changes with depth.

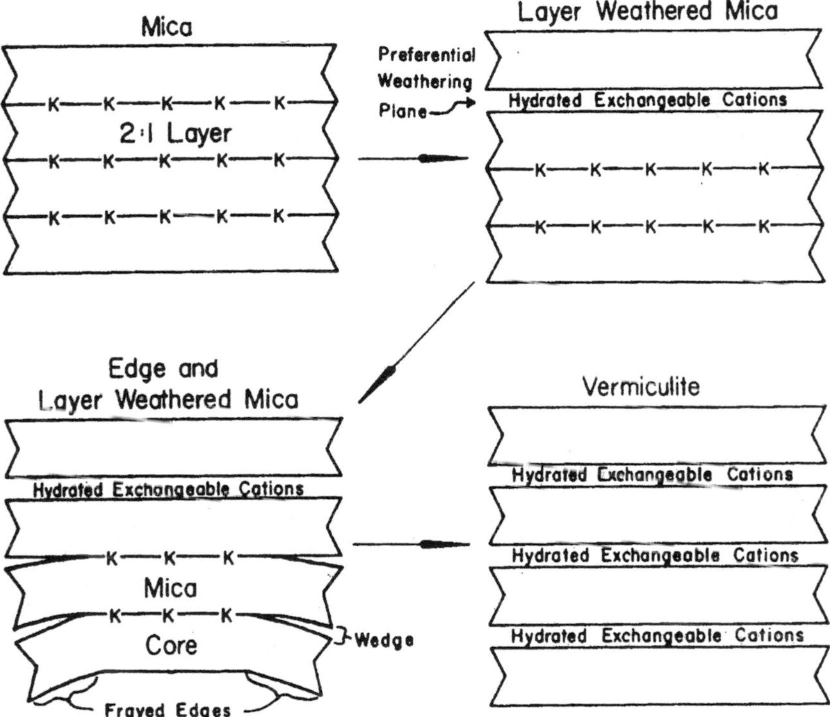

Figure 3.9. Transformation of mica to vermiculite through removal of K at the edges and in planes of weakness. From Fanning and Keramidas, 1977.

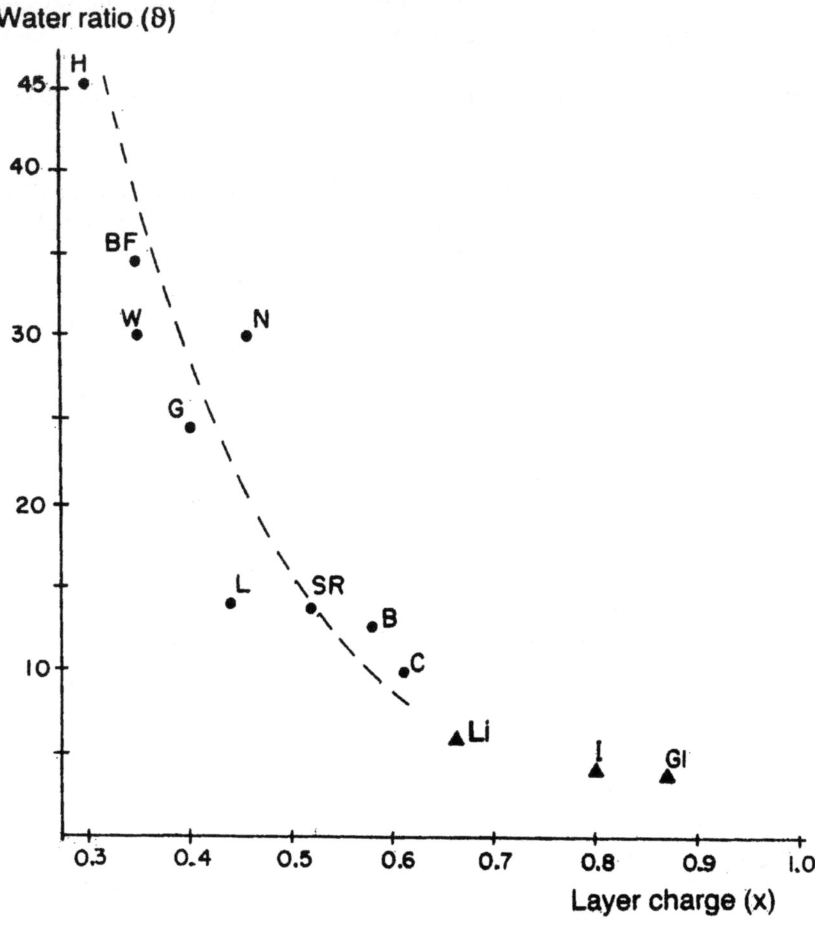

Figure 3.10. Layer charge and water content of 2:1 minerals in 10^{-3} M NaCl at a water potential of -0.32 kPa. The water ratio is the weight of water relative to the weight of clay. For legend, see Figure 3.8. From Wilding and Tessier, 1988.

INTERLAYER WATER AND COLLOIDAL BEHAVIOUR

Removal of interlayer potassium from illites and vermiculites, and of the additional octahedral layer from chlorites, allows the individual TOT-packages to separate, and water (with hydrated cations) to enter between the platelets. Natural micas and vermiculites have many stacked TOT plates. As long as only part of the interlayer potassium has been removed, individual plates do not easily separate, and the clays are relatively rigid. Upon increasing removal of interlayer cations, smaller stacks of TOT plates are formed, and the clays become less rigid. Entry of water is governed both by the removal of interlayer cations and by the charge deficit (Figs 3.8, 3.9). 2:1 Clays with the lowest charge deficit have the highest capacity for hydration, and therefore the greatest shrink/swell potential. The relation between layer charge and water adsorption is illustrated by Figure 3.10 (see also problem 2.7).

Question 3.22. *Why do clays with a low charge deficit have a high capacity for hydration?*

Swell and shrink behaviour of clays (see Chapter 2) is traditionally explained by the variation in the thickness of the double layer (the aqueous phase between the individual clay platelets), in relation to the composition of the exchange complex and the electrolyte concentration outside the double layer (van Olphen, 1963). Research by Tessier (1984), however, suggests that the double layer itself may be less important to swell and shrink behaviour than hydration and dehydration of inter-particle pores with diameters of 0.1-2 μm, that are present within a flexible honeycomb microstructure (see Chapter 2).

In addition to the influence of clay mineralogy, the nature of the exchangeable cations and concentration of the soil solution govern the behaviour of clay in soils. The following gives a recapitulation of the main facts.

Clay particles are attracted to each other by Van der Waals forces and (but probably to a lesser extent) by electrostatic forces resulting from negative (plate-side) and positive (edge-side) double layers. On the other hand, negative plate sides repel each other. Van der Waals forces are mass attraction forces. They strongly decrease with the distance between objects, and are independent of type and concentration of electrolyte.

The thickness of the electrical double layer, on the other hand, is strongly influenced by type and concentration of cations. Monovalent exchangeable cations and low electrolyte concentrations of the surrounding solution tend to cause thick double layers. Di- and trivalent exchangeable cations and high electrolyte concentrations cause thin double layers. When the double layer is expanded, electrostatic forces dominate over mass attraction, clay platelets repulse each other and may disperse. If the double layer is contracted, mass attraction dominates, clay particles attract each other, and flocculation occurs.

SOLUBLE AND EXCHANGEABLE CATIONS

Concentrations of exchangeable cations (cation exchange capacities) vary from a few cmol(+) per kg of clay in kaolinite, to more than 100 cmol(+) per kg in smectites and vermiculites. The most abundant exchangeable cations are those usually released into the soil solution and the groundwater by silicate weathering: Ca^{2+}, Na^{+}, Mg^{2+}, K^{+}, and at low pH, Al^{3+} and H^{+}.

The Kerr equation (for ions of equal valence) and the Gapon equation (for mono-divalent exchange), which are discussed in, e.g., Bolt and Bruggenwert (1976), describe the distribution of cations over the diffuse double layer and the ambient electrolyte solution reasonably well. These models predict that:

- Cations of higher valence are preferentially adsorbed over cations of lower valence. So the ratio of divalent to monovalent cations at the adsorption complex is invariably higher than that in the soil solution.
- This is particularly so at low electrolyte concentrations, so if a soil solution is diluted by addition of rain water, cations of relatively high valence would move to the complex at the expense of equivalent amounts of cations of lower valence. Increasing the concentration e.g. by evaporative removal of water has the opposite effect because this usually causes a dominance of monovalent ions (see Chapter 9).

In most soils, there is a strong connection between pH and the dominant cations at the adsorption complex:

* At pH <5, Al^{3+} is appreciably soluble. On account of its high valence, Al^{3+} can displace other cations at the adsorption complex, even if its concentration in the soil solution is relatively low, as is the case between pH 4.5 and 5.
* At pH values between 5 and 7, electrolyte concentrations are usually low and Ca^{2+} is the dominant cation.
* At pH values between 7.5 and 8.5, the soil usually contains free $CaCO_3$. Presence of this mineral causes a dominance of Ca^{2+} on the adsorption complex, but electrolyte concentrations are usually still low (except in the presence of gypsum).
* pH values above 8.5 are usually caused by the presence of very soluble bicarbonates of monovalent ions (Na^+, K^+), and the concentration of Ca^{2+} is depressed by that of HCO_3^-, which further increases the fraction of the adsorption complex occupied by monovalent cations. This causes a combination of high electrolyte concentration with a high saturation of monovalent ions.

Question 3.23. *Do you expect clay to disperse or to flocculate in each of the following situations:*

a) saline soils due to NaCl;
b) neutral soils with high base saturation;
c) neutral soils with low base saturation;
d) calcareous soils ($CaCO_3$ has a fairly low solubility, so the concentration of the soil solution remains low);
e) non-saline soils with Na^+-dominated clays?

IRON MINERALS

Together with clay minerals, oxyhydrates of iron and aluminium are the most common secondary minerals. Because iron minerals are very strong colouring agents, their presence is usually obvious. Red colours caused by iron oxides may already become apparent at Fe contents of 0.2%. The most common aluminium-hydroxide, gibbsite ($Al(OH)_3$), is white and usually very fine-grained, and therefore not visible to the naked eye.

Five major iron (hydr)oxides are responsible for the yellowish to reddish colours common in soils: goethite, hematite, ferrihydrite, lepidocrocite, and maghemite (Cornell & Schwertmann, 1996).

Goethite (α-FeOOH) is common in soils of all climates and on virtually all parent materials. Goethite has a yellowish brown colour when finely crystalline, but may appear redder when crystals are larger. It forms upon slow hydrolysis of Fe^{+++} ions.

Hematite (α-Fe_2O_3) is abundant in red soils of the tropics and subtropics; it is absent in recent soils of temperate humid climates. Hematite forms a strong red pigment, especially when finely divided through the matrix. When crystals are large, it appears darker: hematite ore has a black appearance. Hematite is frequently found in association with goethite. Its formation is favoured by slightly alkaline pH, as can be found in calcareous soils and saprolites.

Ferrihydrite ($Fe_2O_3.nH_2O$) is a poorly ordered iron hydroxide that is found in fresh precipitates. It is found in drainage ditches and other places where a high concentration of dissolved ferrous iron is exposed to the air. It is frequently associated with organic molecules, which inhibit crystallisation. Consequently, ferrihydrite occurs in lowland peats (bog ore), iron pans, allophanic andosols, and podzol Bs horizons. With time it may recrystallise to one of the other compounds.

Lepidocrocite (γ-FeOOH) forms from precipitated Fe^{2+} hydroxides. As a result it is restricted to hydromorphic soils such as gleys and pseudogleys. It does not form, however, in calcareous soils, where it is replaced by goethite. Lepidocrocite has a yellowish brown colour.

Maghemite (γ- Fe_2O_3) is found in highly weathered soils of the tropics and subtropics. It occurs preferentially on mafic igneous rocks and is usually associated with hematite. Its colour is reddish brown and its form is similar to that of hematite.

Iron compounds found in soils are not always in equilibrium with the present environment. They may have originated in a different environment and can be very persistent. An example of persistent hematite is the occurrence of red paleosols in humid temperate climates. Although hematite may have changed to goethite in the upper part, as witnessed by more yellowish colours, the deeper horizons may still have hematite. In paleosols, hematite and goethite may persist side-by-side for millions of years.

Soils in a seasonal climate with periodic water saturation may have both hematite and goethite, each in equilibrium with conditions prevailing during part of the year in part of the soil.

Question 3.24. *Write a reaction equation for the transformation of hematite into goethite. What environmental conditions should favour the transformation? Do you expect the reaction rate to be low or high?*

3.3. Redox processes

Oxidation and reduction processes of iron, manganese and sulphur, and organic compounds influence many soil properties. You can only understand the brown and grey mottles of very wet soils and the peculiar chemistry of acid sulphate soils (Chapter 7.4) with a good grasp of redox chemistry. This section deals with the basic aspects of redox processes in soils.

REDOX INTENSITY: pe and E_H

The intensity of oxidation or reduction can be expressed by **pe**, the negative logarithm of the electron activity (-log [e]; [e] is expressed on a molar scale; compare pH). The more reduced the higher the electron activity, and the lower **pe**. **pe** is related to the redox potential or E_H, which is the potential of an inert (e.g. Pt) electrode measured relative to the standard H electrode. At 25°C, **pe** = 16.9 * E_H, where E_H is expressed in volts (V). The redox potential or the **pe** of a system is a measure of the ratio of oxidised to reduced substances present. The possible range of **pe** values in soils is dictated by the presence of water. Highest values are found when water and oxygen occur together:

$$O_2 + 4H^+ + 4e^- \leftrightarrow 2\,H_2O \qquad (3\text{-}7)$$

If water, oxygen and H^+ are in equilibrium:

$$pe = 1/4 \log K - pH + 1/4 \log P_{O2} \qquad (3\text{-}7a)$$

in which P_{O2} is the partial pressure of O_2 gas (in bar, 1 bar = 100kPa), and log K= 83.0 at 25°C and 1 bar total pressure.

The lowest values occur where water is in equilibrium with hydrogen gas:

$$2H^+ + 2e^- \leftrightarrow H_2 \qquad (3\text{-}8)$$

$$pe = 1/2 \log K - pH - 1/2 \log P_{H2} \qquad (3\text{-}8a)$$

For this reaction, log K= 0 at 25°C and atmospheric pressure.

Question 3.25*. a) Derive equation 3-7a by applying the law of mass action to equation 3-7. b) The concentration of H_2O does not appear in eq. 3-7a. Why? c) The presence of water in eq. 3-8 is implicit via H^+. Make it explicit by rewriting the reaction with only H_2O, e^-, H_2 and OH^-.*

Any environment at the watery surface of the earth is characterised by **pe** and pH values within the stability limits of water between 1 bar total pressure of the gases O_2 and H_2. The stability limits are usually shown by means of a so-called **pe**-pH diagram.

Question 3.26. *Draw a pe-pH diagram that shows the range of **pe** and E_H where water is stable at a total pressure of 1 bar (=100 kPa) atmospheric pressure (i.e., water does not disappear by complete oxidation to O_2 at P_{O2}=1 bar, or by reduction to H_2 at P_{H2} = 1 bar). Plot **pe** on the y-axis from +18 to -10, and pH on x-axis from 2 to 10.*

OXIC AND ANOXIC CONDITIONS

Most soils have a positive pe. They are "oxic", with a concentration of gaseous O_2 close to that in the atmosphere (20% by volume, $P_{O2} = 0.2$ bar). Under oxic conditions, the yellow, brown, red and and black oxides and hydrous oxides of FeIII and of MnIII or MnIV are stable. Anoxic soils lack O_2, and have greyish, bluish and greenish forms of reduced Fe(II) and Mn(II), minerals that are stable at low pe values. This is illustrated for iron compounds in the pe-pH diagram of Fig 3.11.

Question 3.27 *a) What is the meaning of the dashed line through pe=0 in Figure 3.11? b) Sketch the positions of the solid-solution boundaries c, 3, b and 2 in Fig 3.11 for a higher concentration of dissolved Fe; c) What can you say about the solubility of iron in soils under oxic and anoxic conditions? d) Derive equations defining the phase boundaries 3 and 7; e) What does Fig 3.11 predict about the fate of iron of bicycles and cars?*

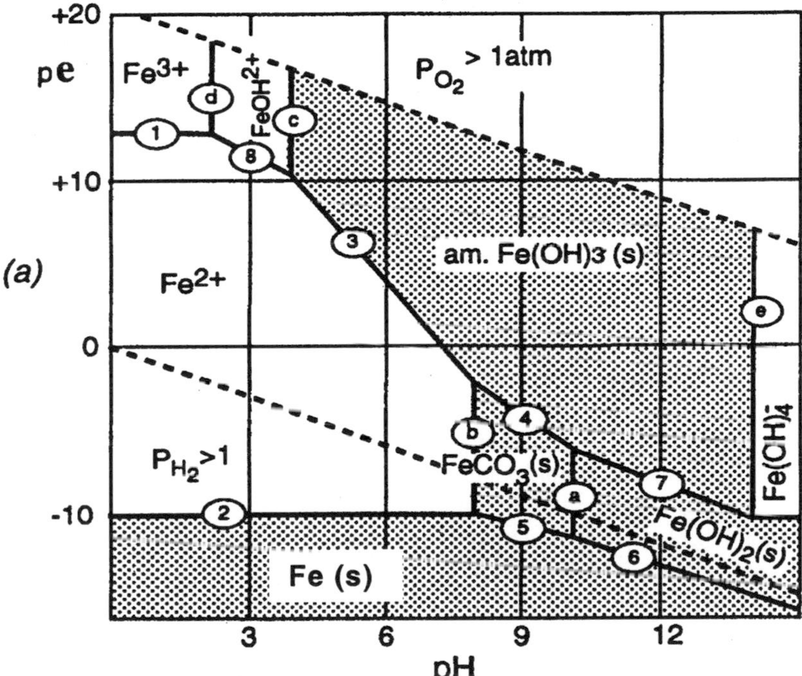

Figure 3.11. pe-pH diagram for the Fe-CO_2-H_2O system at 25°C. The solid phase stability fields are shaded; the non-shaded fields refer to dissolved Fe species at a concentration of 10^{-5} M, and a total dissolved carbonate concentration of 10^{-3} M. $Fe(OH)_3$ stands for ferrihydrite, $FeCO_3$ for siderite, and $Fe(OH)_2$ for a hypothetical Fe(II) hydrous iron oxide. Increasing the concentration of dissolved Fe would expand the solid phase fields parallel to the shown boundaries, and vice versa. The numbers refer to the half-cell redox equilibrium equations used to calculate the phase boundaries. From Stumm and Morgan (1981), reprinted by permission of Wiley and Sons, Inc.

ORGANIC MATTER: THE MOTOR OF SOIL REDUCTION

Organic matter is usually instrumental in changes from oxic to anoxic conditions. By trapping light energy during photosynthesis, plants produce localised centres of strongly reduced conditions (organic matter). Photosynthesis involves reduction of CO_2:

$$CO_2 + 4H^+ + 4e^- \leftrightarrow CH_2O + H_2O \qquad (3\text{-}9)$$

CH_2O, the stoichiometric composition of carbohydrates, stands for living or fresh organic matter.

Equilibrium between CH_2O and CO_2 would be attained if

$$\log K = 4pH + 4\ pe - \log P_{CO2} = 0.8 \qquad (3\text{-}9a)$$

In equation 3-9a, log K = -0.8, so CH_2O can persist only in the complete absence of O_2, very close to the lower stability limit of water, where H_2O will be reduced to H_2. This illustrates how strongly reduced organic matter is.

Question 3.28. *Verify the previous statement by calculating equilibrium P_{O2} in the presence of CH_2O (Combine equation 3-9a with and 3-7a, for a typical value of P_{CO2}, e.g. 0.01 bar).*

During photosynthesis, the electrons are derived from oxidising water to O_2 (write reaction 3-7 from right to left). Combining the two so-called 'half-cell' reactions by balancing for e^-, gives the net overall reaction representative of photosynthesis:

$$CO_2 + H_2O\ (+\ \text{light energy}) \rightarrow CH_2O + O_2 \qquad (3\text{-}10)$$

Question 3.29. *Derive equation 3.10 by combining equations 3-7 and 3-9.*

Equation 3-10 shows how photosynthesis creates a non-equilibrium situation by transforming carbon dioxide and water into a reduced (low **pe**) phase of organic matter, and an oxidised (high **pe**) phase of (atmospheric) oxygen.

Non-photosynthetic organisms tend to restore equilibrium by decomposing (=oxidising) dead organic matter. The bulk of dead organic matter in terrestrial ecosystems is decomposed in the soil during respiration of heterotrophic soil organisms, as will be discussed in Chapter 4. Any free O_2 present is used preferentially as the electron acceptor during respiration (which is the reverse of photosynthesis):

$$CH_2O + O_2 \rightarrow CO_2 + H_2O + \text{thermal or chemical energy} \qquad (3\text{-}11)$$

Note that respiration is the reverse of photosynthesis. If free oxygen is absent, organic matter can be decomposed by anaerobic micro-organisms, using nitrate, Mn(IV), Mn(III), Fe(III) and S(VI) as electron acceptors, in stead of O_2.

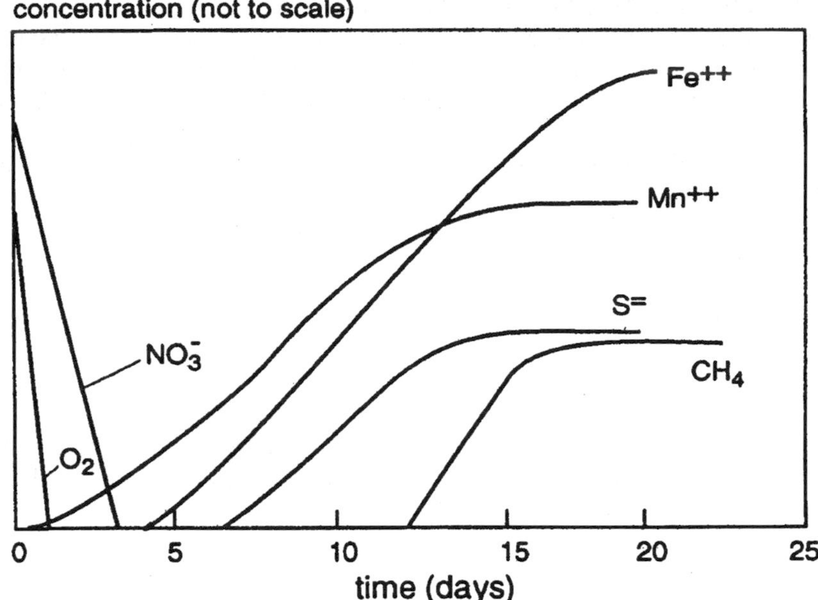

Figure 3.12. The disappearance of O_2 and NO_3^- and the sequential appearance of reduction products during soil reduction. After Patrick and Reddy, 1978.

Moist aerated soils with sufficient decomposable organic matter, placed in a hermetically closed box, will rapidly become anoxic. After disappearance of O_2, any remaining organic matter is oxidised, first by nitrate, followed by Mn(IV), Mn(III), Fe(III) and S(VI), in that order. In that process, MnO_2 will be largely reduced to soluble Mn^{2+}, FeOOH or $Fe(OH)_3$ to a mixture of dissolved Fe^{2+} and solid form of Fe(II) (see later), and SO_4^{2-} to soluble and solid sulphide, S^{2-} (see Figure 3.12).

Question 3.30. *Write coupled redox reactions for the reduction of a) NO_3^- to N_2; b) MnO_2 to Mn^{2+}, and c) FeOOH to Fe^{2+}, with CH_2O as reductant. Start with the half cell reactions of NO_3^-, MnO_2, and FeOOH (Ox) to their reduced counterparts (Red)*: Ox + $e^- \rightarrow$ Red. *Balance them with H^+ and H_2O if needed. Next, combine these halfcell reactions with equation 3.9 proceeding from right to left.*

In the course of the disappearance of O_2 and the appearance of progressively reduced compounds, the E_H of most soils falls from around 600 mV (at pH 4-5), to between 0 and -200 mV (at pH 6 - 7).

Question 3.31. *Trace the changes in E_H and pH described above on Figure 3.11 (first transform E_H into pe). Why would the pH increase (consult also your answer to Question 3.30).*

Reduction of, e.g., Fe(III) in soils, can take place only if the following requirements are met: (1) presence of organic matter, (2) absence of O_2, nitrate and easily reducible manganese oxides, and (3) presence of anaerobic micro-organisms and conditions suitable for their growth.

Well-drained soils normally remain oxic because the O_2 consumed during respiration is replenished by diffusion from the enormous pool of O_2 in the atmosphere. If the soil pores are water-filled, however, gas diffusion is very slow: diffusion of gases in water is approximately 10^4 times slower that in the gas phase. So a water-saturated soil with decomposable organic matter quickly becomes anoxic (see Figure 3.12). A few mm away from the soil-air (or soil-aerated water) interface, the soil tends to become anoxic because oxygen is consumed more rapidly by decomposition of organic matter than it is replenished by diffusion.

In most soils, the content of Fe(III) exceeds that of Mn, NO_3^- and SO_4^{2-} by orders of magnitude. So Fe(III) is usually the dominant oxidant for organic matter in waterlogged conditions, and iron plays a dominant role in the chemistry of periodically flooded soils. If conditions for the reduction of Fe(III) iron are favourable - such conditions occur in relatively fertile surface soils with a microbial population adapted to temporary saturation with water -often between 1 and 20%, and sometimes as much as 90% of the free (= not silicate bound) Fe(III) present in the aerated soil can be reduced to Fe(II) within one to three months of water logging.

FORMS OF Fe IN REDUCED SOILS

Usually only a small fraction of this Fe(II) appears in the soil solution as Fe^{2+}; most of it is in exchangeable or solid form. The nature of solid Fe(II) in reduced soils is still not clear. The well-defined Fe(II) minerals *vivianite* ($Fe_3(PO_4)_2(H_2O)_8$) and *siderite* ($FeCO_3$) are rare in periodically reduced surface soils. Part or most of the reduced solid iron may be present as sulphide in strongly reduced soils with sufficient sulphur: usually in the form of amorphous or tetragonal FeS (mackinawite), or as cubic FeS_2 (*pyrite*). None of the solid Fe(II) forms discussed so far appears to be a quantitatively important constituent of the Fe(II) formed quickly upon soil reduction. The solid Fe(II) typical of reduced soils may be the cause of the greyish to greenish colours that are commonly observed. It can be oxidised by atmospheric oxygen as quickly as it can form in its absence. Mixed Fe(II)-Fe(III) hydroxides that can readily be produced in vitro by partial oxidation of Fe(II) hydroxy salts (carbonates, sulphates, chlorides) have similar properties and may be an important constituent of solid Fe(II) in reduced soils. This so-called *green rust* has the general composition $FeII_6FeIII_2(OH)_{18}$, where Al may substitute for Fe(III), and Cl^-, SO_4^{2-}, and CO_3^{2-} may partially replace OH. So far this substance has rarely been demonstrated in reduced soils (Refait et al., 2001), which may be due to its susceptibility to oxidation and a combination of low concentration and very small particle size.

Figure 3.11 shows where different forms of Fe are stable as a function of pe and pH. Metallic Fe is stable only under very reduced conditions, i.e. outside the stability field of water. $FeCO_3$ and $Fe(OH)_2$ are stable under reduced conditions in soil, but have a high solubility at relatively low pH (<7), as illustrated by the great extension of the dissolved iron (Fe^{2+}) field at low pe. In the presence of sulphide, FeS may replace the carbonates and hydroxides of Fe(II). Ferrihydrite $Fe(OH)_3$ can be present over a wide pe-pH range, but becomes increasingly soluble under slightly acid conditions as pe decreases. Dissolved Fe(III) is observed only under extremely acid, oxidised conditions.

An initially acidic (pH 4-5) soil usually becomes near neutral within a few weeks to months after submersion. This is because the disappearance of NO_3^- or SO_4^{2-} and the appearance of Mn^{2+} or Fe^{2+} upon soil reduction tend to increase the soil pH. This will be explained for the case of iron.

The general reaction equation for reduction of goethite, FeOOH, is:

$$FeOOH + 1/4\ CH_2O + 2H^+ \rightarrow Fe^{2+} + 1/4\ CO_2 + 7/4\ H_2O \qquad (3.12)$$

This reaction shows that 1/4 mole of 'organic matter' (CH_2O) is oxidised to CO_2, along with the reduction of 1 mole of FeOOH to Fe^{2+}. Note that an amount of H^+ is consumed, equivalent to the amount of Fe^{2+} formed. This explains the increase in pH. Actually little or no free H^+ is present, and the H^+ consumed in reaction 3.12 comes from hydrolysis of exchangeable Al^{3+} ($Al^{3+} + 3H_2O \rightarrow Al(OH)_3 + 3H^+$) or from dissociation of carbonic acid ($CO_2 + H_2O \rightarrow HCO_3^- + H^+$). In the first case, exchangeable $2Al^{3+}$ is replaced by exchangeable $3Fe^{2+}$, in the second case Fe^{2+} remains in the soil solution along with HCO_3^-, unless concentrations exceed equilibrium values for $FeCO_3$.

Question 3.32. *Rewrite equation 3.12 with (a) CO_2 as the proton donor, producing dissolved FeII bicarbonate, and (b) exchangeable Al as the proton donor, producing exchangeable FeII and $Al(OH)_3$.*

Figure 3.11 shows that the stability field for Fe^{2+} becomes increasingly smaller at higher pH values. This implies that phenomena of iron redistribution will be less pronounced in, e.g., calcareous soils.

3.4. Problems

Problem 3.1

Data on the release of Mg^{2+}, Fe^{2+}, and H_4SiO_4 from a FeII-containing pyroxene (bronzite) under oxic and anoxic conditions in the laboratory are shown in Figure 3A. The chemical composition of the bronzite used is $Mg_{1.77}Fe_{0.23}Si_2O_6$. The ground mineral sample had a particle size of 100-200μm, and a surface area of 600 cm^2 per gram.

a. Calculate the weathering rate of bronzite under anoxic conditions ($mol\ m^{-2}\ s^{-1}$) from the increase in Si-concentration during the straight- line interval (e.g. 100-500 hour), read from the graph. Compare your result with those for feldspars and with $CaCO_3$, given in the text.
b. Describe and explain the differences in dissolution behaviour in relation to the presence or absence of O_2.

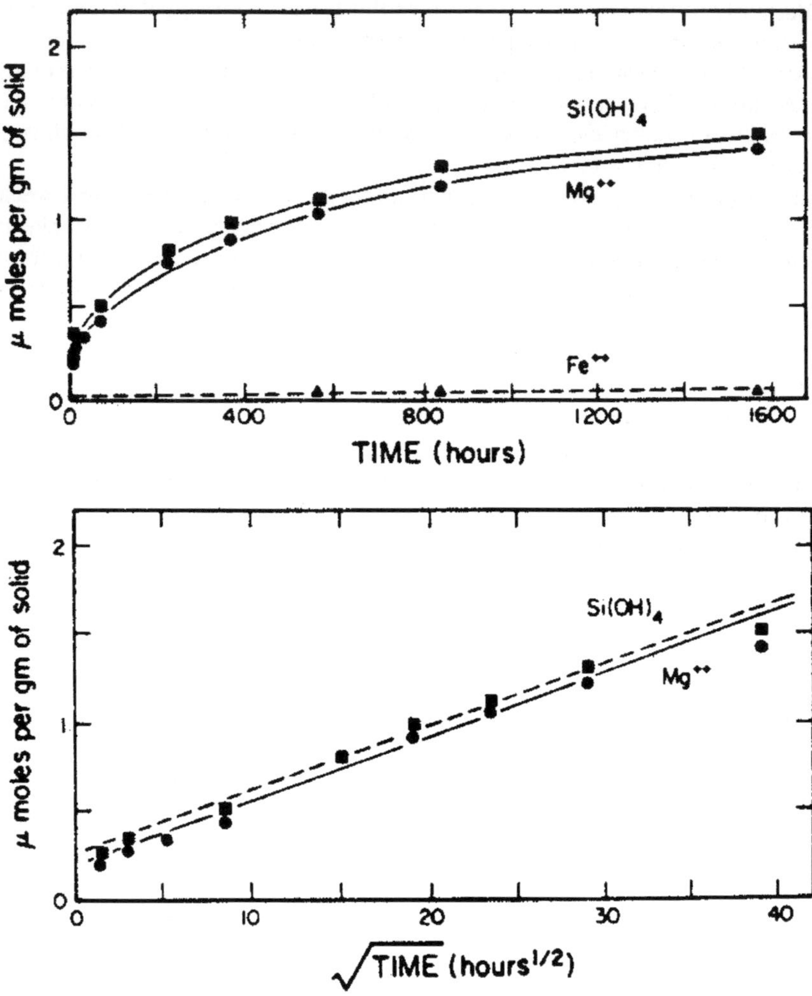

Figure 3A. Mg^{2+}, Fe^{2+} and H_4SiO_4 released as a function of time by the dissolution of bronzite at pH 6 in the absence of O_2 (upper diagram) and at P_{O2} = 0.2 bar. From Schott and Berner (1983), with kind permission of Elsevier Science, Amsterdam.

Problem 3.2

A soil solution with pH 5 is in equilibrium with gibbsite.

a. Using the equilibrium constants from Appendix 2, calculate the concentrations of total dissolved Al_T (= $[Al^{3+}]+[AlOH^{2+}]+[AlL^{2+}]+[AlL_2^{+}]$), in the absence of other compounds, and in the presence of salicylic acid at $[L^-]= 10^{-12}$ mol/l. Assume activity coefficients are equal to 1.
b. What is the concentration of total salicylate needed to maintain the given level of L^-?
c. Salicylic acid can be considered as a model for simple phenolic acids that play a role in e.g. podzolisation. How much could the rate of leaching of Al from a surface soil increase if salicylate were always present in that concentration, other conditions being the same?

Problem 3.3

Figures 3B and 3C give the relation between clay minerals and climate for the topsoils of (3B) acid igneous rocks, and (3C) basic igneous rocks in California. Explain the variation in mineralogy with climate. First consider which clay mineral precursors may have been present in the parent rock, and which minerals could have been formed from primary phyllosilicates and by precipitation from solution.

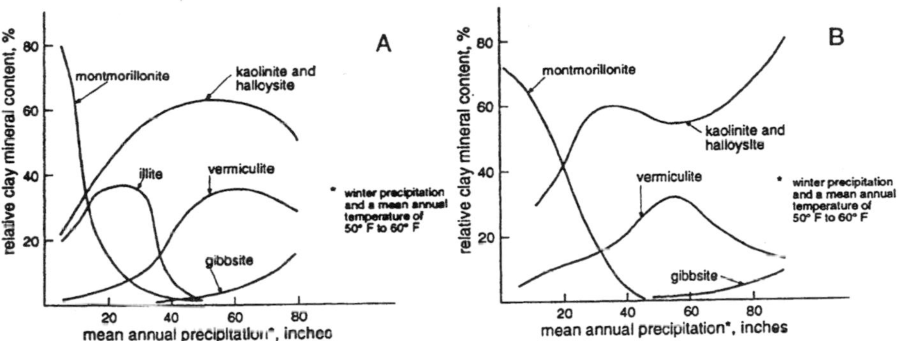

Figure 3BC. Relationship between rainfall and clay mineralogy of surface soils on acid (A) and basic (B) igneous rocks in California. From Singer (1979), reprinted with permission of Elsevier Science, Amsterdam.

Problem 3.4

Table 3D lists rates of clay formation for various parent materials and climates.

a. What is, for each of the cases, the percentage of yearly clay formation with respect to the total soil mass of 1 ha to a depth of 1 metre? Use bulk density = 1 $kg.dm^{-3}$.
b. What are possible causes for the differences between a) loess and pumice in Germany, b) andesite in St Vincent and pumice in Germany, c) sand dunes in Michigan and New Zealand?

Table 3D. Rate of clay formation for various parent materials and climates (various sources).

	Parent material	Mean Temp °C	Mean rainfall mm	clay formation kg clay/ha/yr to 1 m depth
Germany	pumice	8.0	750	55
S. Germany	loess	8.0	600	83
Michigan	sand dunes	6.2	?	17
New Zealand	sand dunes	12.2	?	73
St. Vincent	andesite	27	2300	140

Problem 3.5

Table 3E shows analytical data on soil material from two rice fields (with and without fresh straw matter added) at different times after submergence.

a. Briefly discuss the data in the light of the text of section 3.3

b. What could be the reason for the development of higher concentrations of both dissolved and exchangeable iron in the more acidic soil? Assume that both soils have about the same CEC.

Table 3E. Exchangeable and dissolved Fe in soil material from two rice fields, with and without rice straw added, at various times after the onset of reduction ("weeks submerged"). (IRRI, 1963).

	without rice straw added		with 5% rice straw added	
weeks submerged	exchangeable ½Fe^{2+}, mmol kg^{-1}	dissolved ½Fe^{2+}, mmol l^{-1}	exchangeable ½Fe^{2+}, mmol kg^{-1}	dissolved ½Fe^{2+}, mmol l^{-1}
soil low in exchangeable bases, initial pH 4.8; 2.7% org. matter				
0	1.5	0.02	3	0.02
2	162	3.0	163	5.4
6	170	1.2	180	3.4
12	159	0.46	190	1.6
soil high in exchangeable bases, initial pH 6.6; 2% org. matter				
0	0.2	0.0	1	0.04
2	3.0	0.0	56	0.68
6	2.0	0.02	67	0.76
12	2.7	0.02	67	0.16

Problem 3.6

A soil of pH 6 and bulk density 1.2 contains 10% (mass fraction) albite, present as 200 μm sized cubes. The soil column is always strongly under-saturated with the feldspar.

a. How many years could it take to remove 1% of the albite present in a soil column of 1 m^2 surface and 1 m depth? Use Figure 3.3 and Appendix 2 for additional data.

b. Compare the albite dissolution rate with that of calcite in problem 3.7. Why is the rate found for albite not realistic?

Problem 3.7

Calcite is removed from a soil that has 10% mass fraction calcite, a bulk density of 1.2 Mg/m^3, a constant temperature (20 °C), a CO_2 partial pressure of 1 kPa, an annual rainfall of 1000 mm, and an annual evapo-transpiration of 450 mm. Obtain the solubility of calcite from Figure 3.5.

a. How many years would it take to remove 1% of all calcite present in the top metre of the soil?
b. Compare your answer with that of Problem 3.6 and explain the difference.
c. What would be the calcite dissolution rate in $moles.cm^{-2}.s^{-1}$ if the surface are of the calcite is 10 cm^2/g?

3.5. Answers

Question 3.1

Water 1 is in equilibrium with both the Na feldspar (albite) and with Na smectite, and saturated with amorphous silica. Equilibrium with the primary mineral and, at the same time, with two secondary phases (Na-smectite and amorphous SiO_2), indicates that plenty of time was available to reach saturation with those minerals ("stagnant water"). Water movement in fissures in granite is very slow. The other samples become increasingly dilute from sample 2 to 4 due to increasing rainfall over evapo-transpiration. As a result, under-saturation with albite increases and increasingly base-poor and Si-poor secondary phases can form: smectite (sample 2), kaolinite (sample 3) and gibbsite (sample 4). Sample 4 is even under-saturated with respect to quartz, indicating that this highly resistant mineral can dissolve here.

Question 3.2

a. $Mg_2SiO_4(s) + 4H_2O + 4CO_2(g,aq) \rightarrow 2Mg^{2+}(aq) + 4HCO_3^-(aq) + H_4SiO_4(aq)$

b. $Mg_2SiO_4(s) + 4H^+ + 2\ SO_4^{2-} \rightarrow 2Mg^{2+}(aq) + 2SO_4^{2-}(aq) + H_4SiO_4(aq)$

Question 3.3

$Mg_2SiO_4(s) + 2H_2O + 2CO_2(g,aq) \rightarrow 2MgCO_3(s) + H_4SiO_4(aq)$

Question 3.4

$2KAlSi_3O_8 + 11H_2O + 2CO_2 \rightarrow Al_2Si_2O_5(OH)_4 + 2K^+ + 2HCO_3^- + 4H_4SiO_4$

The reaction equation can be derived as follows:

$KalSi_3O_8 \rightarrow Al_2Si_2O_5(OH)_4$ (start with reactant and product)

$2KAlSi_3O8 \rightarrow Al_2Si_2O_5(OH)_4$ (balance for element conserved in solid phase: Al)

$2KAlSi_3O_8 \rightarrow Al_2Si_2O_5(OH)_4 + 2K^+ + 4H_4SiO_4$ (balance for elements that should appear in solution)

$2KAlSi_3O_8 + 2H^+ \rightarrow Al_2Si_2O_5(OH)_4 + 2K^+ + 4H_4SiO_4$ (balance for charge using H^+)

$2KAlSi_3O_8 + 2H^+ + 9H_2O \rightarrow Al_2Si_2O_5(OH)_4 + 2K^+ + 4H_4SiO_4$ (balance for H and O using H_2O)

$2KAlSi_3O_8 + 11H_2O + 2CO_2 \rightarrow Al_2Si_2O_5(OH)_4 + 2K^+ + 2HCO_3^- + 4H_4SiO_4$

(Replace H^+ by proton donor, $H^+ = CO_2 + H_2O - HCO_3^-$)

Question 3.5

a. $Fe_2SiO_4(s) + 3H_2O + 0.5\ O_2(g,aq) \rightarrow 2FeOOH\ (s) + H_4SiO_4(aq)$

b. $Fe_2SiO_4(s) + 4H_2O + 4CO_2(g,aq) \rightarrow 2Fe^{2+}(aq) + 4HCO_3^-(aq) + H_4SiO_4(aq)$

Question 3.6

a. The primary minerals combine widely different, but not too low (equilibrium) solubilities with moderate to very slow rates of dissolution, giving a weatherability that varies from rather high (forsterite) to very low (quartz, mica). The solubility of secondary clay minerals and gibbsite at normal soil pH values (4.5-7) is orders of magnitude lower that those of the primary minerals, making them much less weatherable (this is of course why clay minerals stay behind in soils, while most primary minerals slowly disappear). Most sulphates, chlorides and carbonates are far more soluble than the silicates; therefore, these secondary minerals can accumulate only in soils of very dry areas.

b. $CaSO_4.2H_2O \rightarrow Ca^{2+} + SO_4^{2-} + 2\ H_2O$: no H^+ appears in the reaction equation because sulphuric acid is fully dissociated.
For the forsterite equilibrium, the law of mass action (see reaction 3-1) gives:

$$[Mg^{2+}]^2[H_4SiO_4]/\ [H^+\ ]^4 = K,$$

so, at constant value of the H_4SiO_4 concentration, the concentration of Mg^{2+} would increase by a factor of 100 per unit pH decrease. The same holds for $[Ca^{2+}]$ in equilibrium with calcite at constant pCO_2: $CaCO_3 + 2H^+ \rightarrow Ca^{2+} + CO_2 + H_2O$. For gibbsite: see equation 3-6 on page 50.

c. The concentration of Al^{3+} increases by $(10^3)^2 = 10^6$ if pH decreases from 5 to 3!

Question 3.7

By removal of solutes released during weathering, the solution remains under-saturated with the mineral in question, so that dissolution can continue.

Question 3.8

At high rates of water percolation the Fe^{2+} released from fayalite could be transported over some distance away from the mineral before it precipitates, so that no protective coating is formed on the mineral and weathering is not hampered mechanically by such a coating. In anoxic conditions no coating will form because no Fe(III) oxide is formed.

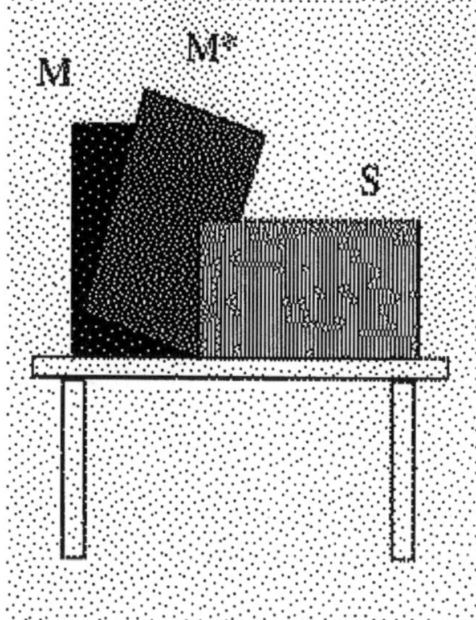

Question 3.9

a. See figure on right.

b. By making the block taller

Question 3.10

The statement holds true when moving from a dry to a somewhat wetter climate, because initially, increasing dilution increases the weathering rate according to equation 3-3 (C_s is lower). At high degrees of under-saturation, further dilution (="a wetter climate") makes no difference because the rate of dissolution depends on the rate of formation of the activated surface complex, which is independent of the degree of under-saturation.

Question 3.11

Because etch-pitting would probably take place at the high degrees of under-saturation, you may expect in a rain forest (high rainfall); micropores would form at low rates of leaching in narrow cracks in the bedrock.

Question 3.12

a. (i) Annual rainfall minus evapotranspiration, (ii) CO_2 concentration in soil atmosphere, (iii) temperature, (iv) pH.
b. The soil solution rapidly equilibrates with calcite, so that the concentration of dissolved $CaCO_3$ is independent of the *amount* of solid $CaCO_3$ present. In case of feldspar weathering, the solution is normally under-saturated with the mineral in question, so that an increasing content (exposed surface) of feldspar in the soil implies that the rate of feldspar weathering (expressed per m^2 of *soil* surface) will increase.

Question 3.13

The brown colour of the decalcified material is probably due to iron liberated from Fe(II)-containing silicate minerals which start weathering faster as soon as $CaCO_3$ is removed, so that soil pH can drop below 7.

Question 3.14

The main differences are concentration (or partial pressure) of O_2, CO_2, and H_2O.

Question 3.15

Calcite forms through combination of Ca^{2+} from silicate weathering with CO_2 from the air. Other salts are combinations of cations with condensed (e.g. volcanic) gases (HCl, SO_2), oxidation products of sulphides, and anions released from complex primary minerals.

Question 3.16

There are two main groups of minerals that may not be stable in the soil environment: minerals that are not hydrated (the variable H_2O component) and minerals of which the stability strongly depends on a cation that is not common in the soil solution (Mg-containing minerals).

Question 3.17

The trioctahedral minerals all have a total of three octahedral cations per unit formula; the dioctahedral minerals have two.

Question 3.18

The charge balance for muscovite is:

positive charges: 2 x Al (3+)+ 3 x Si (4+) + 1 x Al (3+) = 21 +

negative charges: 10 x O (2-) + 2 x OH (1-) = 22 -

The excess negative charge is compensated by 1 K (+) as interlayer cation.

Question 3.19

1:1 Phyllosilicates have one silica tetrahedral layer and one Al/Mg octahedral layer; 2:1 phyllosilicates have two tetrahedral layers. As a result, 2:1 clays have two similar plate sides, while plate sides have different characteristics in 1:1 clays. Chlorite (2:1:1) is usually set apart from the 2:1 clay minerals, because the interlayer is crystalline.

Question 3.20

a. The source of solutes is the weathering of primary minerals, either locally or at some distance (provided by groundwater flow).
b. In case of strong percolation, the necessary increase in concentration of solutes does not happen.

Question 3.21

Removal of interlayer cations increases the surface of contact between the soil solution and the charged mineral surface and it removes charge-compensating interlayer ions.. Both increase the CEC. Chloritisation causes closure of interlayer spaces and has therefore the opposite effect. For the location of the interlayer space, see the water molecules and the second octahedron in Figure 3.7.

Question 3.22

If clays have a low charge deficit, interlayer cations are held with relatively weak force. This means that the plates can be separated fairly easily and water can enter the interlayer space.

Question 3.23

At high electrolyte concentrations in the soil solution, as in saline soils, clays tend to remain flocculated because the electrical double layer is compressed, regardless of the nature of exchangeable cations. At low electrolyte concentrations the double layer will remain compressed (and the clay flocculated) if the exchange complex is dominated by divalent or trivalent cations (Ca^{2+}, Mg^{2+}, Al^{3+}). The double layer will expand, however, and the clay will disperse, if the percentage of exchangeable Na^+ is high (15-20% of the CEC). So: **a, b, d**: flocculation; **c, e**: dispersion.

The relations between pH, exchangeable Na^+, and clay stability are depicted in Figure 8.1.

Question 3.24

$Fe_2O_3 + H_2O \rightarrow 2FeOOH$

The equation contains two solid phases and water, and the reaction can proceed only if water is present. The fact that many soils contain hematite although goethite is stable over hematite + water, indicates that the reaction is very slow.

Question 3.25

a. $2H_2O \leftrightarrow O_2 + 4H^+ + 4e^-$

b. Applying the law of mass action gives $K=[O_2] * [H^+]^4 * [e^-]^4 / [H_2O]^2$, so

$\log K = \log [O_2] - 4pH - 4pe - 2 \log [H_2O]$.

Because $[H_2O]$ is unity (holds true for all dilute solvents and pure solids), and $[O_2]$ is expressed as partial pressure, this is equivalent to:

$pe = -1/4 \log K - pH + 1/4 \log P_{O2}$,

c. $2H^+ + 2e^- \rightarrow H_2$

$2H_2O^- \rightarrow 2H^+ + 2OH^-$

----------------------------------+

$2e^- + 2H_2O \rightarrow H_2 + 2OH^-$

Question 3.26

pe-pH conditions where water is stable against oxidation to O_2 (at P_{O2} of 1 bar (= 100 kPa) and reduction to H_2 (at P_{H2} of 1 bar).

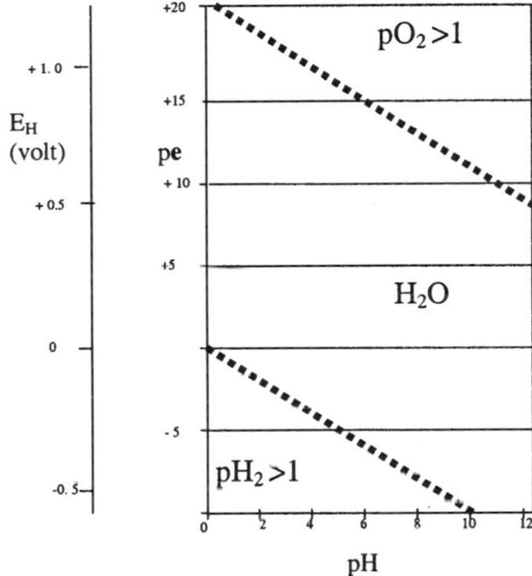

Question 3.27

a. The dashed line indicates the lower stability limit of water (compare answer to Question 3.26). This means that all phases that are only stable below this line will not be stable at the earth's surface.

b. The Fe^{2+} field shrinks at the expense of the grey (solid) fields.

c. Iron is appreciably soluble only at very low pH in oxic conditions and at much higher pH in anoxic conditions

d. Phase boundary 3: the stoichiometry of reduction of $Fe(OH)_3$ can be derived as follows:

$Fe(OH)_3 + 3H^+ \rightarrow Fe^{3+} + 3H_2O$ (dissolve $Fe(OH)_3$ to Fe^{3+})
$Fe^{3+} + e^- \rightarrow Fe^{2+}$ (reduce Fe^{3+} to Fe^{2+})
$Fe(OH)_3 + 3H^+ + e^- \rightarrow Fe^{2+} + 3H_2O$ (sum up and balance for Fe^{3+})

Applying the law of mass action and expressing it logarithmically yields:

$\log K = \log[Fe^{2+}] + 3pH + pe$

Similarly for phase boundary 7:

$Fe(OH)_3 + H^+ + e^- \leftrightarrow Fe(OH)_2 + H_2O$, and $\log K = pH + pe$

e. Bicycles and cars will rust and disappear whatever you do against it.

Question 3.28

pe = $1/4 \log K_{(7\text{-}a)} - pH + 1/4 \log P_{O2} = 1/4 \log K_{(9\text{-}a)} - pH + 1/4 \log P_{CO2}$,
Substituting values for log K and P_{CO2} gives $\log P_{O2} = -85.8$, so $P_{O2} = 10^{-85.8}$. Quite a bit lower than Avogadro's number....

Question 3.29

$2 H_2O \rightarrow O_2 + 4H^+ + 4e^-$
$CO_2 + 4H^+ + 4e^- \rightarrow CH_2O + H_2O +$
$CO_2 + H_2O \rightarrow CH_2O + O_2$

Question 3.30

a. $4 NO_3^- + 5 CH_2O + 4 H^+ \rightarrow 5CO_2 + 2 N_2 + 7 H_2O$
b. $MnO_2 + 1/2 CH_2O + 2 H^+ \rightarrow Mn^{2+} + 1/2 CO_2 + 3/2 H_2O$
c. $FeOOH + 1/4 CH_2O + 2 H^+ \rightarrow Fe^{2+} + 1/4 CO_2 + 7/4 H_2O$

Question 3.31

The values of Eh and pH roughly descend along line 3, but slightly steeper because the concentration of Fe^{2+} tends to increase in the course of soil reduction. The increase in pH is due mainly to the consumption of H^+ during reduction of FeOOH (Q. 3.30, c).

Question 3.32

a. Reaction 3.12 has H^+ as proton donor:

$FeOOH + 1/4 CH_2O + 2 H^+ \rightarrow Fe^{2+} + 1/4 CO_2 + 7/4 H_2O$ (i)

If CO_2 is the proton donor:

$CO_2 + H_2O \rightarrow H^+ + HCO_3^-$ or $2 CO_2 + 2 H_2O \rightarrow 2H^+ + 2 HCO_3^-$ (ii)

Combining equations (i) and (ii) gives:

$FeOOH + 1/4 CH_2O + 7/4 CO_2 + 1/4 H_2O \rightarrow 2 HCO_3^- + Fe^{2+}$

b. If exchangeable Al^{3+} is the proton donor:

$2/3 Al^{3+}_{(exch)} + 2 H_2O \rightarrow 2/3 Al(OH)_3 + 2 H^+$ (iii)

Combining eqs (i) and (iii) gives:

$FeOOH + 1/4CH_2O + 2/3Al^{3+}_{(exch)} + 1/4H_2O \rightarrow 2/3Al(OH)_3 + Fe^{2+}_{(exch)} + 1/4CO_2$

This is a "forced" exchange: reduced Fe *must* take the place vacated by Al^{3+} at the exchange complex, because there is no anion that can compensate the charge of dissolved Fe^{2+}, and no cation that can replace Al at the adsorption complex.

Problem 3.1

a. After an initial parabolic release, Si, Mg and Fe are released linearly with time in the absence of O_2. The initial high dissolution rate can be attributed to dissolution of very fine dust from grinding. From the graph, it can be read that between 100 and 500 h, 2 μmol of Si are released per gram of solid. 1 gram of solid represents 600 cm^2, or 0.06 m^2. So: Dissolution rate is $2*10^{-6}$mol/ 0.06*400*60*60 m^2.s = $2.31*10^{-11}$ mol/m^2s.

b. In the presence of O_2, Fe release is virtually zero and the release of Mg and Si decreases with time. These observations can be attributed to the formation of a Fe(III) (hydrous)oxide precipitate on the mineral surface in oxygenated conditions. The precipitate acts as a barrier causing diffusion-controlled weathering.

Problem 3.2

a. At pH 5, the Al^{3+} concentration in equilibrium with gibbsite follows from the gibbsite equilibrium equation:

$(Al^{3+})(OH^-)^3 = 10^{-34.2}$ $\qquad$ $\log Al^{3+} + 3 \log (OH^-) = -34.2$

substitute $\log (OH^-)$ by (-14 + pH). At pH = 5: $\log Al^{3+} = -7.2$

Also: $\log Al(OH)^{2+} - pH - \log Al^{3+} = -5.02$

So, at pH 5, $\qquad$ $\log Al(OH)^{2+} = -7.2$

Thus, total dissolved Al in the absence of salicylate is $2* 10^{-7.2} = 1.26 * 10^{-7}$

In the presence of salicylate (10^{-12} M L^-):

$\log AlL^{2+} - \log L^- - \log Al^{3+} = 12.9$ $\qquad$ $\log AlL^{2+} = -6.3$

$\log AlL_2^+ - 2 \log L^- - \log Al^{3+} = 23.2$ $\qquad$ $\log AlL_2^+ = -8.2$

So in the presence of salicylate, the total amount of dissolved Al is:

$1.26 * 10^{-7} + 5.01 * 10^{-7} + 6.31 * 10^{-9} = 6{,}34 * 10^{-7}$

b. The total amount of salicylate = (L) + (HL) + (AlL) + 2* (AlL_2)

at pH = 5, and $(L^-) = 10^{-12}$,

$(HL) = 10^{13} * 10^{-5} * 10^{-12} = 10^{-4}$

so $(L)_T = 10^{-12} + 10^{-4} + 10^{-6.3} + 2* 10^{-8.2} = 1.01 * 10^{-4}$

c. If the mineral weathering that produces dissolved Al (the rate-limiting step) is not taken into account:

In the presence of salicylate, the total concentration of Al is about 5 times higher. So, other conditions being similar, the leaching of Al during podzolisation would proceed five times as fast.

Problem 3.3

The relationships are as follows: On acid igneous rocks, the main primary minerals are quartz, potassium feldspars and muscovite. Muscovite gives illite upon weathering, and vermiculite if weathering proceeds further. At low precipitation, all weathering products remain in the soil and recombine to form montmorillonite. When leaching increases, montmorillonite disappears from the topsoil, leading to a relative increase in illite. At the same time, kaolinite and halloysite start to form (relatively low H_4SiO_4 activity). At very high rainfall/percolation, vermiculite and kaolinite become unstable (very low H_4SiO_4) and gibbsite forms at the expense of/instead of these two minerals.

On basic igneous rocks, the main primary minerals are plagioclases, biotite, and amphiboles or pyroxenes. As in acid rocks, smectites are formed at low rainfall. When percolation increases (and pH goes down), the smectites become Al-interlayered and show vermiculite properties. Kaolinite contents increase with increasing leaching and eventually forms at the expense of vermiculite. At very high rainfall and very low H_4SiO_4 concentrations, gibbsite appears and vermiculite (a 2:1 mineral) dissolves.

Problem 3.4

a. The total volume of soil of 1 ha to 1 m depth is $10^4 * 10^3$ dm^3. This is equivalent to 10^7 kg. So the clay formation is 1-15 ppm per year.

b. The difference between loess and pumice in Germany is probably due to the much higher content in weatherable minerals in the pumice. The difference between andesite in St. Vincent and pumice in Germany is mainly a precipitation and temperature effect. The difference between rates of clay formation in sand dunes in Michigan and New Zealand is probably caused by a difference in weatherable minerals and temperature.

Problem 3.5

a. We can distinguish the effect of several factors on the contents of exchangeable and dissolved Fe^{2+} in the soil after submergence: i) time, ii) addition of rice straw and iii) content of exchangeable bases and (initial) soil pH.
i) With time after flooding, the contents of Fe^{2+} increase. Exchangeable Fe^{2+} increases to more or less constant levels, while the concentration of dissolved Fe^{2+} first increases, and than decreases. The increase in Fe^{2+} is clearly due to reduction of Fe(III) oxides by organic substances (under the influence of anaerobic bacteria), in the absence of air. The decrease in dissolved Fe^{2+} following peak concentration after a few weeks of submergence (not discussed in Chapter 3.3) is not well understood, but may be due to precipitation of Fe(II) phases (sulphides, hydroxy carbonates) and their equilibration with the soil solution.
ii) Adding straw strongly stimulates formation of Fe^{2+}, because the fresh, undecomposed organic matter provided with the straw is energy-rich and easily decomposable, forming a strong reductant for Fe(III). The effect of adding straw is much greater in the high-pH soil than in the acid soil. It is not clear to what extent this is caused by the difference in pH or by a higher original content of decomposable organic matter in the acid soil.

b. The much higher levels of dissolved and exchangeable Fe^{2+} in the more acid soil must be attributed to the higher (equilibrium) concentrations of Fe^{2+} in contact with solid Fe(III) and Fe(II). An important extra reason for the high contents of exchangeable Fe^{2+} in the acid soil is the forced exchange (see question 3.31).

Problem 3.6

a. (*Additional data from Appendix 1: 1 mol albite = 262 g; 1m^3 albite weighs 2620 kg*)
1 m^3 of the soil weight 1200 kg. This contains 0.1*1200 kg albite = 120 kg albite, or 120/2620 = 0.0458 m^3 of albite. 1 albite sand grain has a volume of $(200*10^6)^3$ $m^3 = 8*10^{-12}$ m^3, so there are 0.0458 $m^3/8*10^{-12}$ m^3/grain = $5.725*10^9$ albite grains per m^3 of soil. One grain has a surface are of 6 * 0.02^2 cm^2 = 0.0024 cm^2, so the total albite surface are per m^3 of soil is $5.725*10^9$ *0.0024 cm^2 = 1.374 * 10^7 cm^2. At a weathering rate of 10^{-16} mol/cm^2.sec, the 1.374 * 10^7 * 10^{-16} = 1.374 * 10^{-9} mole of albite are removed per second from each m^3 of soil, or 0.0433 mol/year. This

amounts to 0.0433* 262 g = 11.35 g of albite. One percent of all albite present in 1 m^3 of soil amounts to 1200 g, which could be removed in 1200g/11.35 g/yr = 106 years.

b. The rate of dissolution is much higher than that observed in the field. This is in part because laboratory experiments work with fresh and therefore more reactive feldspars, and with relatively diluted, but more aggressive solutions.

Problem 3.7

Figure 3.5 gives for the solubility under the specified conditions about 200 mg $CaCO_3$/L. The amount of water percolating through 1 m^2 of soil is 1000-450 = 550 mm/ m^2 = 550 litre, so 550 L*200 mg/L= 110 g CaCO3 are leached per year. So it takes 1200/110 = 10,9 years to remove 1 % of the calcite present in the top m of the soil profile.

110 g= 110/100 = 1.1 mol calcite. This has a surface area of 110* 10^4 cm^2. So the weathering rate is 1.1 mol/110* 10^4 cm^2 year = 3.2 * 10^{-14} mol/cm^2sec. This is about 30 times faster than the weathering rate (/surface area) of albite.

3.6. References

Anbeek, C., 1994. Mechanism and kinetics of mineral weathering under acid conditions. PhD Thesis, Wageningen, 209 pp.

Berner, R.A. and G.R. Holdren Jr., 1979. Mechanism of feldspar weathering. II. Observations of feldspars from soils. Geochim. Cosmochim. Acta, 43:1173-1186.

Blum, A.E. and L.L. Stillings 1995 Feldspar dissolution kinetics, p.290-351 in A.F.White and S.L. Brantley, Chemical Weathering rates of silicate minerals. Reviews in mineralogy, Vol 31 Mineralogical Society of America.

Bolt, G.H and M.G.M. Bruggenwert, 1976. *Soil Chemistry A*. Elsevier Scientific Publ. Co, 281 pp.

Brindley, G.W., and G. Brown, 1980. *Crystal structures of clay minerals and their X-ray identification*. Mineralogical Society Monograph No. 5, Mineralogical Society, London. 495 pp.

Cornell, R.M., and U. Schwertmann, 1996. *The Iron Oxides*. VCH Verlagsgesellschaft, Weinheim, 573 pp.

Fanning, D.S., and V.Z. Keramidas, 1977. Micas. In J.B. Dixon and S.B. Weed (eds). *Minerals in Soil Environments*. Soil Science Society of America, Madison. pp. 195-258.

IRRI, 1963. International Rice Research Institute, Annual Report for the year 1963, IRRI, Los Banos, Philippines.

Martell, A.E and R.M. Smith, 1974-1977. *Critical stability constants*. Vol.3. Other organic ligands. Plenum press, 495 pp.

Motomura, S, 1962. Effect of organic matter on the formation of ferrous iron in soils. Soil Science and Plant Nutrition, 8: 20-29.

Nordstrom. D.K. 1982. The effect of sulfate on aluminum concentrations in natural waters: some stability relations in the systems Al_2O_3-SO_3-H_2O at

298K. Geochimica et Cosmochimica Acta, 46:681-692.

Patrick, W.H. and R.D. Delaune, 1972. Characterisation of the oxidized and reduced zones in flooded soil. Soil Science Society of America Proceedings, 36: 573-576.

Patrick, W.H., and C.N. Reddy, 1978. Chemical changes in rice soils. p.361-379 in *Soils and Rice*. International Rice Research Institute, Los Banos, Philippines.

Refait, P., M. Abdelmoula, F. Trolard, J.M.R. Génin, J.J. Ehrhardt and G. Bourrié, 2001. Mössbauer and XAS study of a green rust mineral: the partial substitution of Fe^{2+} by Mg^{2+}. American Mineralogist, 86:731-739.

Scheffer, F., and P. Schachtschabel, 1977. *Lehrbuch der Bodenkunde*. 10. durchgesehene Auflage von P. Schachtschabel. H.P. Blume, K.H. Hartge and U. Schwertmann. Ferdinand Enke Verlag, Stuttgart, 394 pp.

Schott, J. and R.A. Berner, 1983. X-ray photoelectron studies of the mechanism of iron silicate dissolution during weathering. Geochim. Cosmochim. Acta, 47: 2233-2240.

Singer, A., 1979. The paleoclimatic interpretation of clay minerals in soils and weathering profiles. Earth Science Reviews, 15:303-326.

Tessier, D., 1984. *Etude experimentale de l'organisation des materiaux argileux*. Dr Science Thesis. Univ. de Paris INRA, Versailles Publ,. 360 pp.

Van Breemen, N., 1976. *Genesis and soil solution chemistry of acid sulfate soils in Thailand*. PhD thesis, Agricultural Univ, Wageningen, the Netherlands, 263 pp.

Van Breemen, N., and R. Brinkman, 1978. Chemical equilibria and soil formation, p. 141-170 in Bolt and Bruggenwert. op. cit.

Van Breemen, N., and R. Protz, 1988. Rates of calcium carbonate removal from soils. Canadian Journal of Soil Science. 68:449-454.

Van Olphen, H., 1963. *An introduction to clay colloid chemistry*. Interscience Publishers, London, 301 pp.

Velbel, M.A., 1986. Influence of surface area, surface characteristics, and solution composition on feldspar weathering rates. In: *Geochemical processes at mineral surfaces*. J.A. Davis and K.F. Hayes. ACS Symposium series 323:615-634.

White, A.F., 1995. Chemical weathering rates of silicate minerals in soils. p. 407-461 in A.F. White and S.L. Brantley, *Chemical Weathering rates of silicate minerals*. Reviews in mineralogy, Vol 31 Mineralogical Society of America,

Wilding, L.P. and D. Tessier, 1988. Genesis of vertisols: shrink-swell phenomena. p. 55-81 in: L.P. Wilding and R. Puentes (eds). *Vertisols: their distribution, properties, classification and management*. Technical Monograph no 18. Texas A&M University Printing Center, College Station, TX , USA.

General and further reading:

Bolt, G.H., and M.G.M. Bruggenwert, 1976. See above.

Deer, W.A., R.A. Howie, and J. Zussman, 1976. *Rock-forming minerals*, Volume3: Sheet silicates. Longman, London, 270 pp.

Drever, J.L., 1982. *The geochemistry of natural waters*. Prentice Hall, New York, 388 pp.

Duchaufour, P., 1982. *Pedology - pedogenesis and classification*. George Allen & Unwin, London, 448 pp.
Garrels, R.M., and C.L. Christ, 1965. *Solutions, minerals, and equilibria*. Harper and Row, New York, 450 pp.
Jenny, H., 1980. *The soil resource*. Springer Verlag, Heidelberg, 377 pp.
Paul, E.A., and F.E. Clark, 1996. *Soil microbiology and biochemistry*, 2d Ed. Academic
Tan, K.H., 1993. *Principles of Soil Chemistry*. Marcel Dekker, New York. 362 pp.

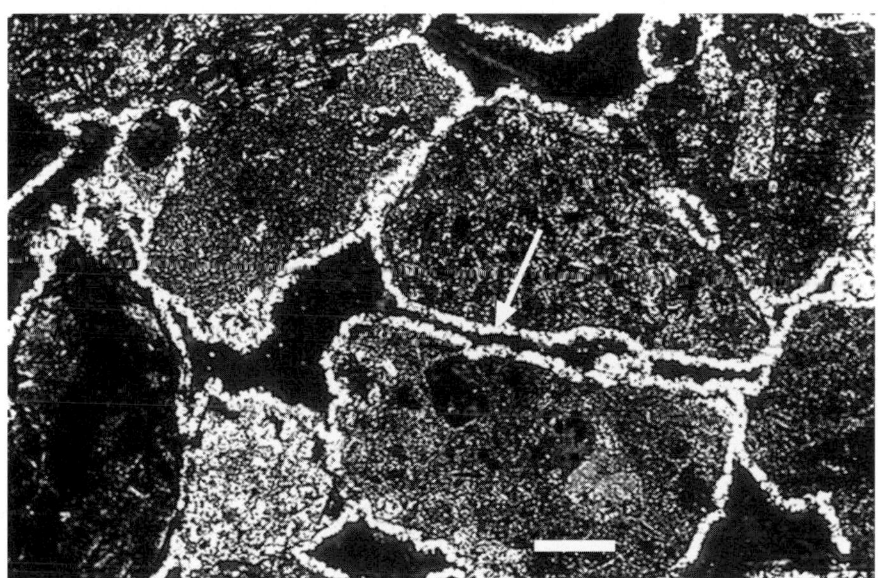

Plate F. Highly birefringent gibbsite coatings (arrow) on sand-size aggregates in an allophanic andosol from Costa Rica. Scale bar is 175 μm. Photograph A.G. Jongmans.

Plate G. Very deep stoneline (arrow) at the bottom of the biomantle in an Oxisol near Lavras, Minas Gerais, Brazil. Note size of person. Underlying layers show sedimentary stratification. Layers above stoneline have horizontal scars due to road works. Photograph P. Buurman.

Plate H. Decaying termitaria in a degraded pasture on an Oxisol. Minas Gerais, Brazil. Distance between mounds is 5-10 metres. Photograph P. Buurman

CHAPTER 4

BIOLOGICAL PROCESSES IN SOILS

This is the last chapter of Part B, the section of this course dealing with basic processes that underlay soil formation. The effects of biological processes on soils can be described in physical or chemical terms (see Chapters 2 and 3), but they are so important and specific that they deserve a separate chapter. After studying this chapter you should be able to (1) specify the soil forming factors that determine the quantity and nature of soil organic matter; (2) apply rate equations that quantify the kinetics of soil organic matter dynamics; (3) describe effects of biota on soil structure, texture, and mineral weathering; and (4) recognise the chemical effects of nutrient cycling on a soil profile.

The motor for biological processes in soils is (a small fraction of the) solar energy harnessed by plants. Plants influence soil directly through the activity of their roots.

Question 4.1. *How do plant roots influence the soil at different time scales (days, years, and millennia)?*

Perhaps more important is the indirect effect of dead plant material on the soil system. Dead plant material serves as an energy source of soil biota, and forms the 'parent material' of all soil organic matter. Both soil biota and soil organic matter have important further effects on soil properties.

Question 4.2. *Give examples of* ***a)*** *important direct effects of biota on soil processes and* ***b)*** *effects of soil organic matter on soil properties.*

4.1. Input of plant litter into soils

ABOVE- AND BELOW-GROUND INPUT OF PLANT LITTER

With our aboveground eyes we tend to underestimate the importance of belowground biotic processes. In some ecosystems, root biomass exceeds aboveground biomass

(Table 4.1). Root biomass is a large proportion of total plant biomass if nutrients or water are scarce, or in climates with very cold or dry seasons.

Question 4.3. *Fill in the empty rows in Table 4.1, assuming that below-ground production is 70 % of above-ground litter fall, and that the system is at steady state Why would the ratio of below-ground to above-ground biomass be high in very cold or dry conditions?*

Table 4.1. Total (above- and belowground) biomass, root biomass, and annual aboveground litter fall in natural ecosystems of major climatic zones from the arctic to the humid tropics. Mass refers to organic matter (about 50% C). The litter layer is equivalent to the organic O-horizon on top of the mineral soil. (After Kononova, 1975; reproduced with permission of Springer Verlag GmbH & Co., Heidelberg).

	tundra		Spruce	oak	dry	semi	subtrop	dry	humid
	arctic	brush wood	Forest	forest	steppes	desert	decid. forest	savan nah	tropical forest
total biomass (10^3 kg/ha)	5	28	260	400	10	4	410	27	500
root biomass (10^3 kg/ha)	4	23	60	96	9	3.5	82	11	90
aboveground litterfall (10^3 kg/ha/yr)	1	2	5	7	4	1	21	7	25
litter layer mass (10^3 kg/ha)	4	83	45	15	2	-	10	-	4
relative root mass (kg/kg)									
belowground annual production (10^3 kg/ha/yr)									

Root biomass, however, is not simply related to annual belowground litter input. In highly productive systems, more carbon is allocated to roots than in low-productive systems, even though root biomass and the fraction of total production allocated to roots may be higher at low production. These points are illustrated in Problem 4.1

In forests, only a small part of the net primary production (NPP) is utilised by grazers and carnivores preying on them and about 95 % of the aboveground NPP eventually enters the soil decomposition food web. In heavily grazed, managed grasslands still more than 75 % of the aboveground NPP eventually reaches the soil and is consumed there. This illustrates the overwhelming importance of soil biological processes in ecosystem dynamics and energy transfers.

4.2. Decomposition of plant litter and formation of soil organic matter

Soil organic matter is generally thought of as all dead organic material present in the soil. It is impossible, however, to separate all living organic matter from a soil sample: fine roots, meso- and micro- soil fauna, bacteria and fungi are closely associated with mineral particles and make up some 10% of the organic fraction. More than 90 % of soil organic matter usually consists of humus: strongly decomposed plant material that has been transformed to dark-coloured, partly aromatic, acidic, hydrophilic, molecularly flexible polyelectrolytic matter. Its properties are described in section 4.3. The present section deals with decomposition of fresh organic matter, derived from plants (roots, stems, leaves) and animals (both faeces and dead macro-, meso- and micro- fauna), and its transformation into humus.

All organic material is thermodynamically unstable in oxic conditions (see section 3.3); it oxidises to water and carbon dioxide:

$$CH_2O \text{ (org. matter)} + O_2 \rightarrow CO_2 + H_2O \qquad (4\text{-}1)$$

This process is the reverse of photosynthesis (eq. 3-10). It can take place abiotically (e.g. during a forest fire), but at normal temperatures it proceeds only under the influence of heterotrophic organisms.

In terms of biomass, fungi and bacteria dominate the decomposer community, but the decomposer chain includes many functional groups and species (Fig. 4.1). During decomposition, part of the carbon stored during photosynthesis is respired, releasing energy; part is converted to cell material of soil organisms, and (a very small) part is transformed to humus compounds which are relatively resistant to further decomposition.

***Question 4.4.** a) Which four groups of organisms in Figure 4.1 live on organic residues (the primary decomposers), and which are predators of the lower order decomposers? **b**) The arrows indicate the flow of useful energy. Do they also indicate the flow of nutrients? What happens ultimately to the nutrients and to the energy present in the plant residues?*

NUTRIENT MINERALISATION AND IMMOBILISATION

Plant litter also contains nutrient elements such as N, P, K, Ca and S. During decomposition, part of these nutrient elements is released in ionic form: e.g., NH_4^+, K^+, Ca^{2+}, SO_4^{2-}, and $H_2PO_4^-$. Such decomposition to inorganic solutes and gases is called mineralisation. A general equation for mineralisation is:

$$C_aN_{(b+k)}P_cS_d...M_gH_{2x}O_x \text{ (fresh org.mat.)} + (a+2b+2d)O_2 \rightarrow$$
$$aCO_2 + bNO_3^- + cH_2PO_4^- + dSO_4^{2-} + gM^+ + kNH_4^+ + xH_2O + (b+c+2d-g-k)H^+ \qquad (4\text{-}2)$$

The remainder is immobilised: it is transformed into humus

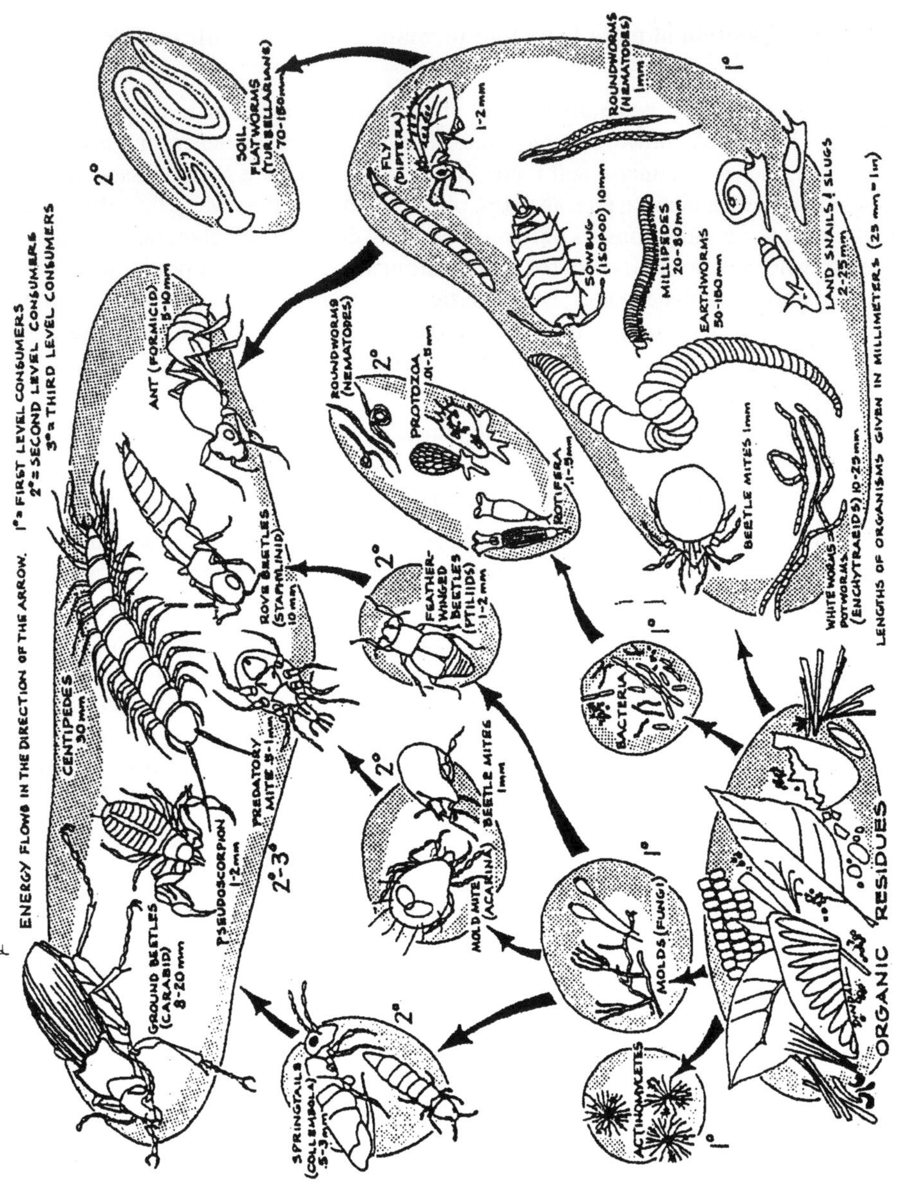

Figure 4.1. The decomposer food web. From Dindal, 1978.

The part transformed into soil microbial biomass is called the assimilation factor, or assimilation efficiency. The C assimilation factor is generally between 0.1 to 0.5. The fraction of C in plant litter that is transformed to humus is generally < 0.05.

Most of the mineralised nutrients are assimilated again by the vegetation. In natural ecosystems, even in high rainfall areas, only a small part of the nutrients released during decomposition of organic matter is removed from the soil by leaching.

Question 4.5. *Equation 4-2 is not only valid for fresh leaf litter, but does realistically portray the overall stoichiometry of mineralization of* ***all*** *soil organic matter under steady state conditions. Explain this statement.*

Question 4.6 *Plant litter with a C/N (mass) ratio of 50 is transformed completely into microbial biomass with a C/N of 10, and mineralised nutrients. The C assimilation factor is 0.2. There is no other N source than organic N.* ***a)*** *What is the N assimilation factor?* ***b)*** *What would happen to N if the C assimilation factor were 0.1?* ***c)*** *What do you expect about the decomposition of the plant litter if the C assimilation factor were 0.5?*

4.3 Kinetics of decomposition and of humus formation

Decomposition rates vary strongly for different materials. Under favourable conditions (moderate to high temperatures, sufficient moisture, oxygen and nutrients), fine roots, deciduous leaves high in nutrients, and fungal and bacterial necromass can be decomposed almost completely within a year. Most coniferous leaves (needles) take a couple of years to decompose. Depending on their size and on the climate, pieces of wood may take decades (branches) to centuries (tree trunks) to disappear.

PHYSICAL AVAILABILITY AND CHEMICAL NATURE OF SUBSTRATES

Differences in decomposition rate relate to the ease by which constituent organic compounds can be attacked by fungi and bacteria. Both the physical availability and the chemical nature play an important role here. Fresh organic matter is decomposed more quickly if it is first made finer by soil fauna, so that it is more easily accessible to fungi and bacteria. Soil fauna such as termites, earthworms, woodlice, mites and springtails, respire only 5 to 10 % of the decomposable organic C, and their contribution to the total biomass of decomposers is even smaller. The main contribution of soil fauna to decomposition is chewing up coarse organic particles (Figures 6.1 and 6.2) and providing favourable conditions for micro-organisms in their gut and faeces. Leaf-cutter ants and many species of termites consume fungi that they supply with fresh plant litter in well-managed fungal gardens. Mites and springtails stimulate fungal activity by grazing hyphen.

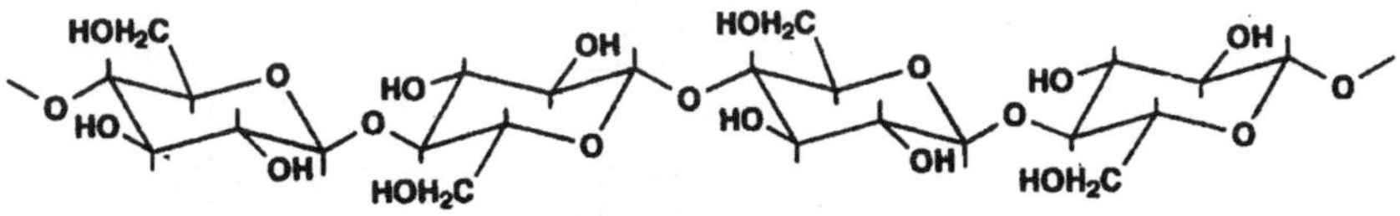

Figure 4.2. Structure of cellulose. From Engbersen and De Groot, 1995.

	Coniferyl alcohol	Sinapyl alcohol	*p*-Coumaryl alcohol
Conifers	+	0	0
Deciduous trees	+	+	0
Graminees	+	+	+

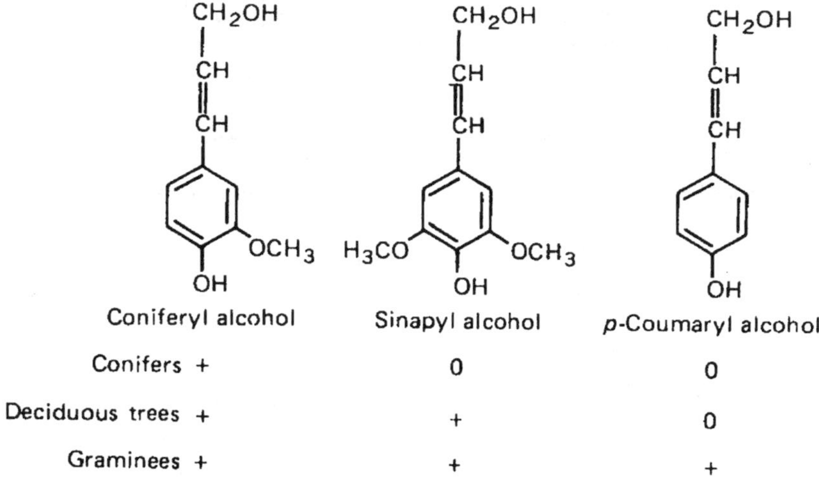

Figure 4.3. Lignin building blocks and their occurrence (+) or absence (-) in three groups of vascular plants. From Flaig et al., 1975. Reproduced with permission of Springer Verlag GmbH & Co., Heidelberg.

Of the organic constituents of fresh plant litter, amino acids and sugars can be taken up directly by soil organisms, and hence can be decomposed very quickly. The bulk of plant litter, however, consists of polymeric macromolecules, which must be depolymerised into smaller units by a host of extra-cellular enzymes before they can be ingested. The most important polymers are:

- cellulose, a glucose polymer (Figure 4.2) which constitutes about 10 to 15 % of leaf material, 20 % of deciduous wood, and 50 % of coniferous wood.
- hemicellulose, which mainly consists of branched chains of glucose-xylose polymers, making up 10 to 20 % of most plant materials, and
- lignin, a polymer of phenyl propane units (Figures 4.3 and 4.4), making up 10 to 20 % of leaves and up to 30 % of wood.
- plant and microbial lipids and aliphatic biopolymers, such as cutin, suberin, that protect roots and leaves.

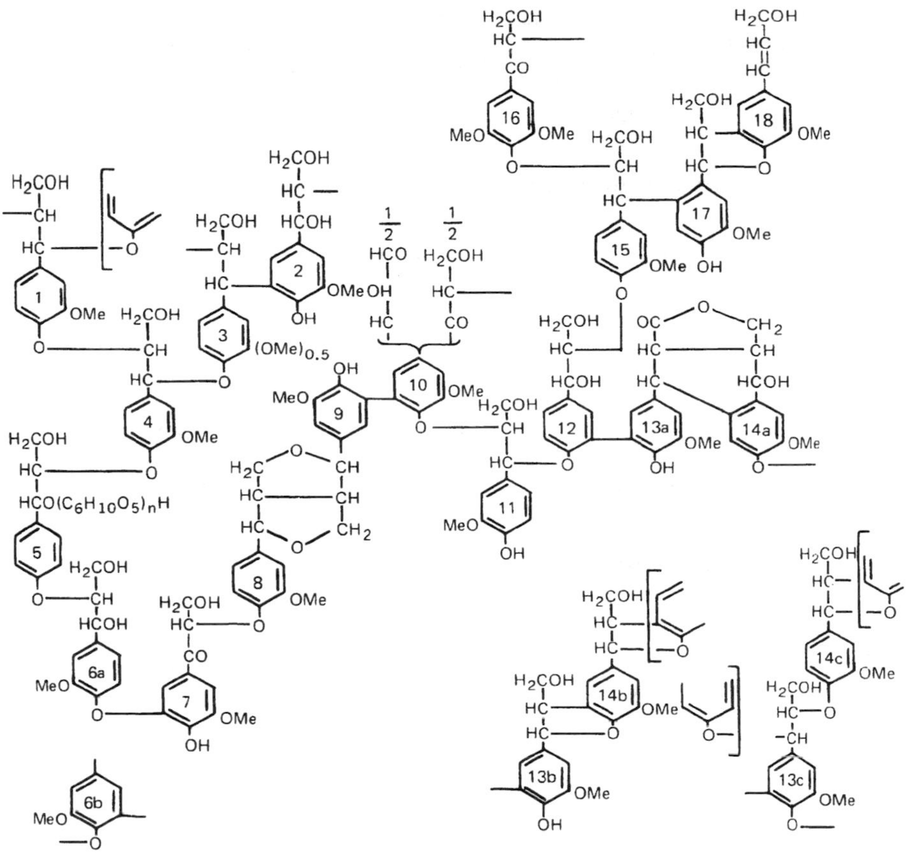

Figure 4.4. Example of a coniferous lignin molecule. From Flaig et al., 1975. Reproduced with permission of Springer Verlag GmbH & Co., Heidelberg.

For oak leaf litter in the Netherlands, the rate of decomposition of these and other constituents is illustrated in Fig. 4.5. Each solid straight line shows a logarithmic decline with time of the concentration of a certain constituent added at time 0. Sugars decompose most quickly (99% decomposed after 1 year), followed by hemicellulose (90% decomposed after 1 year), cellulose (75 %), lignin (40 %), waxes (25 %) and non-lignin phenolic acids such as tannins (10 %).

The lower dotted curve S shows the decomposition of total fresh litter if it were composed of a simple mixture of its constituents. The actual rate of decomposition is slower (upper dotted curve, M), because part of the cellulose is shielded from microbial attack by more resistant lignin, and can only be decomposed after lignin is broken down.

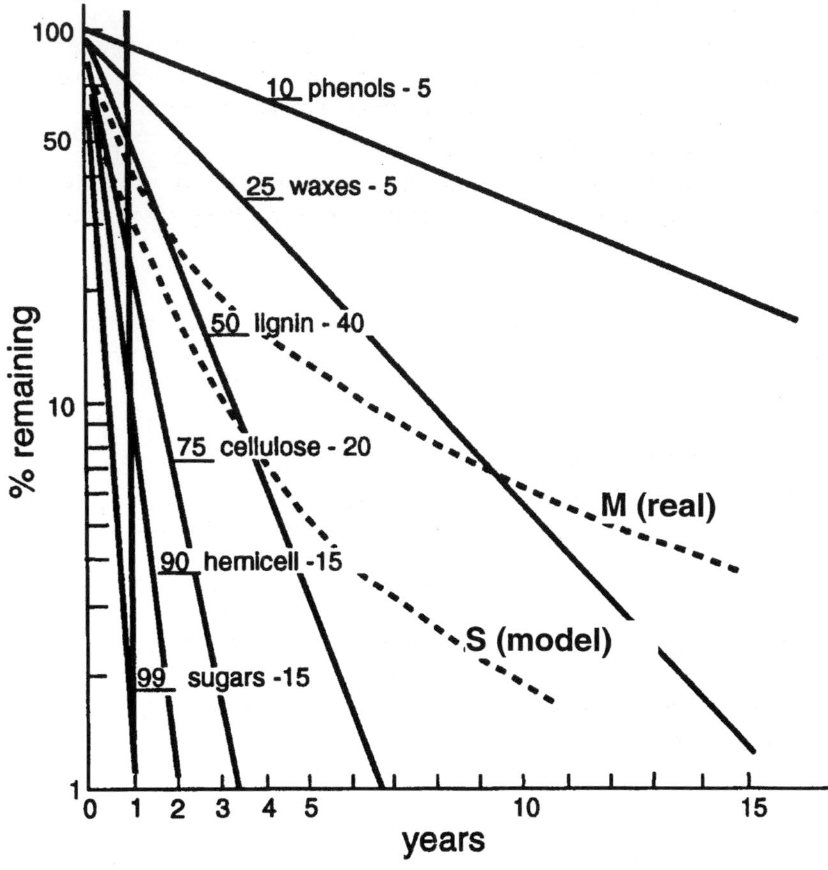

Figure 4.5. Rate of decomposition of constituents of oak litter in Hackfort, the Netherlands. The number in front of the name of the constituent indicates the percentage of that constituent lost by decomposition in one year. The number after the name represents the constituent's mass fraction (%) in the original litter. M and S, see text. From Minderman, 1968; cited by Swift et al., 1979.

Decomposition of lignin, mainly by so-called white-rot fungi (in contrast to brown-rot fungi that can only break down celluloses), may cost as much or more energy than is gained. The main advantage of decomposing lignin apparently is the exposure of hitherto unaccessible cellulose for use by the fungi.

RATE OF DECOMPOSITION

As illustrated in Fig. 4.5, the rate of decomposition of individual constituents can be approximated by a logarithmic decay rate, corresponding to first order kinetics (first order because the rate depends on the concentration of the substrate):

$$dX/dt = -k.X \qquad (4\text{-}3)$$

where X is the amount ("pool size") of the substrate (e.g. in t/ha or g/kg soil), t is time in days or years, and k is the rate constant in day^{-1} or $year^{-1}$. Integration of equation 4-3 yields

$$X_t = X_o.\, e^{-kt}, \quad \textit{or} \quad \ln X_t/X_o = -kt \qquad (4\text{-}4)$$

In steady state conditions, when input equals output, the mean residence time, or turnover time, is defined as MRT ≡ amount in reservoir/ sum of input or removal rates. So, according to equations 4-3 and 4-4:

$$MRT\ (= X/(-dX/dt)) = 1/k \qquad (4\text{-}5)$$

While decomposition rates for individual constituents in fresh plant material vary from about 5 to 0.1 $year^{-1}$, decomposition rates of humus are much lower, in the order of 0.01- 0.0001 $year^{-1}$.

Question 4.7 a) *Use Figure 4.5 to estimate the rate constant k (in units of* $year^{-1}$*) and the residence times (in years) for sugars, cellulose, and phenols.* ***b)*** *The apparent residence time of cellulose in oak litter in the field is 1.1 year. Why is that different from the value you found?*

4.4. Environmental factors influencing decomposition and humification

Good drainage favours decomposition by providing aeration. Aerobic respiration (involving O_2 as oxidising agent or electron acceptor) by soil organisms is relatively energy-efficient: the assimilation factor is 0.1 to 0.5. As has been discussed in more detail in Chapter 3, oxygen is absent in a water-saturated soil, and other electron acceptors must be used in respiration.

Question 4.8 *Mention a number of those electron acceptors. Which are quantitatively important in most soils?*

The energy yield of this so-called anaerobic respiration is lower than that of aerobic respiration, and decreases as reduction proceeds and new oxidants that function at successively lower **pe** values come into play (see Figure 3.12). Consequently, the assimilation factor in anoxic conditions is lower, about 0.05. So more organic carbon is oxidised per unit of microbial biomass formed. Since the N content of anaerobic and aerobic microorganisms is similar, less N is immobilised under anoxic than under oxic conditions.

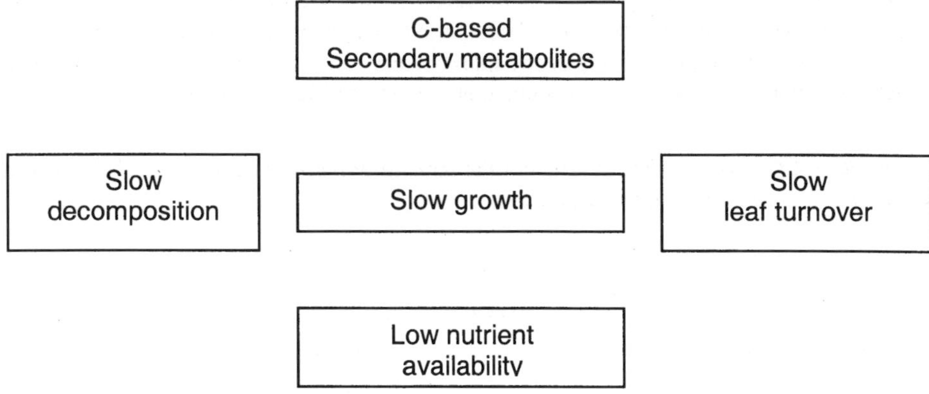

Figure 4.6. Feedbacks between low nutrient availability, plant growth, and substrate quality - see Question 4.10. After Chapin, 1991.

Question 4.9 *Rice is generally better supplied with nitrogen in flooded than in dry-land fields. Explain why.*

Under anoxic conditions, decomposition of intermediate products such as low molecular weight organic acids, is slowed down. As a result, high concentrations of such acids may develop. Decomposition of lignin requires much oxygen, and comes to a complete standstill under waterlogged conditions. Under continuously anoxic conditions, decomposition is very limited and cell structures of plant material remain intact for millennia: peat formation.

Also low soil pH slows down decomposition. First, low pH by itself may affect soil microorganisms and soil fauna, and so decreases the rate of decomposition. Second, high concentrations of dissolved Al^{3+} at low pH may be toxic to soil biota.

Another factor is the quality of the plant litter, the so-called 'substrate quality'. On infertile acidic soils, plants invest much energy in the production of C-based secondary metabolites such as polyphenols, and in unpalatable structural materials such as lignin. Plants probably do so to deter grazers (mammals, insects, which also use microbes for ingestion), but these products are also difficult to decompose by soil organisms. As a result, on infertile soils, leaves tend to live longer, and plant litter decomposes more slowly than litter produced on more fertile soils.

Question 4.10. *Slow turnover of plant material at infertile sites is a nutrient conserving strategy by plants, which also tends to keep availability of soil nutrients at a low level.*

a) Illustrate this by drawing "cause-and-effect" arrows between the boxes in Figure 4.6.

b) Why would secondary metabolites in infertile soils be C-based? (Hint: some plants deter grazers by N-based metabolites).

Climate influences decomposition mainly through effects of temperature on the activity of soil micro-organisms, and of soil fauna able to fragment litter. Litter fragmentation is almost negligible in tundra systems, appreciable in temperate systems, and very high (termites, ants, earthworms) in savannah and equatorial forest ecosystems.

Question 4.11 *Discuss the size of the litter layer in different ecosystems of Table 4.1 in terms of climate and soil faunal activity.*

EFFECTS OF AGRICULTURE AND OF TEXTURE

Land reclamation involving removal of natural vegetation, and ploughing of land for arable crops stimulates decomposition of organic matter. The effects of cultivation on soil organic matter content are caused by differences in (1) fresh organic matter inputs and (2) decomposition rates. Inputs are often lower in arable land, and higher in heavily fertilised pasture, than under natural vegetation. By regular ploughing, the availability of organic substrates to micro-organisms is increased by stirring up the soil. Clay tends to protect organic matter against decomposition by (1) organic matter adsorption on clay particles and (2) entrapment of organic matter in small pores or in soil aggregates, unaccessible to microbes. Fig. 4.7a/b illustrates how organic matter decomposes more slowly as it is more closely associated with mineral matter. Increasing association of coarse organic matter (density about 1 g/cm^3) with mineral particles (density about 2.6 g/cm^3) is reflected by increasing density fractions (see also paragraph 4.5). The three curves refer to different density fractions: <1.1 (L = light), via 1.1-1.4 (I = intermediate) to >1.4 g/cm^3 (H = heavy).

Question 4.12 *Estimate the value of k (in year^{-1}) for the five fractions of organic matter for which decomposition is shown in Fig. 4.7a b. Assume that decomposition follows first order kinetics (see Equation 4-4) until at least 1000 days (fractions L, I, and H), or 6000 days (C in micro-aggregates).*

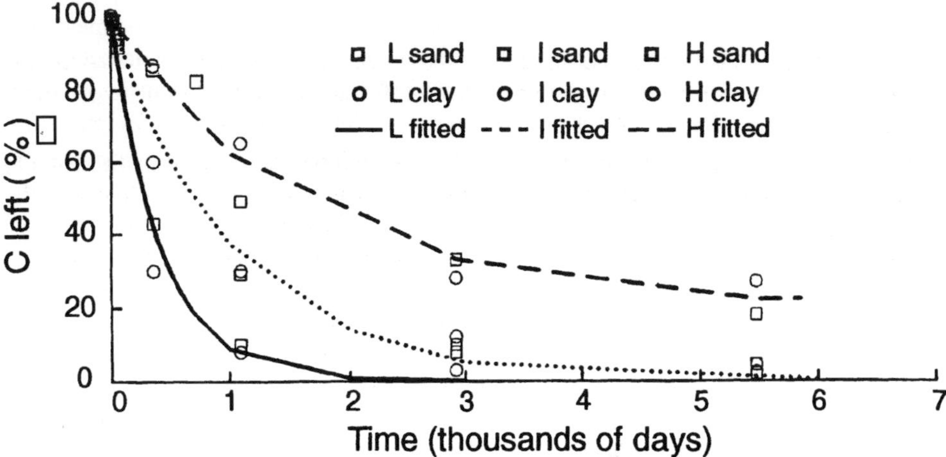

Figure 4.7a. Change with time in the content of coarse (>150μM) organic matter fragments (expressed as C in % of amount at t=0) in a soil kept bare for 15 years. L = light fraction; I = intermediary fraction; H = heavy fraction. From Hassink, 1995.

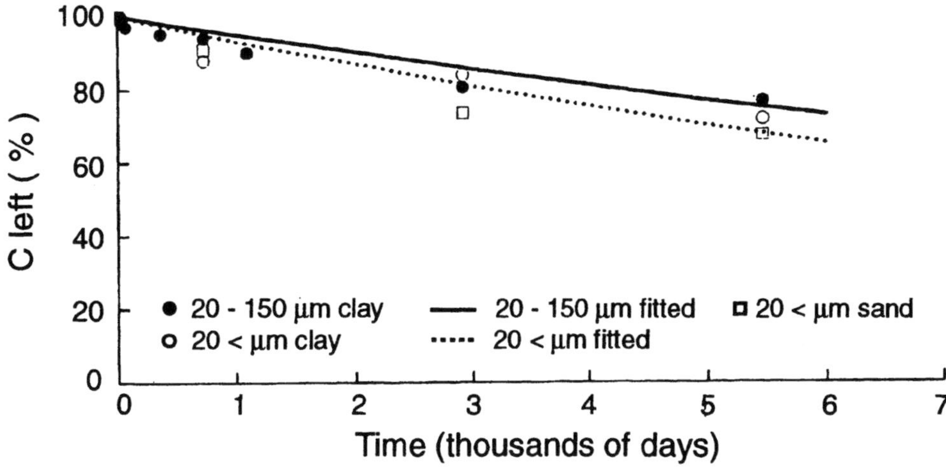

Figure 4.7b. As Figure 4.7a, for more humified organic matter in soil micro-aggregates with diameters of 20-150μM (solid line) and < 20μM (dotted line). From Hassink, 1995.

4.5. Formation of humus

Most of the biomass that reaches the soil as litter and animal material is mineralised and only a minor part is transformed into humus. Humification consists of three major steps: a) fragmentation of large polymers such as lignin, cellulose, and aliphatic biopolymers; b) partial oxidation of the fragments, and c) combination of the oxidised material into new configurations of relatively small molecules. Humus does not have a specific chemical structure, and it contains a large number of different chemical components and bonds in varying spatial relations. Schnitzer (1986) described humus as follows:

> *Humic substances, which constitute 70 to 80% of organic matter in mineral soils, are dark-coloured partly aromatic, acidic, hydrophilic, molecularly flexible polyelectrolyte materials. They can interact with metal ions, oxides, hydroxides, minerals and other inorganic and organic substances, forming associations of widely differing chemical and biological stability. The interactions include formation of water-soluble ligand complexes, adsorption and desorption, dissolution of minerals and adsorption on external mineral surfaces.*

Schnitzer's definition under-emphasises hydrophobic, aliphatic components, which form an important portion of soil organic matter. Another point of discussion in the definition of humus is that in most classical definitions, recognisable chemical compounds, such as polysaccharides and proteins, are considered not to be part of 'humus' as such. The 'polymeric' model is now being replaced by a model of stacked smaller units (conformational structure), with an *apparent* large molecular weight. Recognisable chemical compounds participate in the conformational structure and, although chemically better defined, such components play an essential role in the behaviour of soil organic matter as a whole.

The building blocks of humus are fairly well known: phenolic components formed upon degradation of lignin; polysaccharides formed by degradation of cellulose, hemicellulose and starches; long aliphatic chains formed from biopolymers; N- and S-containing compounds from proteins; and a large variety of smaller degradation products (see e.g. Beyer, 1996). The absence of a specific organisation of the building blocks makes the study of humus extremely difficult. Many models for humus structure (e.g. Figure 4.8 for humic acid), are not realistic. They are based on a polymeric view of humic substances and, in addition, systematically underestimate aliphatic components, do not include elements of small molecular size, only depict a stochastic composition based on general chemistry, and do not use structural information from, e.g., analytical pyrolysis (see later). Nevertheless, we have chosen to include Figure 4.8 because it contains a number of recognisable building blocks.

For practical purposes, humic substances are characterised according to their behaviour, e.g. in interaction with metals, by titration curves, or with respect to standardised extraction techniques. Properties can be expressed as, e.g., content of aromatic or aliphatic C, of polyphenols, of contents of functional groups (groups that interact with other soil components, e.g. carboxyl groups, alcoholic groups, phenolic OH, N-containing groups, etc.).

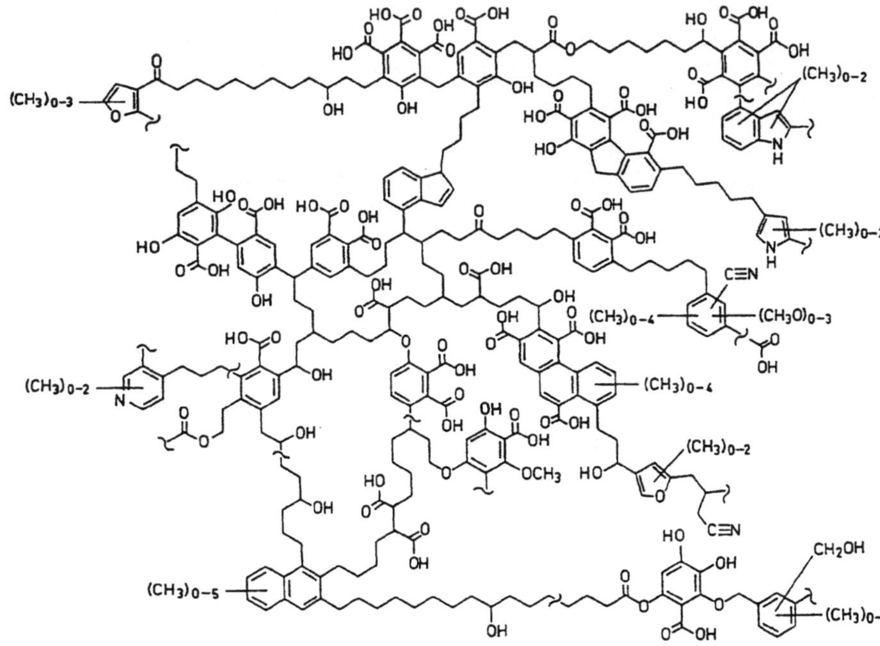

Figure 4.8. Polymeric model structure for humic acid, depicted with various combinations of building blocks. From Schulten, 1995. Reproduced with permission of Srpinger Verlag GmbH & Co., Heidelberg.

***Question 4.13** a) Try to find in Figure 4.8 some degradation products of lignin, polysaccharides (compare Figures 4.2-4.4), and aliphatic biopolymers (aliphatic biopolymers have long -CH_2- chains). b) What evidence is there that the source materials underwent partial oxidation?*

For analysis, humus is commonly extracted from soils by NaOH or $Na_4P_2O_7$ (sodium pyrophosphate). Both extractants have a high pH, which disrupts H-bonds. In addition, pyrophosphate forms complexes with Al and Fe, so that these metals are removed from organic complexes.

Extraction changes the organic matter. Operationally defined humus fractions, such as Humic Acid and Fulvic Acid, are very broad groups (Figure 4.9). Their proportion may be different in various soils (e.g. 15-20% fulvic acid in chernozems and up to 70% in podzols), but also their chemical properties change with soil type. Such fractions, therefore, are very rough and only serve a general purpose.

Functional groups in humus are extremely important for its properties. In the range of normal soil pH values, between 3 and 8, many of the acidic groups are dissociated, which causes a negative charge (CEC). This negative charge may be as large as 4000 cmol(+)/kg C at pH 4 and 8000 cmol(+)/kg at pH 8. This is 4 to 8 time the CEC of the same mass of smectite or vermiculite.

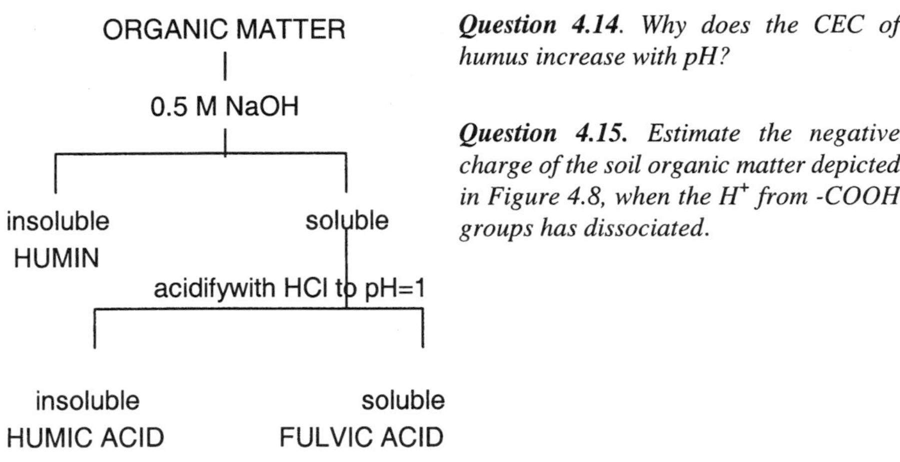

Figure 4.9. Classical separation of soil organic matter into humin, humic acid, and fulvic acid.

Question 4.14. *Why does the CEC of humus increase with pH?*

Question 4.15. *Estimate the negative charge of the soil organic matter depicted in Figure 4.8, when the H^+ from -COOH groups has dissociated.*

Functional groups also influence the solubility of humus in water. Charged (dissociated) molecules are more soluble than neutral ones. Consequently, organic molecules with acidic groups become less soluble when the acidic groups are neutralized through binding of cations (Ca, Mg), complexation (Fe(III), Al), or binding to mineral surfaces. The interactions between humus and the mineral fraction of the soil is extremely important for soil forming processes, as we will see in Chapters 6 (surface horizons), 12 (andisols), and 11 (podzols). The organisation (arrangement) of humus molecules may change upon wetting or desiccation (seasonal climates), upon addition of small molecules (acids, alcohols), upon changes in pH (change in dissociation), upon change of cation adsorption, and upon metal or mineral binding.

BLACK CARBON

Burned vegetation remains, as charcoal and soot (char) are an important part of the 'humin' fraction in various soils (e.g., Haumaier and Zech, 1995; Schmidt and Noack, 2000; Schmidt et al., 1999). Although charcoal is inert when it first forms, it gradually oxidises and develops exchange capacity for cations and anions. Because charred material is highly aromatic, it considerably changes the humus chemistry. Charred material is relatively stable, and may constitute an important pool of stable organic matter in the soil. The amount of charred material is considerable in soils with grassy vegetation that are subject to regular burning, such as savannah soils and black (melanic) Andosols. It is a very important component in soils under traditional cultivation, such as the *Terra preta dos Indios* (black Indian soils) of the Amazon region (Glaser, 1999).

4.6. Effects of soil fauna on soil properties: bioturbation, homogenisation and formation of aggregates.

BIOTURBATION

Plant roots and burrowing soil animals create large and small channels to various depths. By this so-called bioturbation they alter the original structure of sediments and of weathered rock (Plate H, p.122; Plate J, p.158). The channels can be filled again later. In this way the original rock structure or sedimentary layering disappears with time. Soil animals important in perforating soils and sediments include moles, rodents, earthworms, beetles, termites and ants.

Question 4.16. *How do you expect the intensity of bioturbation to vary with soil depth?*

VERTICAL HETEROGENISATION AND HOMOGENISATION

A result of burrowing is biological homogenisation of the soil profile, which counteracts the vertical heterogenisation associated with the formation of distinct soil horizons by other soil forming processes. Homogenisation is mainly caused by animals that move soil material downwards or upwards. E.g. earthworms ingest soil at one depth but may excrete it at another depth. If worms are highly active for a long time, as for instance in the black earths (Chernozems) or in old orchards as in southern Limburg, the Netherlands, very deep (up to about 1 m), homogenous Ah horizons may develop.

On a small scale, homogenisation takes place if biopores are filled in from above, creating a so-called 'pedotubule' consisting of generally more humose and porous material than that surrounding the former biopore (Plate H, p.122). Soil material can be slowly incorporated in the material of the upper soil horizons. This process also tends to decrease chemical or mineralogical differences between soil materials at different depths.

Moles and many rodents (e.g. ground squirrels or *gophers*) in N. America, sousliks in Russia, mole rats in Africa), dig mainly horizontal tunnels in the upper soil horizons and deeper sloping tunnels to a depth of several metres, where they build their nests. The soil material removed during digging is partly placed in abandoned tunnels, but part is deposited on the soil surface. After such a system of tunnels has been abandoned and is partly collapsed or filled in, a network of channels and highly porous, loosely packed soil material persists. The rodents are particularly active in steppe areas. The traces of their filled-in burrows are called *krotovinas*.

In the Netherlands, earthworms are mainly confined to fertile, loamy to clayey (calcareous) soils under permanent vegetation (e.g. grass). In some temperate areas (e.g. Black Forest) and in many tropical soils earthworms also occur in acidic soils.

Ploughing greatly disturbs earthworms, both directly (wounding) and indirectly (e.g. by increasing soil temperature, decreasing soil moisture, and removing crop residues from the soil surface), so they are rare in most tilled land. Earthworms produce mainly vertical pores, between < 1 mm to 1 cm in diameter, sometimes to several metres deep. Earthworm channels are formed by consumption of the soil and by lateral compression of soil material. The consumed soil is excreted, often at the soil surface, in the form of worm casts (Darwin, 1881). Soil aggregates in soils with many earthworms can often be recognised as former worm casts. On the surface of seasonally flooded soils, worm casts are often found as a kind of chimneys or as cylindrical blocks consisting of many individual casts with a channel in the centre.

Scarab beetles dig macro-pores with their specially adapted front legs. Their burrows are generally vertical and can be found to a depth of about 1 metre, especially in densely packed sandy soils. These channels are filled up with material from shallower depth. The fillings are more loosely packed than the original, undisturbed soil. Roots can easily penetrate such loosely filled channels.

In the tropics, termites (insects of the Isoptera order) are extremely important in moving soil. Most species have their nests inside the soils, but some build metres-high termite mounds (termitaria; Plate H, p.82) of soil material; in which many maintain fungal gardens that they supply by bringing in plant material. They move soil upwards from deep horizons, create soil aggregates, galleries, and channels, increase pore space, and stimulate decomposition of organic matter by growing fungi on litter in special chambers. Leaf cutting ants (Atta) in South America do the same in enormous below-ground nest areas. Because they only use specific particle sizes (e.g. silt, very fine sand), termites and ants also influence particle size distribution (see Chapter 8).

The large channels of moles and rodents are rather unstable. Worm channels are generally more stable because of their smaller size, but also because of the stabilising effect of excreta (organic matter intensively mixed with mineral material) smeared along channel walls.

Biota plays a great role in the formation and stabilisation of soil aggregates: clusters of sand and silt particles linked with clay and organic matter (Plate J, page 140). Many small soil aggregates are faecal pellets are enriched in clay size material relative to the bulk soil because the preference of many soil ingesting animals for fine material. The water stability of aggregates and the stability of pores often depend on organic materials. Tisdall and Oades (1982) classified the organic binding agents into: (a) transient, mainly polysaccharides, often produced by micro-organisms as exocellular mucilages or gums, (b) temporary, such as root and fungal hyphae, and (c) persistent, such as humic and phenolic compounds associated with polyvalent metal cations, and strongly sorbed polymers. The effectiveness of various binding agents at different stages in the structural organisation of aggregates is illustrated in Figure 4.10. Roots and hyphae stabilise macro-aggregates, defined as >250 μm diameter.

Soil organic matter fractions

Humus can be separated into fractions in various ways. Here we will only mention the classical fractionation into *Humic Acid, Fulvic Acid, and Humin*, the grain-size and density separation, and the fractionation of dissolved organic C.

Humic acid, fulvic acid and humin

This subdivision is based on the solubility in acid and base of various humus fractions. The extraction scheme is given in Figure 4.9. Humin includes both organic matter strongly bound to minerals and in complexes, charcoal, apolar components, and undecomposed plant remains. The differences between fulvic acid and humic acid are only gradual:

	Fulvic acids	*Humic acids*
colour	yellow to yell.brown	brown to black
apparent molec. weight	2000-9000 D	up to 100.000 D
C content	43-52%	50-62%
O content	higher	lower
solubility in water	higher	lower
functional groups	higher	lower

(High molecular weights in Humic Acids are only apparent; see p.95).

Particle-size and density fractionations

In particle-size fractionation, sand, silt, and clay-size fractions of organic matter are separated by sieving and sedimentation of the bulk sample, without other pre-treatment than sonication (the mineral-organic fractions are supposed to behave as pure mineral fractions). Sand-size organic matter is usually undecomposed plant litter; clay-size material is strongly humified and may be strongly bound to clay minerals.

Density fractionation uses the different density of fresh, humified, and mineral-bound organic matter. Density fractions are similar to those of the grain-size separation. Clay size = heavy, mineral-bound fraction; sand size = undecomposed, light fraction.

These fractionations are useful for roughly separating fresh from strongly humified material, but the separations are necessarily coarse and the separates do not have predictable chemical properties.

Fractionation of dissolved organic carbon (DOC).

Dissolved organic carbon in natural water is sometimes fractionated according to its charge and solubility. For separation, charged resins are used. The six fractions are hydrophobic and hydrophilic acids, bases, and neutrals. The fractions are used to predict the interaction of dissolved organic matter with metals.

Studying humus

We can use various ways to obtain information on humus, depending on the kind of information we are interested in.

Chemical components

Chemical components of humus (e.g. sugars, lipids, phenols, etc.) can be studied after extracting them one by one, by using different reagents. General information on the total chemistry of the humus can be obtained by pyrolysis - gas chromatography - mass spectrometry (Py-GC-MS). A small sample of purified organic matter is quickly heated to 600-650°C in an inert atmosphere, causing breaking up of large molecules into smaller units of <50 carbons. These smaller units are separated by gas chromatography and analysed by mass spectrometry. They can be identified, and their relative abundance gives insight into the total humus chemistry and into the components from which the humus was derived.

Environment of C atoms

Nuclear magnetic resonance (NMR) and infra-red absorption (IRA) give an indication of the relative numbers of C atoms in different chemical environments. In principle, amounts of CH, COOH, C-OH, C H and other groups can be determined, but because of the many ways in which carbon atoms in humus are attached to other carbons and to H, O, N, and S, quantification of solid-state NMR is still debated. NMR can be applied directly, both to humus concentrates and to soil samples, without the necessity of extracting the organic matter, so that changes due to extraction procedures are avoided. Paramagnetic elements (Fe), however, may cause bad spectra. For IRA measurements, only one milligram of sample is required; NMR needs much more.

Acidity and dissociation

Humus contains a large variety in acidic and phenolic OH groups, with or without bound metal ions, which dissociate at different pH. Information on such groups can be obtained through titration curves.

Stable isotopes

The ^{13}C isotope, which has a natural abundance of about 1% of total C, is used as a tracer in studies of C dynamics. ^{13}C is partially excluded during photosynthesis. Plants with a C3 photosynthetic pathway exclude ^{13}C more strongly than C4 plants. This isotopic signature is also found in the decomposition products. Rates of change in soil organic matter fractions can be measured when a C4 vegetation replaces a C3 one, or vice versa, by studying the change in ^{13}C abundance (Balesdent & Mariotti, 1996).

Question 4.17. *The number of stable macro-aggregates, but not micro-aggregates (<250 μm), decreases quickly when land under natural vegetation is reclaimed for arable farming. Why?*

CHEMICAL EFFECTS OF BIOTA

During decomposition of 'low-quality' litter in acid, infertile soils, higher concentrations of soluble low-molecular weight (LMW) organic acids are observed than in more fertile soils. The same trend is observed when moving from soils of warmer to soils of colder regions. The most common of these acids are: citric, p-coumaric, galacturonic, glucuronic, p-hydroxybenzoic, lactic, malonic, oxalic, succinic, and vanillic acid (Figure 4.11). High concentrations of dissolved LMW acids in acid, infertile soils have often been attributed to (1) a higher proportion of precursors of those substances in plant material (e.g. polyphenols) from unproductive sites, and (2) lower rates of breakdown of such acids. More important, however, may be exudation of such acids by (plant-symbiotic) mycorrhizal fungi, which are particularly abundant in infertile soils. Roots infected by mycorrhizal fungi have a much larger surface for uptake of nutrients and water than non-mycorrhizal roots. The LMW acids form strong complexes with dissolved Fe and Al, a property that is utilised by mycorrhizal hyphae on (nutrient- poor) podzols to dissolve narrow channels inside weatherable minerals to obtain nutrients for their host plant (Fig.4.12).

Question 4.18. *Plants provide sugars for their symbiotic mycorrhizal fungi. Mycorrhizal fungi are particularly common on infertile soils. Can you think of a simple mechanism related to lack of nutrients for new growth, why plants put more energy to the mycorrhiza when growing on infertile than on fertile soil?*

Soil fauna strongly influences the soil chemical composition, by mixing soil horizons, incorporating organic matter into the mineral soil, and (in case of earthworms and termites) promoting decomposition of plant material.

4.7. Nutrient cycling

During decomposition of plant material, part of the nutrient elements are liberated in inorganic dissolved form, for instance as the ions Ca^{2+}, Mg^{2+}, K^+, NH_4^+, SO_4^{2-}, $H_2PO_4^-$. In soils of humid areas, dissolved nutrients tend to migrate down the soil profile. Under vegetation, the larger part of those nutrients is usually taken up again by plant roots and assimilated by the vegetation, and only a small fraction is lost with the drainage water. An actively growing, closed vegetation may take up several hundred kg of nutrient elements per ha per year from the root zone, which often extends to several metres below the soil surface.

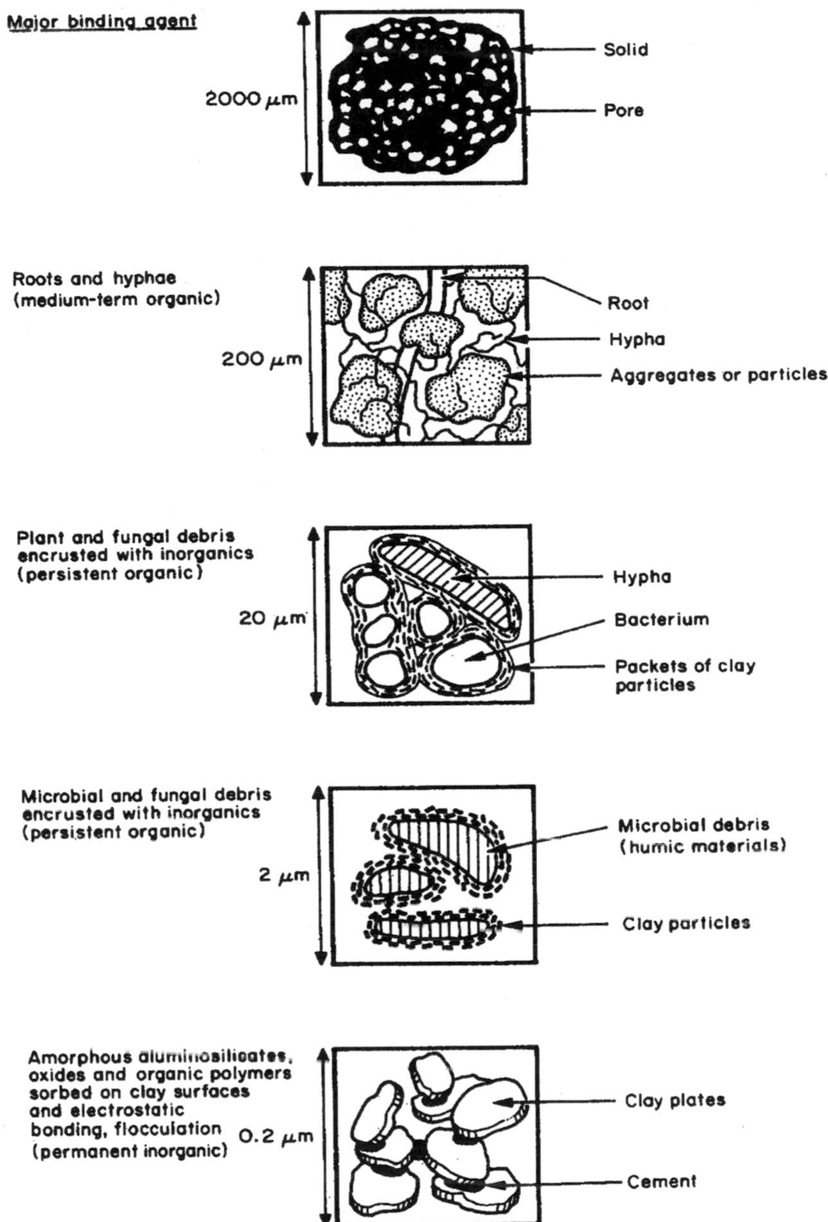

Figure 4.10. Model of aggregate organisation with the major binding agents involved. From Tisdall and Oades, 1982. Reprinted with permission of Blackwell Science Ltd., Oxford.

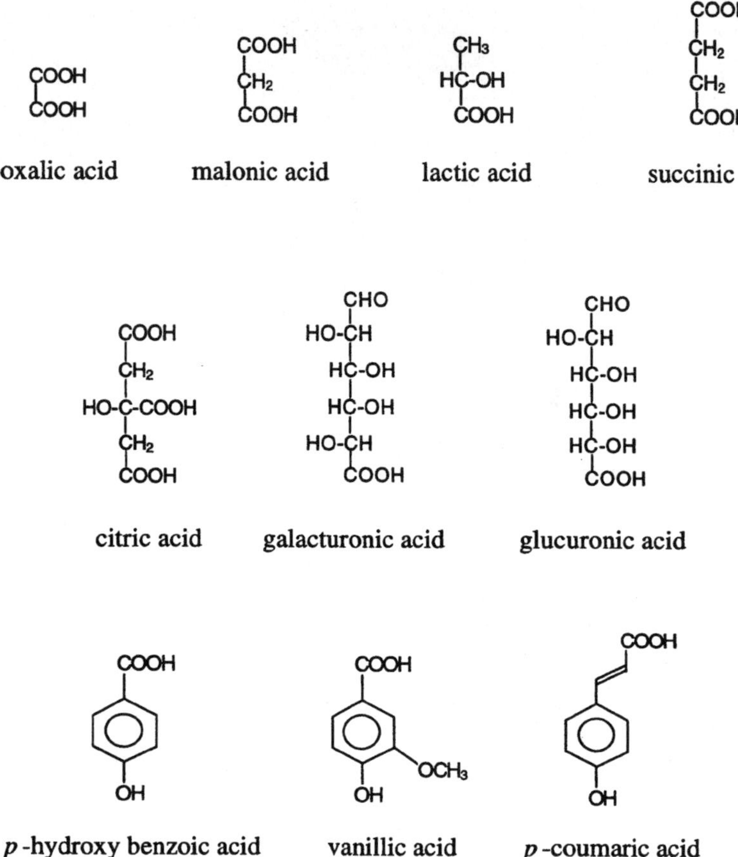

Figure 4.11. Structural formulas of common low molecular weight (LMW) organic acids.

Recent work showed that in many ecosystems, a small fraction of the roots of trees and shrubs extends to very great depths (10-40 m; Canadell et al., 1996). It is not clear if these roots are mainly for supply of water during droughts, or if they also play a role in nutrient supply. Most nutrients in plants are eventually returned to the soil surface by litter fall. So the net effect of uptake by plants is that nutrients from deeper soil layers are transported to the soil surface. Part of the nutrients in the surface soils is easily available to plants (at the exchange complex or temporarily stored in microbial biomass), but part may be locked up in slowly available mineral form (especially P). This process of "nutrient pumping", which counteracts the leaching of nutrients, is especially important in mature, undisturbed forests.

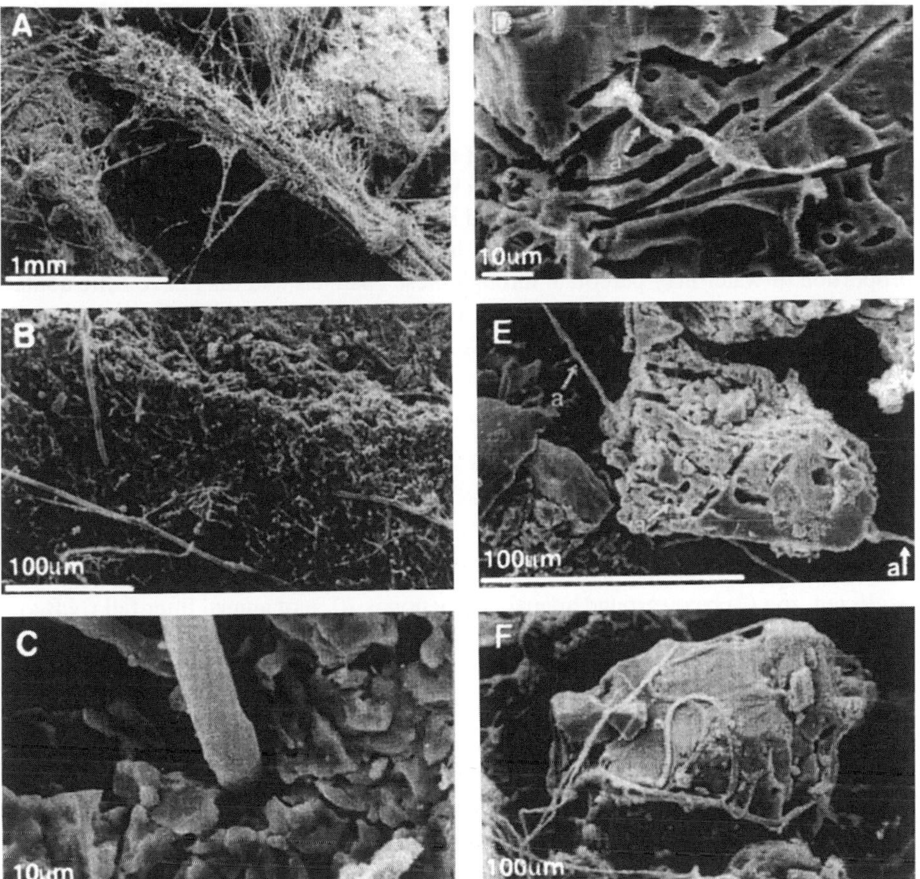

Figure 4.12. Scanning electron micrographs of fungal hyphae in soils. **A.** Mycorrhizal root and associated mycelial mat on granite rock; **B.** Hyphae entering a weathered Ca-rich feldspar at the granite surface; **C.** Ditto, detail; **D.** Interior of a crushed Na-Ca feldspar grain with interconnected tubular pores and associated fungal hyphae (a); **E.** Hyphae partly inside a Ca-feldspar grain; **F.** Hyphae on a quartz grain. **D-F** from a podzol under *Pinus sylvestris*. Photographs A.G. Jongmans.

The effect of nutrient cycling can be seen by vertical profiles of 'available' nutrients, e.g. mineralisable N, exchangeable Ca, Mg, K, or extractable P, which often show markedly higher concentrations in the rooting zone than at greater depth.

Certain broad-leaf trees (e.g. *Acer, Tilia)* have deep root systems compared to coniferous trees (*Picea, Tsuga*), and pump up more Ca^{2+} from greater depth. This results into a higher pH and exchangeable Ca concentrations in the soil surface, with better conditions for decomposition and thinner O-horizons (See Chapter 6; Dijkstra, 2000).

Question 4.19. *Why is the effect of nutrient "pumping" on the concentration profiles of nutrients seen more clearly in nutrient-poor than in nutrient-rich soils, and more clearly as the climate becomes wetter?*

Earthworms contribute to conservation of nutrients in the soil (even though they may distribute nutrients over a relatively thick surface horizon), both by incorporating decomposed organic matter in the mineral horizons, and by forming vertical channels. Especially in climates characterised by high rainfall intensities, a relatively large proportion of water may percolate through these channels and thus have little contact with the soil material. In this way earthworm activity tends to decrease the rate of nutrient leaching.

4.8. Problems

Problem 4.1

Consult table 4.1. Assume that (1) the biomass is in steady state, (2) grazing can be neglected, and (3) belowground annual production is 70 % of annual litter fall (N.B.: (3) holds for temperate forests (Nadelhoffer and Raich, 1992), but not necessarily elsewhere).

a. Explain why NPP equals belowground plus aboveground litter production at steady state.
b. Draw a graph of relative root mass (root/total biomass) on the y-axis, as a function of net primary production (NPP). What is the relationship between NPP and the fraction of total biomass that occurs below ground?
c. How does climate affect (1) NPP and (2) relative root mass? Explain the relationship between NPP and relative root mass from the viewpoint of plant strategy to allocate energy above and below-ground?
d. If the climate is favourable for plant growth, how would chemical soil fertility affect the relative root biomass?
e. Draw a graph of root biomass (y-axis) as a function of annual root litter (x-axis), and discuss the relationship between the two.

Problem 4.2

A soil contains *Y* kg/ha of humus-C. Every year *X* kg/ha of plant litter-C enters the soil. Assume: 1) soil humus is in a steady state, 2) every year all new plant litter decomposes at a rate k_f, and the remaining fraction $1-k_f$ (also called the humification coefficient) is directly transformed to humus; 3) humus mineralises completely at a rate of k_h year^{-1}.

a. Write an equation of the form $input_{humus} = output_{humus}$, expressed in Y, X, k_f, and k_h.

b. Calculate the humus pool if the annual input is 5000 kg/ha.yr, $k_f = 0.01$, and $k_h = 0.05$; check if your answer is realistic by recalculating it to an organic C content (mass fraction, %) in the 0-20 cm surface soil at a bulk density of 1.2 kg/dm^3
c. List at least two reasons why this approach is simplistic.

Problem 4.3

In 1970, trees in a pine forest shed 3500 kg leaves (needles) per ha. The freshly fallen leaves contained 50% C and 0.38% N. In 1974, 27% of the mass of leaf litter fallen in 1970 still remained. The N content of this partly decomposed material increased to 1.2%, while the C content was still 50%. (Sources: Berg and Staaf, 1981; Gosz, 1981).

a. Calculate the C/N ratio of fresh and 4-year old pine leaf litter.
b. How much (kg ha^{-1}) of the N and C present in the fresh leaf litter has disappeared in four years?
c. Explain the difference in the mineralization rates of C and N.

Problem 4.4

Table 4A refers to an experiment in which the decomposition of straw from alfalfa (*Medicago sativa* L., a leguminous plant; C/N = 13) under anaerobic and aerobic conditions was compared. Assume that all water-soluble C, CO_2 and CH_4 produced came from the 500mg of C in alfalfa that was added to the soil.

a. Plot ln X_t/X_o against time for the aerobic and anaerobic case. X_o is the initial amount of C added (500mg), and X_t is the amount left after t days (=500 mg C minus C released as CO_2 + CH_4 + org.C).
b. Indicate three main stages in aerobic and anaerobic decomposition. Estimate the rate constants (day^{-1}) of each stage. Can you explain these different stages? Discuss the differences between aerobic and anaerobic decomposition.
c. Compare the results for alfalfa with rate constants for oak litter (Question 4.7). Explain the differences.
d. Compare the amounts of N mineralised per amount of C mineralised under aerobic and anaerobic conditions. How did the C/N ratio of the residue of the alfalfa straw change as a result of decomposition in anaerobic and aerobic conditions? The initial amount of N present follows from the C/N ratio, in combination with the 500 mg C added. To get an impression of the changes in C/N during decomposition, calculate these ratios for 2 and 30 days after the start of the experiment, by quantifying the composition of the residue.

Table 4A. Redox potentials (Eh), pH, forms of mineralised C, and mineralised N from aerobic and anaerobic flow-through incubations for 0.5g carbon from alfalfa straw applied to 100 g soil. Results were corrected for substances released when the same soil was incubated without alfalfa. Taken from Gale and Gilmour, 1988.

day	Eh (mV)	pH	mineralised C (mg)			new N mineralised (mg)
			water soluble C.	CO_2	CH_4	
aerobic						
1	-	-	0	25.4	-	-
2	-	6.1	0	44.1	-	1.3
5	-	-	0	69.5	-	-
7	-	6.2	0	82.4	-	1.7
10	-	6.1	0	91.5	-	-
14	-	6.0	0	95.9	-	2.0
22	-	6.2	0	102.9	-	3.5
30	-	6.2	0	110.8	-	4.5
anaerobic						
1	-	-	34.0	0.9	-	-
2	59	5.9	41.2	4.4	-	1.9
5	-	-	46.0	9.7	-	-
7	-35	6.1	50.4	11.7	-	4.3
10	-105	6.7	54.0	15.6	0	-
14	-101	6.6	67.5	17.4	0	7.0
22	-131	6.8	43.4	24.9	15	6.6
30	-139	7.0	18.5	38.4	27	6.3
LSD($p<$ 0.05)	23	0.1	5.5	2.5	1	0.6

Problem 4.5

Table 4B shows the contents of organic carbon in the surface soil of three pastures in land reclaimed from the IJsselmeer (East Flevoland, The Netherlands). Worms are absent in soil nr. 1; soil nr. 2 has worms since two years, and soil nr. 3 since eight years.

a. Plot the organic carbon contents against depth.
b. Briefly explain the changes in the soil as function of the period of time the worms have been active.

Table 4B. Organic carbon contents (mass fraction, %) in the surface soil of three pasture soils in East Flevoland. The (-2 to 0 cm) layer on top of the mineral soil largely consists of undecomposed plant remains; this so-called *litter layer* is absent in profiles 2 and 3 (Hoogerkamp et al., 1983).

profile nr.	1	2	3
depth (cm)			
-2 - 0	16.5	---	---
0 - 2	1.8	5.9	3.3
2 - 5	1.2	1.9	2.9
5 - 10	1.1	1.3	1.8
10 - 20	1.1	1.2	1.2

Problem 4.6

Table 4C shows some physical and chemical differences between surface soil material (0-10 cm) and worm casts in four West African soils. All differences between S and W are statistically significant (P<1%).

Table 4C. Properties of surface soil material (S) and worm casts (W) in four West African soils. Source: De Vleeschauwer and Lal, 1981.

soil series	clay (%)		organic C (%)		CEC ($mmol_c$/kg)	
	S	W	S	W	S	W
Eketi	10.3	12.7	1.43	3.05	29	89
Ibadan	4.4	10.6	0.80	3.12	31	161
Apomu	4.3	4.8	0.50	2.83	19	122
Matako	2.4	4.3	0.70	3.15	34	155

a. Briefly describe the nature of the changes that the soil material has undergone by passage through the digestive system of the worms.
b. Calculate the contribution of clay and organic C to the CEC by assuming these are the same per gram for S and W in a given soil. What is more important in changing the cation exchange capacity of the soil material during passage through the worms: changes in the clay content or changes in the content of organic carbon? The assumption about equal specific CECs of organic matter and of clay in S and W

implies that worms do not select certain clays when feeding, and that clay and "quality" of organic matter do not undergo changes during passage of the worm gut. These assumptions are not completely correct, which is the reason for impossible negative values of the CEC of clay that you find.

Problem 4.7

Lumbricus terrestris, one of the most common earthworms in Western Europe, daily consumes an amount of soil material equivalent to about 10% of its own weight. Assume (1) that a soil contains 200 g of *L. terrestris* per m^2, (2) that these worms are active for 300 days per year and (3) that the bulk density of the 20 cm thick surface soil is 1.3 kg/dm^3. Calculate the amount of time it takes for the worms to transform all of the soil material over 0-20 cm depth into worm casts. Assume that worms never take in worm casts and only eat "fresh" soil; this, of course, is a gross simplification. (Sources: Paton, 1978; Hoeksema, 1961)

Problem 4.8

The following analyses (mass fraction, %) refer to a soil from a termite mound (A) and from its surroundings (B). Briefly describe the data in the light of the text of this chapter (data from Leprun, 1976).

	A	B
Clay	27	13
Sand	25	39
Organic matter	3.6	0.8

4.9. Answers

Question 4.1

Plant roots influence the soil i) via uptake of water and nutrients: soil moisture and soluble nutrients contents (effect measurable within hours to days), ii) ditto, for nutrients bound in minerals: increasing the rate of mineral weathering (effects noticeable after centuries to millennia) , iii) via penetration and growing thicker: formation of pores of different sizes (effects measurable after years).

Question 4.2

a) Direct effects of biota: i) bioturbation (soil macro- and meso- fauna: moles, earthworms, termites, ants), ii) comminution of plant litter (all detritivore soil fauna), iii) consuming plant litter and soil organic matter (bacteria, fungi, soil fauna).

b) Effects of soil organic matter: i) increasing structural stability, ii) increasing water holding capacity, iii) increasing cation exchange capacity, iv) structural storage of plant nutrients (organic N, P and S).

Question 4.3

In dry or infertile soils, plants need to invest a relatively larger fraction of their primary production below-ground in order to obtain sufficient nutrients and water. In seasonally dry or cold soils, plants store energy below-ground to survive the inhospitable season.

	tundra		spruce forest	oak forest	dry steppe	semi desert	subtropical deciduous forest	dry savan-nah	humid tropic. forest
	arctic	brush wood							
relat. root mass kg/kg	0.8	0.82	0.23	0.24	0.9	0.875	0.20	0.41	0.18
annual belowground production	0.7	1.4	3.5	4.9	2.8	0.7	14.7	4.9	17.5

Question 4.4

a) Primary decomposers include actinomyces, saprotrophic fungi ("molds"), bacteria, earthworms, potworms, certain nematodes, millipedes, slugs and sowbugs. Fungi are grazed by springtails and mites, bacteria by protozoa and by certain nematodes. All of the larger detritivores, and fungivores and bacterivores are prey of certain beetles, pseudoscorpions, centipedes and ants.

b) Part of the nutrients tend to follow the arrows, but ultimately nutrients are excreted by organisms at various trophic levels, and are recycled via root uptake by plants. Energy is degraded from chemical energy to heat, which is lost, and never recycled!

Question 4.5

Ultimately, microbial litter and humic substances are decomposed too, and in a steady state condition (input of litter into the soil organic carbon pool = output of CO_2 , water and nutrients from that pool), net immobilisation is zero, and equation 4-2 holds for the bulk of all soil organic matter.

Question 4.6

a) For every 100 g of plant litter decomposed, 1/50 * 100 = 2 g of N is released. At a C assimilation factor of 0.2 , 0.2* 100 = 20 g of microbial C is formed, containing 1/10* 20= 2 g N. So all N must be transformed into microbial tissue. The N assimilation factor is 1.0.

b) At a C assimilation factor of 0.1, only 10 g of microbial C and 1 g of microbial N are formed. The remaining N is mineralised; the N assimilation factor is 0.5.

c) If the C assimilation factor is 0.5, not enough N is present to transform plant litter completely to microbial biomass. Any inorganic N present in the environment may be utilised to transform more plant litter into microbial biomass: N is immobilised.

Question 4.7

a) See table.

Component	Xt/Xo after 1 year	ln Xt/Xo	k $(year)^{-1}$	MRT (year)
sugar	0.01	-4.6	4.6	0.22
cellulose	0.25	-1.39	1.39	0.72
phenols	0.9	-0.105	0.105	9.5

b) The higher residence time of cellulose in the field is because in plant litter cellulose is partly protected against decomposition by the more recalcitrant lignin associated with it. So pure cellulose decomposes faster.

Question 4.8

NO_3^-, Mn(IV, III) oxides, Fe(III) oxides, and SO_4^{2-}. Manganese and iron are common in most soils; nitrate is an important electron acceptor in heavily fertilised soils; sulphate is common in acid sulphate soils (and gypsiferous soils).

Question 4.9

In an anoxic flooded rice field, the C assimilation factor is lower than under oxic conditions, so that a larger fraction of the mineralised N is available for plants (see answer to Question 4.6).

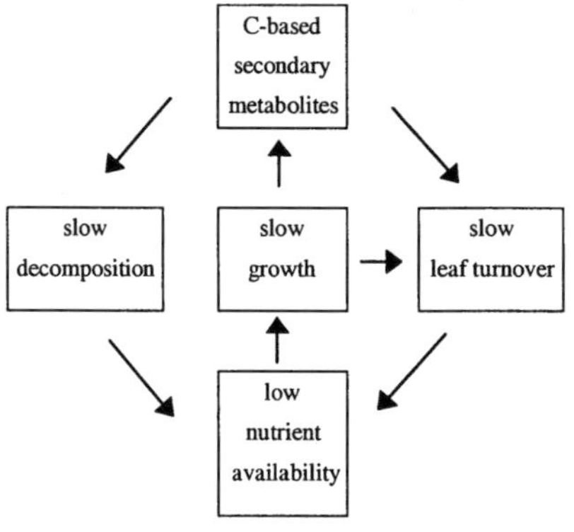

Question 4.10

a) See drawing.

b) C-based secondary metabolites are made up only of C (together of course with O and H). Examples are phenols and polyphenols. They are typical of plants of nutrient-poor sites. By contrast, plants of nutrient-rich soils often deter grazers with N-rich metabolites, e.g. HCN or nicotine. The advantage to a plant of using C-based deterrents in nutrient-poor sites is obvious.

Question 4.11

Sizeable litter layers are found under forest systems, where net primary production (and annual litter fall) is much higher than under grass or herb vegetation. Because decay is slowest at low temperatures, litter layers are thickest under such circumstances. In the tropics, decay is fast, but yearly litter production is high enough to maintain a litter layer.

Question 4.12

ln Xt/Xo = -kt. We can calculate k in year^{-1} by: k = ln Xt/Xo x 365/t

	Xt/Xo	ln Xt/Xo	t (days)	k
Heavy	0.6	-0.51	1000	0.19
Intermediate	0.4	-0.92	1000	0.34
Light	0.1	-2.30	1000	0.84
20-150 microns	0.75	-0.29	6000	0.018
< 20 microns	0.68	-0.39	6000	0.024

Question 4.13

a) Phenolic remnants, which may be partly due to lignin, are characterised by aromatic rings, and materials derived from aliphatic polymers by long 'zig-zag' C-chains; remnants of sugar are not found in this structure (because the authors of this model are convinced that sugars are not part of the 'structure').

b) Comparison with the lignin building blocks shows increased content of O, especially as -COOH groups.

Question 4.14

CEC of organic matter (at a given pH) depends on dissociation of acidic groups. The relative amount of dissociated acidic group increases with pH.

Question 4.15

The number of C atoms in figure 4.8 is about 280; 28 of these are in -COOH groups. So the charge is 28 moles COOH/280*12 g C = 833 cmol(-)/kg C.

Question 4.16

Bioturbation is intense in the upper horizons but proceeds much more slowly in the subsoil because supply of food for burrowing organisms (leaf and root litter) decreases with depth.

Question 4.17

(i) Large aggregates are more easily disrupted by physical disturbance (ploughing), (ii) organic binding agents are decomposed (and generally not replenished as fast in arable land as e.g. under grassland),), and (iii) inorganic binding agents, which probably are little affected by the change to arable land, play a role only in the smallest aggregates (heavy fraction); see Fig 4.7).

Question 4.18

On infertile sites, plants have not enough nutrients to translate their photosynthetic production (basically carbohydrates) into new tissue (= carbohydrates + N, P, K etc). So the carbohydrates 'spill over' to the mycorrhizal fungi, the growth of which increases the uptake capacity of the plant for nutrients.

Question 4.19

In a nutrient-rich soil, e.g. a young soil with high contents of weatherable minerals throughout the soil profile, differences in available nutrient contents between the surface soil and the subsoil are likely to be small. Therefore, the relative effect of nutrient addition to the surface soil by nutrient cycling is small. In a wet climate where leaching rates are high, the soil solution is very dilute and temporary precipitates (e.g. $CaCO_3$, Ca phosphates) can hardly form. In these conditions, nutrients stored at exchange sites and in soil organisms are best protected against leaching. This storage capacity is mainly in a "forest floor" (= the O horizon) and in surface horizons.

Problem 4.1

a) At steady state the plant biomass is constant from year to year, so an amount of dead biomass equal to annual net primary production must be returned to the soil.

b) See figure A below; the belowground fraction of total biomass decreases with increasing NPP.

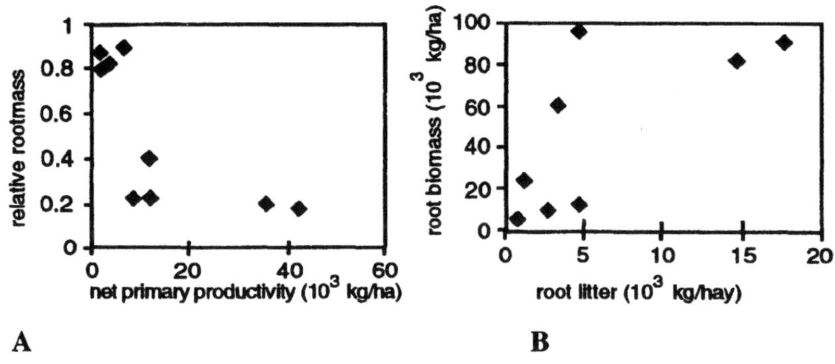

A B

c) NPP increases and relative root biomass decreases as the climate becomes more favourable for plant growth. Under unfavourable climatic conditions, plants can survive below-ground; in dry climates they need a large root system to obtain sufficient water.

d) In an infertile soil, plants also must invest in their root system to obtain sufficient nutrients. e) See graph B above. Root biomass tends to increase with annual root production, but in highly productive systems, annual inputs are high relative to the standing root biomass: apparently in such systems roots grow quickly and die quickly.

Problem 4.2

a) $input_{humus} = (1 - k_f)*X$; $output_{humus} = k_h * Y$, so $Y = (1 - k_f)*X/k_h$.

b) Y= 99.000 kg/ha; this would give an org. C content of 4%, rather normal!

c) Not all plant litter decomposes in one year to humus (the humification rate is not equal to $1\text{-}k_f$); plant litter is composed of many different compounds, that all decompose at different rates. Some of this turns into humic substances, and some is mainly respired.

Problem 4.3

a) C/N of fresh litter: per hectare: C:3500kg leaves * 50%C = 1750 kg C
N:3500 kg leaves * .38%N = 13.3 kg N, so C/N = 131
C/N of 4-year old leaves: per hectare: C: 0.27* 3500* 50%C kg= 472.5 kg C
N: 0.27*3500*1.2%N = 11.34 kg N, so C/N = 41.7

b) C and N disappeared in 4 year: C:1750-472.5 = 1277.5 kg/ha; N: 13.3-11.3= 2 kg/ha

c) During decomposition, plant litter is partly transformed to microbial biomass with a much lower C/N. So N is largely retained, while most of the C is lost (as CO_2).

Problem 4.4

Day	Aerobic decomposition			Anaerobic decomposition		
	C mineralised	X_t/X_0	$\ln X_t/X_0$	mg C mineralised	X_t/X_0	$\ln X_t/X_0$
1	25.4	0.94[1]	-0.05	34+0.9=34.9	0.93	-0.072
2	44.1	0.91	-0.092	45.6	0.91	-0.096
5	69.5	0.86	-0.15	54.7	0.89	-0.116
7	82.4	0.83	-0.18	62.1	0.88	-0.133
10	91.5	0.817	-0.202	69.6	0.86	-0.150
14	98.3[2]	0.803	-0.22	84.9	0.83	-0.18
22	103	0.794	-0.23	83.3[3]	0.83	-0.18
30	111	0.778	-0.25	83.9	0.83	-0.18

[1]) (500-25.4)/500; [2]) 95.9 (CO_2)+2.4(water soluble org. C) = 98.3; [3]) including 15 mg C in CH_4

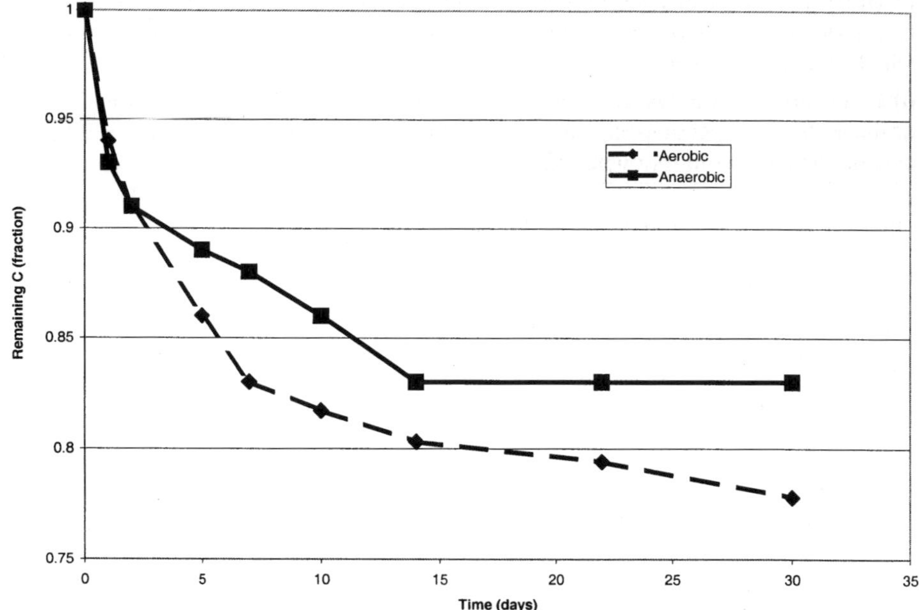

Roughly three linear stages can be distinguished (see graph above). These phases are clear in anaerobic decomposition, but rather arbitrarily in case of aerobic decomposition.

Calculation of k: $\ln X_t/X_0 = -kt$

1st order decomposition rate (day^{-1})	stage I	stage II	stage III	over-all
aerobic	day 0-5; $\ln X_t/X_0$= -0.15, k= 0.15/5=0.03	day 5-14; $\ln X_t/X_0$= -0.07[1], k=0.07/9=0.008	day14-30; k=0.002	k= 0.008
anaerobic	day 0-2; $\ln X_t/X_0$=-0.09, k=0.09/2=0.04	day 2-14; $\ln X_t/X_0$= -0.084, k = 0.064/12 = 0.007	day 14-30; k=0	k= 0.006
[4]) -0.22- (-0.15)=-0.07				

If k values would indeed "jump" from higher to lower values, these might represent decomposition of specific substrate fractions. In the last stage (III) of anaerobic decomposition, not the original substrate decomposes (k=0), but organic acids ferment to CO_2 and CH_4. Except in the very first stage, aerobic decomposition is faster than anaerobic decomposition.

c) The overall rate constants, recalculated from day^{-1} to $year^{-1}$, are resp. 3 $year^{-1}$ (aerobic) to 2 $year^{-1}$ (anaerobic). This is appreciably quicker than decomposition of oak litter, which is about 0.5 $year^{-1}$.

d) Initial amount of N present = mg C/(C/N)= 38.5 mg N. So after 2 days, the C/N ratio of the aerobically decomposed material would be (500-44.1)/(38.5-1.3) = 12.2. Similar calculations yield the following table:

time	C/N aerobic	C/N anaerobic
2	12.2	12.4
30	11.4	12.9

Clearly, in spite of slightly higher C/N ratios, more N is liberated under anaerobic conditions than when the soil is aerobic.

Problem 4.5

a, b) the plot of carbon content versus depth indicates that, in profiles 2 and 3, organic matter is gradually mixed with deeper soil. The litter layer has already disappeared in profile 2. In profile 3, the high carbon content in the upper horizon has decreased, and that of the deeper horizons increased, because of increasingly deep mixing.

Problem 4.6

a. Worm casts have higher contents of clay and organic matter, and hence a higher CEC than the surface soil material. This can be attributed to the preferential consumption of clay by worms, as well as uptake (followed by partial digestion) of organic matter.

b. Assuming that the specific CEC of clay (**a**) and of SOC (**b**) are the same in worm casts and in soil, they can be calculated from two equations and two unknowns of the type:

CEC_{soil} = **a** * clay content + **b** * O.M. content

= **a** mmol/g clay*g clay/g soil + **b** mmol/g C *C/g soil

E.g., for Eketi: solve two unknowns from these two equations.

CEC of S: **a** *0.103 + **b** *0.00143 = 0.029

CEC of W: **a** *0.127 + **b***0.0305 = 0.089

soil series	Eketi	Ibadan	Apomu	Makato
a mmol(±)/g clay	-0.3	-0.6	-0.08	-0.03
b mmol(±)/g org.C	4.1	7.2	4.0	5.0

The negative values of **a** indicate that the assumption about similar properties of clay and organic carbon in worm casts and bulk soil is incorrect. It is likely that is this mainly due to changing properties of the organic fraction as the humus becomes older. At any rate, clay clearly contributes little to the CEC, and organic C a lot.

Problem 4.7

The bulk density of the soil is 1.3 kg/dm^3. This gives 260 kg of soil per m^2 for the upper 20 cm of the soil profile. At an ingestion rate of 10/100 * 200 g of soil per m^2, it would take 13000 days to ingest all soil once. This would take 13000 days/ 300 days per year, or 43.3 year.

Problem 4.8

Soil from the termite mound contains more clay and organic matter than the surrounding soil, because termites preferentially collect and transport fine material and organic-rich soil for building their nests.

4.10. References

Andriesse, J.P., 1987. *Monitoring project of nutrient cycling in soils used for shifting cultivation under various climatic conditions in Asia.* Final report of a Joint KIT/EEC project. No. TSD-A-116-NL. Royal Tropical Institute, Amsterdam.

Balesdent, J., and A. Mariotti, 1996. Measurement of soil organic matter turnover using ^{13}C natural abundance. In: T.W. Boutton anf S. Yamasaki (eds.): *Mass Spectrometry of Soils.* Marcel Dekker, New York, pp. 81-111.

Berg, B., and H. Staaf, 1981. Leaching, accumulation and release of nitrogen in decomposing forest litter. Ecological Bulletin (Stockholm), 33:163-178

Beyer, L., 1996. The chemical composition of soil organic matter in classical humic compounds and in bulk samples - a review. Zeitschrift für Pflanzenernährung und Bodenkunde, 159:527-539.

Canadell, J., R.B. Jackson, J.R. Ehleringer, H.A. Mooney, O.E. Sala, and E.D. Schulze, 1996. Maximum rooting depth for vegetation types at the global scale. Oecologia, 108:583-595.

Chapin, F.S., 1991. Effects of multiple environmental stresses on nutrient availability and use. In: H.A. Mooney, W.E. Winner and E.J. Pell (Eds.): *Response of plants to multiple stresses.* Academic Press, San Diego, p. 67-88.

Darwin, C., 1881 (13th thousand, 1904). *The formation of vegetable mould through the action of worms with observations on their habits.* John Murray, London. 298 p.

De Vleeschauwer, D., and R. Lal, 1981. Properties of worm casts under secondary tropical forest regrowth. Soil Science, 132: 175-181.

Dijkstra, F., 2000. *Effect of tree species on soil properties in a forest of the northeastern United States.* PhD Thesis, Wageningen University.

Dindal, D.L., 1978. Soil organisms and stabilizing wastes. Compost Science, 19:8-11.

Engbersen, J.F.J., and Æ. De Groot, 1995. *Inleiding in de bio-organische chemie.* Wageningen Pers, Wageningen, 576 pp.

Flaig, W., H. Beutelspacher, and E. Rietz, 1975. Chemical composition and physical properties of humic substances. In: J.E. Gieseking (Ed.): *Soil components, Volume 1: Organic Components*, pp. 1-211. Springer Verlag, Berlin.

Gale, P.M. and J.T. Gilmour, 1988. Net mineralization of carbon and nitrogen under aerobic and anaerobic conditions. Soil Science Society of America Journal, 52: 1006-1010.

Glaser, B., 1999. *Eigenschaften und Stabilität des Humuskörpers in 'Indianenschwarzerden'.* Bayreuther Bodenkundliche Berichte, Band 68.

Gosz, J.R. 1981. Nitrogen cycling in coniferous ecosystems. Ecological Bulletin (Stockholm), 33:405-426.

Hassink, J., 1995. *Organic matter dynamics and N mineralization in grassland soils.* PhD Thesis, Wageningen Agricultural University, 250 pp.

Haumaier, L., and W. Zech, 1995. Black carbon - possible source of highly aromatic components of soil humic acids. Organic Geochemistry, 23:191-196.

Hoeksema, K.J., 1961. Bodemfauna en profielontwikkeling. p. 28-42 in: Voordrachten B-cursus bodemkunde 14-18 september 1959. Dir. Landbouwonderwijs.

Hoogerkamp, M., H. Rogaar, and H. Eijsackers, 1983. The effect of earthworms (Lumbricidae) on grassland on recently reclaimed polder soils in the Netherlands. In: J.E. Satchell (ed.) *Earth worm ecology:* 85-105. Chapman & Hall, London.

Johnson, D.L., D. Watson-Stegner, D.M. Johnson, and R.J. Schaetzl, 1987. Proisotropic and proanisotropic processes of pedoturbation. Soil Science, 143, 278-292.

Jongmans, A.G., N. Van Breemen, U. Lundström, P.A.W. van Hees, R.D. Finlay, M. Srinivasan, T. Unestam, R. Giesler, P.A.Melkerud and M. Olsson, 1997. Rock-eating fungi. Nature, 389:682-683.

Kononova, M.M., 1975. Humus of virgin and cultivated soils. In: J.E. Gieseking (ed): *Soil Components. I. Organic Components*, p 475-526. Springer, Berlin.

Leprun, J.C., 1976. An original underground structure for the storage of water by termites in the Sahelian region of the Sudan zone of Upper Volta. Pedobiologia, 16, 451-456.

Nadelhoffer, K.J., and J.W. Raich, 1992. Fine root production estimates and below-ground carbon allocation in forest ecosystems. Ecology, 73:1139-1147.

Paton, T.R., 1978. *The formation of soil material.* Allen & Unwin, London.

Persson, H. 1990. Methods of studying root dynamics in relation to nutrient cycling.

Schlesinger, W.H. 1977. Carbon balance in terrestrial detritus. Annual Review of Ecological Systems, 8:51-81.

Schmidt, M.W.I., and A.G. Noack, 2000. Black carbon in soils and sediments: analysis, distribution, implications, and current challenges. Global Biogeochemical Cycles, 14:777-793.

Schmidt, M.W.I., J.O. Skjemstad, E. Gehrt, and I. Kögel-Knabner, 1999. Charred organic carbon in German chernozemic soils. European Journal of Soil Science, 50:351-365.

Schnitzer, M, 1986. Binding of humic substances by soil mineral colloids. p 77-101 in: P.M. Huang and M. Schnitzer (eds.): *Interactions of soil minerals with natural organics and microbes*. SSSA Special publication 17, Madison, Wisc. USA.

Schulten, H.R., 1995. The three-dimensional structure of humic substances and soil organic matter studied by computational analytical chemistry. Fresenius Journal of Analytical Chemistry, 351:62-73.

Tisdall, J.M. and J.M.Oades, 1982. Organic matter and water-stable aggregates in soils. Journal of Soil Science, 33: 141-163.

General and further reading:

Anderson, J.M., and J.S.I. Ingram, 1989. *Tropical soil biology and fertility: a handbook of methods.* C.A.B. International, Wallingford, 171 pp.

Huang, P.M., and M. Schnitzer (Eds.), 1986. *Interactions of soil minerals with natural organics and microbes*. SSSA Special Publication No. 17. Soil Science Society of America, Madison, 606 pp.

Piccolo, A. (ed.), 1996. Humic substances in terrestrial ecosystems. Elsevier, Amsterdam, 675 pp.

Swift, M.J., O.J. Heal, and J.M. Anderson, 1979. *Decomposition in terrestrial ecosystems.* Studies in Ecology, Vol. 5. Blackwell Science Publishers, 372 pp.

PART C

SOIL PROFILE DEVELOPMENT

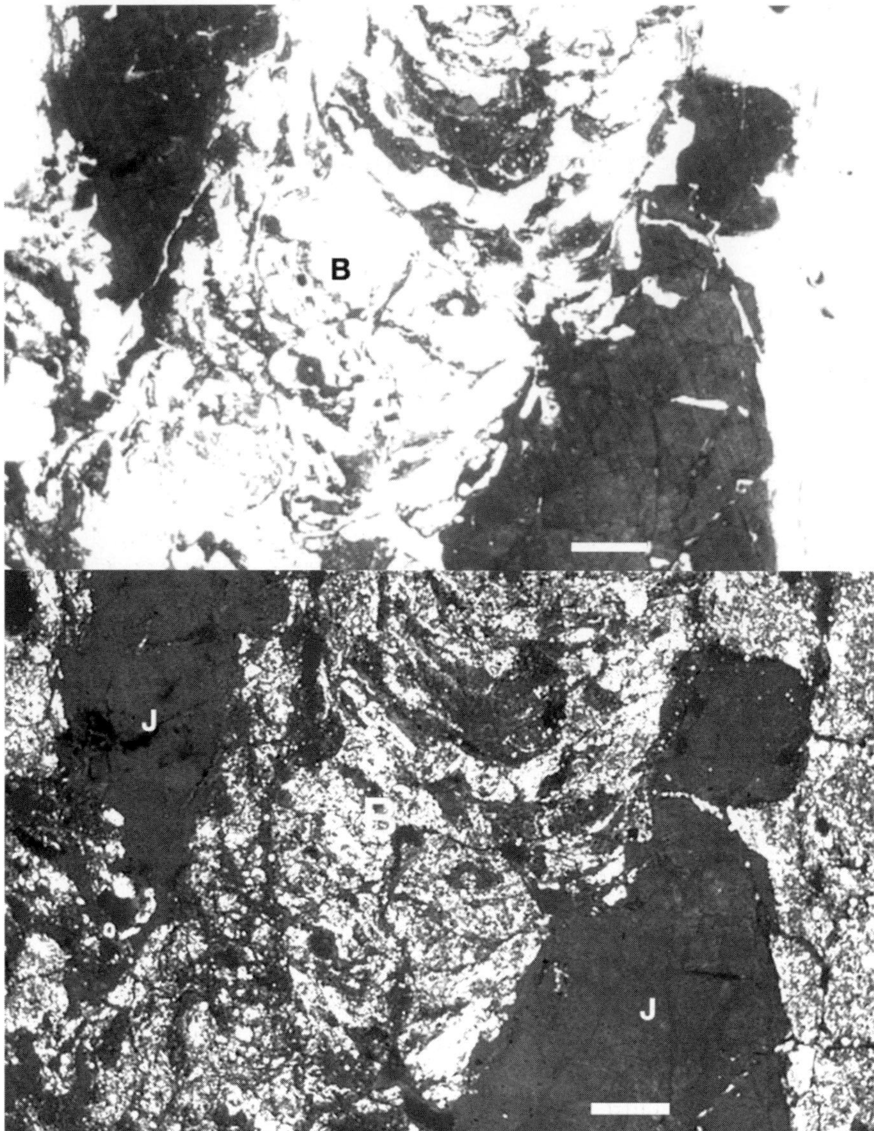

Plate I. An animal burrow or *pedotubule* (B) dissecting a previously formed jarosite accumulation (J) in an acid sulphate soil from Guinee Bissau. Top, plain light; bottom: crossed polarisers. Note striations in burrow; the animal moved from bottom left to top right. Scale bar is 585 μm. Photographs A.G. Jongmans.

CHAPTER 5

STUDYING SOIL PROFILES

5.1. Introduction

In part C we will treat the formation of the main genetic soil horizons that are used as diagnostic criteria in the international soil classification systems (FAO-Unesco and Soil Taxonomy). You will be able to describe the formation of these diagnostic horizons, and thereby, of the different genetic soil types, in terms of the physical, chemical and biological processes you studied in Part B.
A specific soil forming process (made up of a number of sub-processes) is usually responsible for the formation of a particular genetic soil horizon or a particular set of genetic horizons that are usually observed together.

Question 5.1.*Give examples of a soil forming process that causes formation of one genetic soil horizon, and of a soil forming process responsible for 'twin' sets of diagnostic horizons.*

CAMBIC HORIZON
Many of the constituent processes underlying several different specific soil-forming processes take place in the same pedon. One example is complexation of Al^{3+} and organic matter, which dominates in podzols and some andisols, but which takes place in a less extreme form in practically all acidic surface soil horizons. Another example is silicate weathering, which is the dominant soil forming process in ferralitisation, but proceeds in practically all soils in humid areas.
The Cambic horizon is typically intermediate between a number of more outspoken diagnostic horizons (Figure 5.1). It usually has undergone moderate weathering, because of young age or low intensity of soil formation. Often, it has weakly expressed characteristics of some other diagnostic horizon, towards which it may develop. Its properties grade naturally into those of the other diagnostic horizons, although it is defined by strict, but arbitrary, classification boundaries. Sometimes the cambic horizon is strongly weathered and very stable, remaining virtually unchanged for thousands of years. In Soil Taxonomy (SSS, 1990), the cambic horizon can be poorly drained, but in FAO (1988) and WRB (Deckers et al., 1998), it is restricted to aerobic circumstances

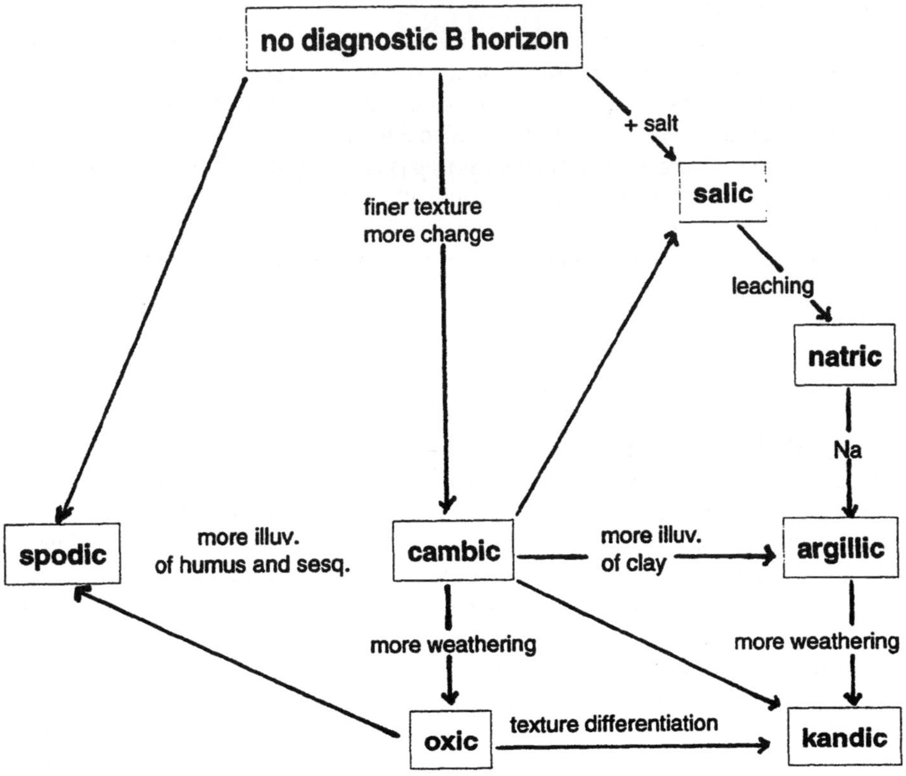

Figure 5.1. The cambic horizon and its relation to some other diagnostic horizons (USDA terminology).

and excludes the influence of gley. In short, it is a wastebasket' kind of diagnostic horizon, not caused by a specific set of processes. Therefore, we will not treat the cambic horizon in a separate chapter. At the same time, when studying the material in Part C, you should be aware that the definition of other genetic soil horizons has arbitrary aspects as well. Moreover, horizons due to different soil forming processes that overlapped in time or space may be found in one and the same soil profile. In Chapter 15, temporary and spatial overlaps of soil forming processes will be illustrated.

Before discussing the formation of the various diagnostic soil horizons, you should be familiar with the tools to identify and quantify the physical, chemical and biological changes that took place during the formation of a soil from its parent material. These will be provided in this chapter.

HISTORIC APPROACH

A soil profile is a vertical sequence of soil horizons that is the result of a set of soil forming processes that have affected the parent material on a particular spot. The soil profile is a reflection of these soil forming processes, and is therefore the primary source of information about the conditions that prevailed during soil formation (= soil forming factors). When you study a soil profile you take a historic approach. The genetic history of a profile does not always reflect its ongoing soil formation. Some of these ongoing soil-forming processes can be studied in real-time. After studying this chapter you should be able to quantify, in principle, changes in a soil profile according to the historic and according to the 'here and now' approach.

5.2. The historic approach to characterise and quantify effects of soil formation

REFERENCE STATE = PARENT MATERIAL

The result of soil formation can be assessed *only* by reference to some former state. This may be the same soil at an earlier stage of soil formation. The ultimate reference is the original, unaltered material: the parent material. There are several ways to obtain information about the parent material, or about earlier stages of soil formation.
Sometimes, soil development has been largely or completely arrested locally at some time in the past, because the soil surface has been covered, e.g. by a sediment or a man-made object such as a dike or a building. Comparison of the soil below and adjacent to such a cover can yield much information about the changes that have taken place since the object was put in place. If, moreover, the age of the cover is known, the mean rate of change can be determined, provided that processes do not continue under a cover. S. Matsson made use of stone walls separating agricultural fields in south Sweden to determine the effect of fertilisation and liming on the chemical properties of podzols. Sometimes, old samples (or analytical data on old samples, plus a detailed description of the analytical methods applied) are available, together with precise data on the sample locations. More often, the soil scientist has to rely on assumptions about earlier stages, from a soil profile or from a chronosequence of soils. Then, the C-horizon, with little or no signs of soil formation is a good first approximation of the parent material.

Question 5.2. *Give a number of morphological field criteria a C-horizon has to meet in order to qualify for parent material in terms of absence of signs of soil formation. Is the absence of such signs a guarantee that the C material* ***is*** *the parent material?*

HOMOGENEITY OF PARENT MATERIAL

Field observations on the landscape around the site and on the morphology of the soil profile often provide a preliminary answer to the question whether or not the soil horizons near the surface (where the influence of climate and biological agents are strongest) developed from material with the same properties as that found now at greater depth. Sharp textural changes at some depth often indicate a boundary between two different sediments. To test whether a soil profile developed from vertically homogenous parent material, we can use textural, mineralogical, or chemical clues. Soils on Holocene or late Pleistocene sediments are usually too young to have undergone much weathering of silicate minerals. Therefore the original size distribution of the Si-containing minerals has hardly changed, and the ratio of the sand and silt fractions can be used to test for vertical homogeneity of the parent material.

Question 5.3. *Why should you not use grain-size fraction ratios to test for the vertical homogeneity of the parent material in case of very old soils?*

In old soils, studying the distribution with depth of minerals that are very resistant to weathering and that are not transported easily by physical processes can test vertical homogeneity. Examples of such minerals are zircon (ZrO_2), rutile (TiO_2), tourmaline, and, except in very old soils, quartz. If a mass fraction ratio of such minerals, e.g., % zircon/ % rutile, does not change with depth, the parent material is probably homogenous vertically.

Question 5.4. *Why do we consider the mass ratio between two minerals and not simply the mass fraction of one of these minerals?*

INDEX COMPOUND

The absolute mass in g of these weathering-resistant minerals has not changed with time, and therefore other changes (e.g. the loss of weatherable minerals such as feldspars) can be calculated with reference to that absolute mass. Such a reference compound, that neither weathers nor moves, is called an **index** compound.

STRAIN

Brimhall and Dietrich (1987) presented a systematic way to calculate mass balances of weathering profiles. First they consider the relative change in thickness ('strain', ε) if a certain thickness of parent material (d_p) is converted to a layer of soil with a thickness d_s (d_p and d_s can be expressed in any unit of length). The strain is a dimensionless fraction, defined as:

$$\varepsilon = (d_s - d_p)/d_p \qquad (5\text{-}1)$$

or

$$\varepsilon = (d_s/d_p - 1) \qquad (5\text{-}1a)$$

Question 5.5 *What has happened to the thickness of a certain layer of parent material if $\varepsilon = 0$? What if $\varepsilon > 0$? What processes could be responsible for a positive value of ε, what for a negative value? Do you expect positive or negative strain to prevail in young soils derived from fine-textured sediments without easily weatherable minerals?*

To calculate ε we need an index compound. By definition, the mass fraction of an index compound or element **i** present in a certain layer of parent material does not change upon expansion or collapse of that layer in the course of soil development.

Question 5.6 *Does the concentration (mass fraction of total mineral soil) of an index compound change in the course of soil development?*

The amount of index compound (kg/m^2) follows from its concentration C_{is} (kg/kg soil), the thickness of the soil layer d_s (m), and the soil bulk density ρ_s (kg/m^3) :

$$C_{i,s} * d_s * \rho_s = \text{constant} \qquad (5\text{-}2)$$

From equations 5-1 and 5-2 it can be shown that for each soil layer considered,

$$\varepsilon = \rho_p C_{i,p} / \rho_s C_{i,s} - 1 \qquad (5\text{-}3)$$

CALCULATION OF GAINS AND LOSSES

The subscript (p) stands for parent material, the subscript (s) for the soil derived from that parent material. Equation 5-3 is used to calculate the strain. Once ε is known, we can calculate, for each horizon, the fraction of element j added to or removed from the amount originally present in the parent material during soil development, $\tau_{j,s}$, as outlines on the following page:

$$\tau_{j,s} = (\rho_s C_{j,s} / \rho_p C_{j,p})\,(\varepsilon_i + 1) - 1 \qquad (5\text{-}4)$$

Like ε, $\tau_{j,s}$ is a dimensionless fraction.

Question 5.7. *What can you say about the loss of an element j since the start of soil formation in a soil horizon s, for which $\tau_{j,s} = 0.5$*

If we know both $\tau_{j,s}$ and the amount of the substance j in the (present-day) soil horizon (= $C_{j,s} * d_s * \rho_s$, e.g. expressed in kg/m^2), we can calculate the amount of j that was originally present ($x_{j,p}$) from:

$$x_{j,p} + \tau_{j,s} * x_{j,p} = C_{j,s} * d_s * \rho_s \qquad (5\text{-}5)$$

From the mass loss of j per horizon, $\tau_{j,s} * x_{j,p}$, the total mass flux of each element can be found by adding up the values of $\tau_{j,s} * x_{j,p}$ for all horizons of the soil profile.

Derivation of ε and τ

Brimhall and Dietrich (1987) give the following derivation of their formulas. We have substituted 'depth' (d) for the original 'volume' (v), because we suppose that weathering does not affect the surface area of any given profile.

In equations 5.1, 5.2, and 5.3 the strain and the change in concentration of an index compound in relation to strain are given. If we define the loss or gain of a none-index compound as *flux*, the relation between content of the element before and after weathering is:

$$(d_p \times \rho_p \times C_p)/100 + \text{flux} = (d_s \times \rho_s \times C_s)/100 \qquad \text{A}$$

or

$$\text{flux} = (d_s \times \rho_s \times C_s)/100 - (d_p \times \rho_p \times C_p)/100 \qquad \text{B}$$

division of (B) by d_p gives:

$$\text{flux}/d_p = (d_s \times \rho_s \times C_s)/100\, d_p - (\rho_p \times C_p)/100 \qquad \text{C}$$

now substitute (ε + 1) for (d_s / d_p) from Eq. 5-1a into C:

$$\text{flux}/d_p = \{(\rho_s \times C_s)/100\} \times (\varepsilon + 1) - (\rho_p \times C_p)/100 \qquad \text{D}$$

if the flux is expressed as fraction τ of the original amount:

$$\tau = 100 \times \text{flux}/(d_p \times \rho_p \times C_p) \qquad \text{E}$$

solving flux, and substitution in E gives:

$$\tau = (\rho_s \times C_s/\rho_p \times C_p) \times (\varepsilon + 1) - 1 \qquad (5.4)$$

ISOVOLUMETRIC WEATHERING

At a certain depth below the soil surface, outside the range of activity of plant roots and burrowing soil fauna, chemical weathering of rock often proceeds without disruption of the original rock structure. In that case, absolute mass transfers from the 'rotten rock' or saprolite can be calculated by expressing concentrations of minerals or components on a volume basis: the isovolumetric technique of studying weathering (Millot, 1970).

Sometimes one need not worry about the reference situation. E.g. changes in organic matter contents with depth or time in profiles with parent materials that did not contain significant amounts of organic matter can usually be calculated with reference to the mass of inorganic soil components.

Question 5.8. *Give an argument for the preceding statement. Can you suggest any situations where it is not valid?*

5.3. The "here and now" approach

In the 'here and now' approach we try to measure the nature and rates of soil forming processes that take place now. A general problem with this approach is that most soil forming processes proceed so slowly that their effects are hardly measurable within the few weeks to months that we might call 'now'. However, some soil forming processes proceed relatively fast, and can be followed by repeated measurements or observations over a few months or years, or so-called 'monitoring'. Several soil properties may fluctuate seasonally. Only oscillations that leave a net effect over time are included in 'soil formation'.

Question 5.9. *List a few processes that can be considered 'soil formation', and that could be observed to proceed within months to years. Also list some soil processes that could be monitored but that are not considered 'soil formation'.*

Very slow processes such as weathering of silicate minerals cannot normally be studied by monitoring soil properties that are related directly to the content of the weatherable minerals. Even the decrease in the content of $CaCO_3$ (which weathers more quickly than most silicates) in a few years of weathering is usually much smaller than the spatial variability in the $CaCO_3$ content and than the analytical precision. So, repeated (destructive) sampling and soil analyses provide little information on the rate of mineral weathering. However, such information can be obtained by studying the composition of drainage water.

Question 5.10. *Water draining from a calcareous surface soil (bulk density 10^3 kg/m^3) contains 200 mg/l of dissolved $CaCO_3$. Annual drainage is 500mm. a) What is the decalcification rate (g/m^2 soil surface/ year). b) In 1990 the upper 20 cm of the soil contains 20 $\pm$ 5 % (mass fraction) of $CaCO_3$, the uncertainty being due to a combination of spatial variability and analytical errors. In what year would you come back to sample and analyse the soil if you want to ascertain if the $CaCO_3$ content actually decreases?*

Question 5.11. *What would be a better way than repeated destructive sampling and analysis to quantify the rate of decalcification?*

Question 5.12. *Why is the method of Question 5.11 so much more accurate than the repeated sampling and analysis of the soil?*

The chemistry of drainage water or of the soil solution can also give insight in the degree to which minerals are in equilibrium with the surrounding soil solution. Through thermodynamic calculations, the so-called saturation index can be determined, which shows whether a given mineral will dissolve (weathering) or will precipitate instead.

Question 5.13. *The soil solution of a calcareous soil with a P_{CO_2} of 0.5 kPa and a temperature of 20 °C contains (a) 50, (b) 142, and (c) 200 mg $CaCO_3$. Can you decide in which case $CaCO_3$ forms or dissolves? Consult Figure 3.5 on page 49.*

So, in conclusion, we can study soil profile development by two approaches. First by considering the distribution of solid soil components in the profile, we can try to obtain quantitative information about mass transfers out of (or into) the whole soil and between different horizons over long, but not over short, time scales. Second, by analysing drainage fluxes we can quantify mass transfers that take place during the time of the study. Comparing the two can provide useful insights as well. For example, annual losses of dissolved Al from most European forest soils now are 10 to 100 times higher than integrated losses in the course of the Holocene. The reason for the discrepancy is the greatly increased rate of Al dissolution as a result of acid atmospheric deposition.

5.4. Problems

Problem 5.1

Table 5A gives texture data on a loess soil which apparently has a textural B-horizon (Brinkman, 1979). The question is how much clay has been washed from the surface soil into the textural B-horizon. We will assume that the soil profile developed from vertically homogenous parent material, represented by the C1 horizon.

a. Calculate for each horizon the strain (ε), using two different indexes: (1) the contents of sand plus silt, and (2) the content of TiO_2 present in sand plus silt-sized material.
b. Explain the somewhat different results obtained with the two indexes.
c. Calculate the amounts of clay (expressed in kg/m^2), (1) present now in each horizon and in the whole soil, (2) originally present in each horizon and in the whole soil, and (3) lost or gained by each horizon and by the whole soil. Why would the TiO_2 content of the whole soil be a less suitable index as that in the sand and silt fraction?

d. Can clay migration alone explain the outcome of d. (the "clay balance")? What possible reasons are there for a discrepancy?

Table 5A. Particle size data (expressed as mass fractions of the fine earth (= soil sieved through a < 2 mm sieve)) of a soil with clay illuviation developed in loess, Nuth, South Limburg, The Netherlands (From Brinkman, 1979).

Depth cm	Horizon	bulk density $10^3 kg/m^3$	sand	silt	clay	TiO_2 in sand +silt
			mass fraction (%)			
0 - 25	Ap	1.40	10	79	11	0.74
25 - 41	Bt1	1.44	7	78	15	0.75
41 - 80	Bt2	1.47	4	76	20	0.71
80 -130	Bt2	1.56	6	73	21	0.72
130 -180	Bt3	1.57	7	73	20	0.71
180 - 223	C1	1.47	5	76	19	0.71

Problem 5.2

Table 5B gives elemental analyses and the bulk density of the parent material and of selected soil horizons from two soils developed in andesitic material in the Atlantic zone of Costa Rica (present day rainfall 3500-5500 mm/year). Assume that TiO_2 is immobile during weathering and soil development.

a. Calculate for each horizon the value of ε_{TiO2}
b. Calculate for each horizon the values of τ for each element
c. Discuss the results: what processes were involved in dilation and compaction, and in losses of elements.

Problem 5.3

Table 5C gives annual drainage fluxes (mm/yr) and (flux-weighted) concentrations of selected solutes in the soil solution at different depths in an acidic sandy forest soils, with (A) and without (B) a calcareous subsoil. Values are three-year means (from Van Breemen et al., 1989). The water at 0cm depth percolates from the organic forest floor. This problem illustrates the use of hydrological data (water fluxes) in combination with chemical composition of soil solution samples, to quantify present-day processes of solute transfer. E.g. Al moves from the surface soil to the subsoil (soil A) or is completely lost from the soil profile (soil B). The high mobility of Al is unusual, and typical for sandy forest soils heavily impacted by "acid rain".

Table 5B. Chemical analysis (mass fraction, %) and bulk density (g/cm^3) of selected horizons from two soils developed in andesitic material, Costa Rica (From Nieuwenhuyse and Van Breemen, 1996)

depth/cm	SiO_2	TiO_2	Al_2O_3	MgO	CaO	ρ
	Profile AT 7, 5000 yrs old, developed in sandy beach ridge					
0-8	55.6	1.66	15.8	6.13	5.57	0.3
28-33	52.9	1.50	19.0	6.46	5.66	0.4
90-100	53.9	1.39	18.0	6.41	6.18	0.85
150-160 (parent material)	55.03	1.30	17.0	6.52	6.93	1.10
	Profile RC, 450.000 years old, developed in lava					
0-5	25.3	3.41	48.8	0.17	0.02	0.92
25-35	24.0	3.44	49.9	0.11	0.00	0.99
150-160	13.1	3.89	58.1	0.09	0.00	1.14
375+ (parent material)	55.5	1.17	18.1	4.01	7.82	2.60

Table 5C. Soil solution composition and soil water fluxes at different depths in two forested sandy soils in the Netherlands. From Van Breemen et al., 1987.

Soil	depth (cm)[1]		concentration in soil solution (mmol/l)					water flux (mm/yr)
		pH	Ca^{2+}	Al^{3+}	NH_4^+	NO_3^-	HCO_3-	
A	0	3.6	0.24	0.03	0.38	1.0	0	551
	10	3.6	0.39	0.2	0.24	1.4	0	448
	60	4.6	1.2	0.2	0.0	2.1	0	191
	90	7.1	2.4	0.01	0.0	2.1	2.1	193
B	0	4.2	0.15	0.01	1.1	0.9	0	551
	10	3.5	0.20	0.19	0.2	1.2	0	486
	60	4.0	0.25	0.64	0.1	1.7	0	198
	90	4.0	0.23	0.71	0.0	1.4	0	206

[1]depth below the boundary between O horizon and mineral soil

a. What is the rooting depth of the trees? Be aware that the change in the flux of percolating water with soil depth is caused by water uptake by roots.
b. Calculate the flux of each solute for each horizon (solute flux in mmol/l * mm/yr = $mmol/m^2.yr$)
c. The following processes may take place
   (i) percolation of solutes,
   (ii) mineralization of organic matter to $NH_4^+ + OH^-$ and to $Ca^{2+} + 2OH^-$
   (iii) nitrification of NH_4^+ to $NO_3^- + 2H^+$,
   (iv) uptake by trees of NO_3^- and Ca^{2+},
   (v) dissolution of $Al(OH)_3 + 3H^+$ to Al^{3+}
   (vi) precipitation of Al^{3+} to $Al(OH)_3 + 3H^+$ and
   (vii) dissolution of $CaCO_3 + H^+$ to Ca^{2+} and HCO_3^-

Assuming that over one year the amount of dissolved substances within any one soil layer (0-c, 10-c, 60-c) did not change appreciably, the net release to (or net uptake from) the soil solution in a layer is indicated by the differences in fluxes at its upper and lower boundary. By comparing calculated solute fluxes at different depths, deduce which of the processes listed above could take place in which soil at what depths.

5.5 Answers

Question 5.1

Accumulation of humus (giving an A-horizon) and gleying (giving a cambic gley horizon) are examples of processes responsible for the formation of one genetic soil horizon. Coupled eluviation/illuviation as in podzolisation and vertical clay movement are responsible for 'twin sets' of diagnostic horizons, an E (eluvial) horizon depleted of certain substances, overlying an illuvial B horizon.

Question 5.2

No signs of chemical weathering (e.g. brown colours due to liberation of iron from primary silicates), undisturbed sedimentary layering, no root activity (and, therefore, no input of organic matter). The absence of such signs is no guarantee that the C-horizon is the parent material: there can be a lithological discontinuity above the C-horizon.

Question 5.3

In old soils, weatherable sand- and silt-sized minerals, e.g. feldspars, may have been weathered, in part to clay minerals, so the texture itself is affected by soil formation.

Question 5.4

The mass fraction of a weathering-resistant mineral increases with time because other minerals disappear. The mass ratio of two resistant components does not.

Question 5.5

If $\varepsilon = 0$, the thickness of a certain layer of parent material has not changed in the course of soil formation. If $\varepsilon > 0$, a certain layer of parent material has expanded to a thicker layer of soil e.g. due to formation of pores or addition of organic matter. If $\varepsilon < 0$, the parent material has collapsed, e.g. by removal of weatherable minerals. In all soils with low to moderate weathering, the increase in volume caused by porosity is larger than the loss of material, so usually $\varepsilon > 0$.

Question 5.6

The concentration of an index compound will increase if other substances are removed by weathering.

Question 5.7

If $\tau_{j,s} = 0.5$, half of the mass of j that was present in the original parent material from which s was formed, has been added in the course of soil formation.

Question 5.8

Accumulation or loss of organic matter due to soil formation or changes in land use are usually rapid relative to changes in the mineral part of the soil, so quantifying changes in soil organic matter, the total mineral mass can be used as an index. If there is a considerable change in the mineral mass (e.g. accumulation of calcite), the total mineral mass cannot be used as an index.

Question 5.9

Salinization, gley mottling and extreme acidification as a result of pyrite oxidation are examples of soil formation that may produce marked changes in a few years. Seasonal changes in contents of water and nutrients, soil pH and temperature are not considered "soil formation". However, the boundary between "soil formation" and "not soil formation" is somewhat arbitrary for properties such as gleying, salinization and acidification, which may oscillate on a seasonal basis.

Question 5.10

a) The annual loss of $CaCO_3$ per m^2 of soil surface is 0.5 m (drainage)*1 m^2 (soil surface) * 200 g (dissolved $CaCO_3$ per m^3 of drainage water) = 100 g $CaCO_3$.

b) The total amount of $CaCO_3$ in the 20 cm surface soil is 0.2 * 1 (m^3 of soil) * 10^3 (kg soil per m^3) * 200 (g $CaCO_3$ per kg soil) = $4*10^4$ g $CaCO_3$. So, only 0.25 % of the $CaCO_3$ initially present is lost by drainage every year. Given the uncertainty in the $CaCO_3$ content, at least 10 % of the $CaCO_3$ initially present must have disappeared to establish the change with confidence. So you would have to wait 10 % / 0.25 % /year = 40 years before resampling. To establish the rate exactly, more time would be needed, or more (replicated) samples would have to be taken to determine the uncertainty in the determination of the $CaCO_3$ content.

Question 5.11

Question 5.10 implicitly provides the answer: establish the annual drainage rate (estimate precipitation and annual evapotranspiration) and measure, during the year, the concentration of dissolved $CaCO_3$ in drainage water.

Question 5.12

In most solid-water mixtures (= most soils) the ratio of the mass of an element present in solid form to that in solution is very large. As a result, a very small change in the mass of the element in the solid phase (by precipitation or dissolution) is translated into a very large change in the concentration of the element in solution.

Question 5.13

(a) Undersaturated: $CaCO_3$ will dissolve (b) in equilibrium and (c) supersaturated: $CaCO_3$ will precipitate.

Problem 5.1

a) The following table shows the calculated values for the strain (ε) using the two indexes. However, note that the TiO_2 content in the sand and silt fraction must be expressed as mass fraction to the whole soil in any horizon, before it can be used as an index!

depth	thick-ness	ρ	s+s	TiO_2(s+s)	TiO_2(s+s)	ε_{s+s}	ε_{TiO2}
cm	cm	g/cm^3	fraction in soil	% in s+s	fraction in soil	by equation 5-3	
0-25	25	1.4	0.89	0.74	0.0066	-0.044	-0.077
25-41	16	1.44	0.85	0.75	0.0064	-0.027	-0.079
41-80	39	1.47	0.8	0.71	0.0057	0.013	0.013
80-130	50	1.56	0.79	0.72	0.0057	-0.034	-0.047
130-180	50	1.57	0.8	0.71	0.0057	-0.052	-0.052
180-223	43	1.47	0.81	0.71	0.0058	0.000	0.000

b) Except in the B2t horizon, both indexes indicate a collapse, which must be attributed to loss of minerals by weathering. Because weathering took place, the amount of s+s must have changed, so s+s is not a suitable index. Weathering of minerals in the s+s fraction explains the lower degree of collapse indicated by s+s as an index.

c) See table below.
d) TiO_2 in the whole soil would include TiO_2 present in the clay fraction, which is clearly mobile, and is therefore unsuitable as an index.
e) Clearly, the amount of clay lost from the surface soil (38+14= 52 kg/m^2) is not balanced by the amount gained in the illuvial horizons (7+17+10 = 34 kg/m^2). Possible reasons are a) weathering of clay in the surface soil, or b) loss of very fine clay with the drainage water. Possibility a) is the most likely explanation (see Chapter 7.3 (Ferrolysis).

depth (cm)	$\tau_{s+s, clay}$	$\tau_{TiO2, clay}$	clay (kg/m^2)		
			now	original	loss (cf eq.5-5)
	by equation 5-4			based on τ_{TiO2}	
			$C_{is} * d_s * \rho_s$	x_j	$\tau_{j,s} * x_j$
0-25	-0.47	-0.49	39	76	38
25-41	-0.25	-0.29	35	49	14
41-80	0.07	0.07	115	108	-7
80-130	0.13	0.12	164	147	-17
130-180	0.07	0.07	157	147	-10
180-223	0.00	0.00	120	120	0
		Total	629	646	18

Problem 5.2

a) and b): See table on next page.
c) The soil of AT 7 has expanded (ε >0) and undergone some weathering of Ca and Mg in the surface soil, but less in the subsoil (τ only slightly negative). Expansion must be due to addition of organic matter and pores. The soil of RC has collapsed due to very extensive weathering.
Practically all weatherable Ca- and Mg- minerals (consisting wholly of aluminium silicate minerals) have disappeared, as well as most of the Si that has been originally present (τ practically equal to -1). Most of the Al that was present in the parent material, on the other hand, has remained in the soil, just as Fe (data not shown here).
These changes are typical of strongly weathered, old soils in the tropics. The relatively high value for τ_{Al2O3} in the surface horizon of the young soils is remarkable, and must perhaps be attributed to inhomogeneities in the parent material.

depth/cm	ε_{Ti}	τ				
		SiO_2	TiO_2	Al_2O_3	MgO	CaO
	Profile AT7, 5000 yrs old, developed in sandy beach ridge					
0-8	1.87	-0.21	0.0	-0.27	-0.26	-0.37
28-33	1.38	-0.17	0.0	-0.03	-0.14	-0.29
90-100	0.21	-0.08	0.0	-0.01	-0.08	-0.17
150-160	0	0	0	0	0	0
	Profile RC, 450.000 years old, developed in lava					
0-5	-0.03	-0.84	0.0	-0.07	-1.0	-1.0
25-35	-0.10	-0.85	0.0	-0.06	-0.99	-1.0
150-160	-0.31	-0.93	0.0	-0.03	-0.99	-0.99
375+	0	0	0	0	0	0

Problem 5.3

See tables on next page.

a. The bulk of the roots penetrate to anywhere between 10 and 60 cm: in that zone the water flux decreases with depth (from about 450 mm at 10 cm depth to about 200 mm at 60 cm depth) due to transpiration. Field observation showed that the rooting depth is in fact about 50 cm.

b. Flux calculation: see table below

- solute flux in mmol/l * mm/yr = mmol/m^2.yr
- water flux in mm/yr

c. Compare superscripts with following list: (i) percolation of solutes (not indicated, always where flux is positive), (ii) mineralization of organic matter to Ca^{2+} + $2OH^-$, (iii) nitrification of NH_4^+ to NO_3^- + $2H^+$, partly from NH_4^+ derived from mineralization of organic matter to NH_4^+ + OH^-, (iv) uptake by trees of NO_3^- and Ca^{2+}, (v) dissolution of $Al(OH)_3$ + $3H^+$ to Al^{3+} , (vi) precipitation of Al^{3+} to $Al(OH)_3$ + $3H^+$ and (vii) dissolution of $CaCO_3$ + H^+ to Ca^{2+} and HCO_3^-, (viii) transpiration.

soil	depth	H	Ca	Al	NH_4	NO_3	HCO_3	water
	fluxes at each depth							
A	0	138	132	16	209	551	0	551
	10	112	175	89	107	627	0	488
	60	4.8	229	38	0	401	0	191
	90	0.01	463	1.9	0	405	405	193
	release (+) or uptake (-), per compartment							
	0-10	-26	43 [ii]	73[v]	-102[iii]	76[iii]	0	-63[viii]
	60-10	-108	54	-51[v]	-107[iii]	-226[iv]	0	-289[viii]
	90-60	-4.8	234[vii]	-36[v]	0	4	405[vii]	2
fluxes at each depth								
B	0	34.8	83	5.5	606	496	0	551
	10	154	97	92	97	583	0	486
	60	20	49.5	127	20	337	0	198
	90	21	47	146	0	288	0	206
	release (+) or uptake (-), per compartment							
	0-10	149[iii]	14	86 [v]	-509[iii]	87 [iii]	0	-65[viii]
	60-10	-134 [iv]	-47.5[iv]	35 [v]	-77	-246[iv]	0	-288[viii]
	90-60	1	-2.5	19 [v]	-20	-49 [iv]	0	2

5.6. References

Brimhall, G.H. and W.E. Dietrich, 1987. Consecutive mass balance relations between chemical composition, volume, porosity, and strain in metasomatic hydrochemical systems: results on weathering and pedogenesis. Geochimica et Cosmochimica Acta, 51: 567-587.

Brinkman, R., 1979. *Ferrolysis a soil forming process in hydromorphic conditions.* Agricultural Research Reports 887:1-105. PUDOC, Wageningen.

Deckers, J.A., F.O. Nachtergaele, and O. Spaargaren, 1998. World Refernce Base for Soil Resources. Introduction. ISSS/ISRIC/FAO. Acco, Leuven/Amersfoort, 165pp.

Nieuwenhuyse, A., and N. van Breemen, 1996. Quantitative aspects of weathering and neoformation in volcanic soils in perhumid tropical Costa Rica. Pp 95-113 in A. Nieuwenhuyse: *Landscape formation and soil genesis in volcanic parent materials in humid tropical lowlands of Costa Rica.* PhD Thesis, Wageningen University.

Millot, G., 1970. *Geology of clays.* Springer Verlag, New York, 429 pp.

Van Breemen, N., W.F.J. Visser and Th. Pape, 1989. *Biogeochemistry of an oak-woodland ecosystem in the Netherlands affected by acid atmospheric deposition.* Agricultural Research Reports 930:1-197. PUDOC, Wageningen.

Plate J. Very finely aggregated B-horizon material in an iron-rich Oxisol from Minas Gerais, Brazil. Aggregation is probably due to ants and termites. Coin is 3 cm across. Photograph P. Buurman.

CHAPTER 6

ORGANIC SURFACE HORIZONS

6.1. Introduction

Organic matter reaches the soil as litter at the soil surface and as decaying roots, root exudates, and microbial biomass (dead microbes, hyphae of fungi) in the soil. Breakdown of litter to CO_2 and humic substances, and mixing of the organic material with mineral soil are affected by soil biota (Chapter 4). Between 1 and 5% of the organic matter in the soil consists of living microorganisms.

Organic matter in Ah and O horizons is the result of the balance of production, decomposition and mixing of organic substances. Decomposition and mixing of organic matter are mainly due to soil micro- and meso fauna, and are enhanced by a good availability of oxygen and nutrients, a high temperature, and low amounts of trivalent cations (Al^{3+}, Fe^{3+}) bound to the organic matter.

If, in an environment where litter is produced, one or more of these factors are less than optimal, decay and mixing are hampered. As a result, organic matter contents in the soil will be higher.

PEAT AND O-HORIZONS

In conditions of continuous water saturation, as in swamps, anoxia causes slow organic matter decay and thick organic deposits can be formed: peat.

In well-drained, non-arable soils, low temperature and low nutrient status can cause the accumulation of organic layers on the soil surface. These so-called litter layers, also called *ectorganic horizon* or *forest floor*, consist almost entirely of organic matter. The litter layer on well-drained soils is not recognised as a diagnostic surface horizon in international soil classification systems, but it has been studied in detail with regard to dynamics of forest soils (Klinka et al., 1981; Emmer, 1995). If well developed, several subhorizons can be recognised; from top to bottom:

L layer Consisting of almost undecomposed litter (L).

F layer Consisting of recognisable, but fragmented (F) organic remains, and

H layer Consisting of humified (H) organic matter, in which plant remains are not recognisable. **N.B.** *This H layer is <u>not</u> the Histic H horizon of the FAO classification, which is a peat layer.*

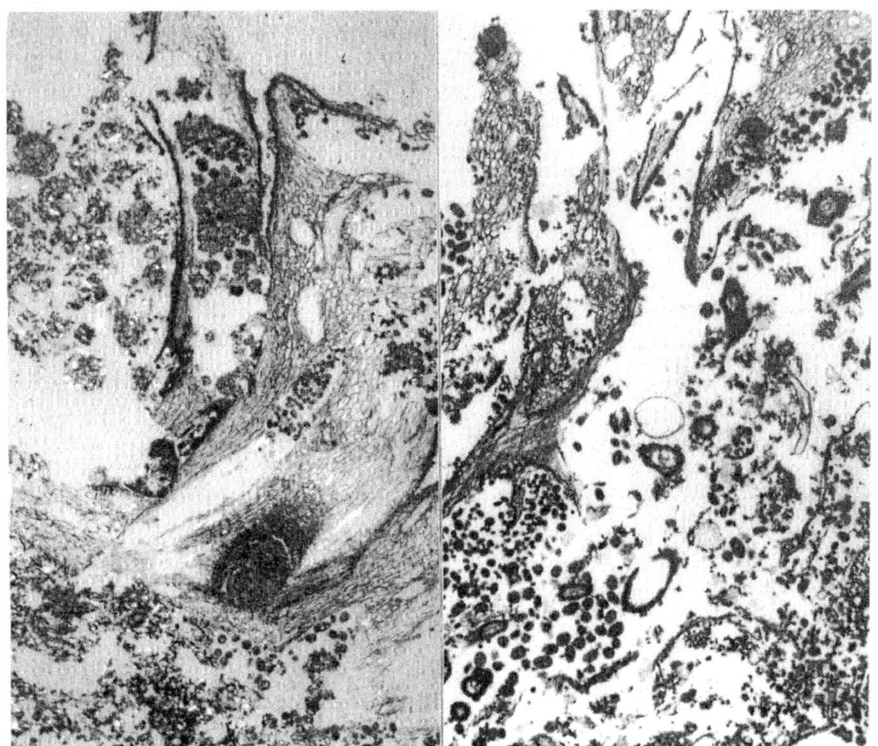

Figure 6.1. Decaying litter in an F horizon under beech. Tissue structure, moder pellets, and decayed moder. Width of picture 5 mm. Photo A.G. Jongmans.

Figure 6.2. Moder formation in an F horizon under beech. Note predominance of moder pellets. Width of picture 5 mm. Photo A.G. Jongmans

Question 6.1. *a. What are endorganic horizons? b. Why are ectorganic horizons restricted to non-arable soils?*

MOR, MODER AND MULL

Thick litter layers that have stratified F horizons with plenty of saprotrophic fungal hyphens and relatively few faeces of soil organisms indicate particularly poor conditions for decomposition. Such layers are called *mor*. In addition to saprotrophs (litter decomposers), ectomycorrhizal fungi are particularly common in soils with mor. Ectomycorrhizal fungi do not decompose litter, but supply their host plants with nutrients, in exchange for sugars. Gadgil and Gadgil (1975) provided field evidence that ectomycorrhizal fungi depress the activity of saprotrophs by competing with them for nutrients, and thus contribute to formation of mor.

Humus that consists largely of excrements of mesofauna (see figures 6.1 and 6.2) is called *moder*. It is typical of well-drained soils of slightly acid pH. Moder occurs both

in the F and H layers and in the mineral soil.
Humus that is intimately mixed with, and bound to clay minerals, without recognisable plant remains, is called *mull*. Calcium-saturated soils invariably have mull humus, but relatively fertile acid soils can also have mull. Mor, moder, and mull are morphological humus types that have little to do with humus organic chemistry.

In the foregoing a distinction is already made between surface litter and humus, and humus in the soil. Because of the presence of mineral compounds in the latter, the dynamics of the two systems are completely different.
In the litter layer, decomposition of organic matter is almost fully governed by nutrient status (nitrogen and basic cations), moisture availability, and temperature. In the mineral soil, there are various mechanisms that protect organic matter from decay:

- Inherent chemical recalcitrance: some components are difficult to break down anyway. Examples are aliphatic biopolymers and some aromatics.
- If the organic matter is locked up (occluded) in dense and stable aggregates, microbes cannot reach it, and it is physically protected against decay.
- Substances that are toxic to microbes, such as ionic aluminium, slow down decomposition when bound to organic matter. Mineral surfaces may strongly bind part of the organic matter. Especially organic matter bound to amorphous silicates (allophane) and clay can be strongly protected against decay.
- Organic molecules may organise in such a way that they become water repellent. This is common in soils with low clay contents. Water repellent units are not accessible for microbes and are therefore also protected against decay. In addition, such units may be extremely stable thermodynamically.

Question 6.2. *Which of the five protection mechanisms do you expect in a) a calcareous clay soil; b) an acid sandy soil?*

In Chapter 4 we have shown that various components of litter have different decay rates. Such differences also exist in humified material, but the time scale is very different. Breakdown of humified components in soils takes hundreds to thousands of years. Long-chain aliphatic components appear to be most stable of all, both in aerobic and anaerobic environments. Phenolic compounds are less stable, but also remain in the soil for a considerable time.

6.2. Humose mineral horizons

A-HORIZONS
Organic matter content and distribution vary widely among soils. We have already discussed the factors that contribute to organic matter accumulation: high production (forests, grasslands), and unfavourable conditions for decomposition (low nutrients contents, low pH, low temperature, excessive wetness, and high Al contents). Figure 6.3 illustrates various climate/vegetation combinations and the resulting organic matter profile.

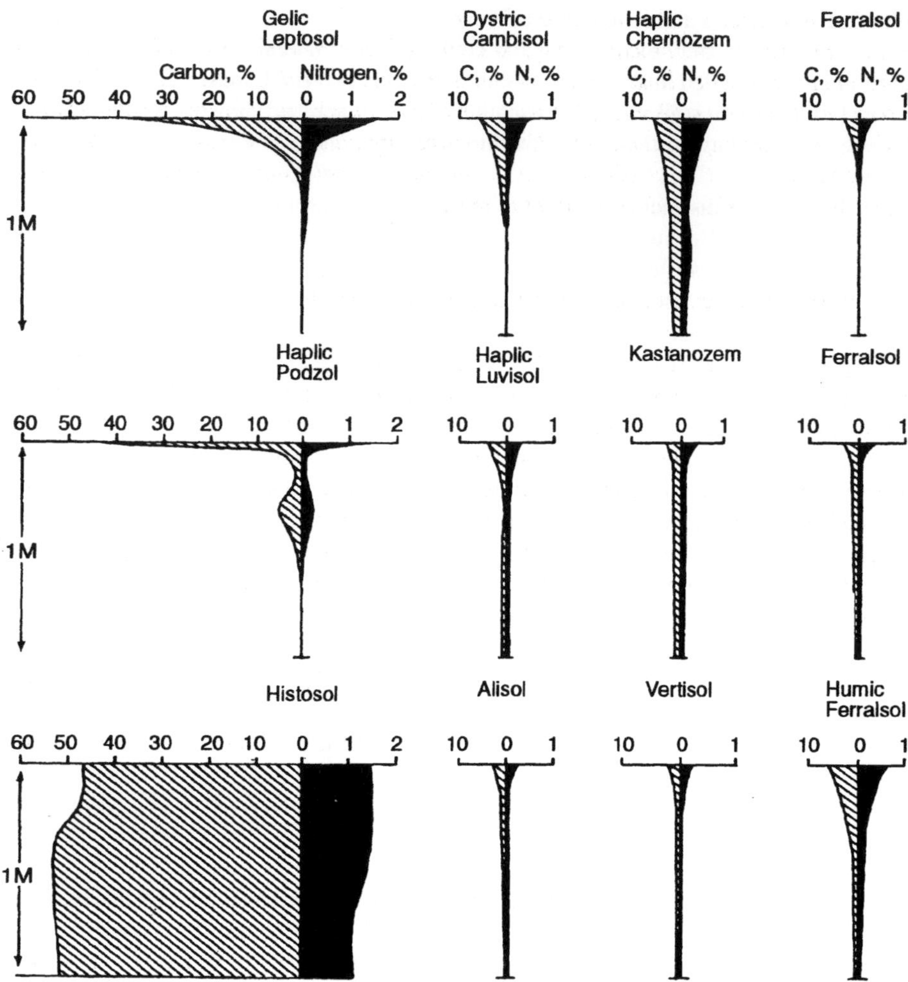

Figure 6.3. Organic carbon profiles of selected soils. For climatic regimes, see Question 6.3. From Parsons and Tinsley, 1975. Reproduced by permission of Springer Verlag GmbH & Co., Heidelberg.

Question 6.3. *Explain the humus profiles of Figure 6.3 in terms of organic matter production, decay, biological mixing, and above/below ground production. The soils occur in the following climatic zones: Gelic Leptosols, boreal climate; Haplic Podzols, boreal and temperate climates; Dystric Cambisols, humid, temperate climate; Haplic Luvisols, humid temperate climate; Chernozems, humid steppe climate; Kastanozems, dry steppe climate; Ferralsols and Alisols, humid tropical climate; Vertisols, monsoon tropical climate; Histosols, all humid climates.*

6.3. Mean residence time of organic matter in topsoils

Soil organic matter content reflects the balance between additions and decay. A simple model can describe the accumulation of soil organic matter under stable vegetation, where additions are equal each year and the decay speed is constant. If the litter deposition is F_l, and the first order decomposition rate of fresh plant litter is k_f, the fraction that remains after one year equals $(1\text{-}k_f)$. Assume that this fraction represents humified material, which decomposes further at a different rate, which we will call k_h. If a bare soil (e.g. volcanic ash, drift sand) is invaded by vegetation at time 0, and if the rate of litter production is constant, we can write the following formula for the pool size after *n* years:

$$\text{pool size} = F_l(1\text{-}k_f)\{1+(1\text{-}k_h)+(1\text{-}k_h)^2...+(1\text{-}k_h)^{n-1}\} \qquad (6.1)$$

The power series $1 + x + x^2 + ...+ x^n$ equals $1/(1\text{-}x)$ for $x<1$ and $n \rightarrow\infty$. Because k_h is always larger than zero, in the final situation:

$$\text{pool size} = F_l\,(1\text{-}k_f) / k_h \qquad (6\text{-}1a)$$

The graphic representation of this model is given in Figure 6.4 for two systems. Each system has a k_f of 0.9 $year^{-1}$, but k_h is 0.005 and 0.002 $year^{-1}$, respectively.
The mean age of the organic matter in the soil also depends on the decomposition rate. Using the same parameters as in the pool equation, and setting the age of the humus that is formed in year 1 at one year, and that of year *n* at n years, the following equation gives the mean age of the humus pool at time *n*:

$$\text{mean age} = F_l(1\text{-}k_f)\{1+(1\text{-}k_h)*2+(1\text{-}k_h)^2*3...+(1\text{-}k_h)^{n-1}*n\} / \text{pool size} \qquad (6.2)$$

or, after a large number of years, for $k_h > 0$:

$$\text{mean age} = \{F_l(1\text{-}k_f) / k_h^2\} / \{F_l(1\text{-}k_f) / k_h\} = 1 / k_h \qquad (6\text{-}2a)$$

Question 6.4. *For both curves in Figure 6.5, estimate the equilibrium value for organic matter content, and the time needed to reach this value.*

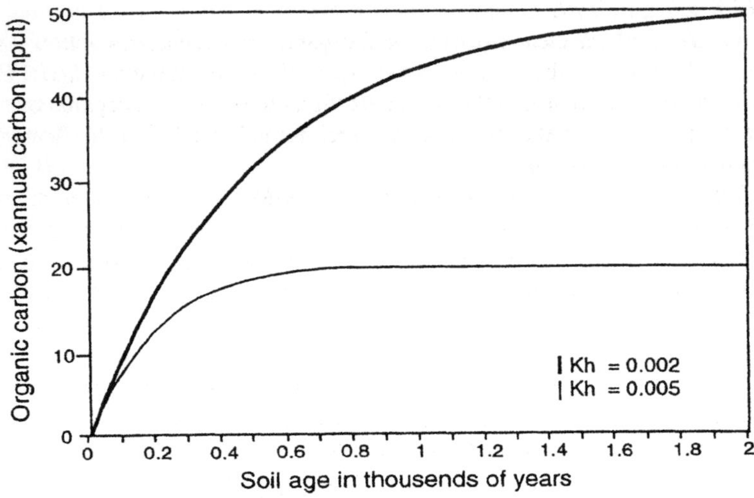

Figure 6.4. Calculated change in organic C content in a soil that is invaded by vegetation at time = 0. The organic C content is expressed as its ratio to annual input.

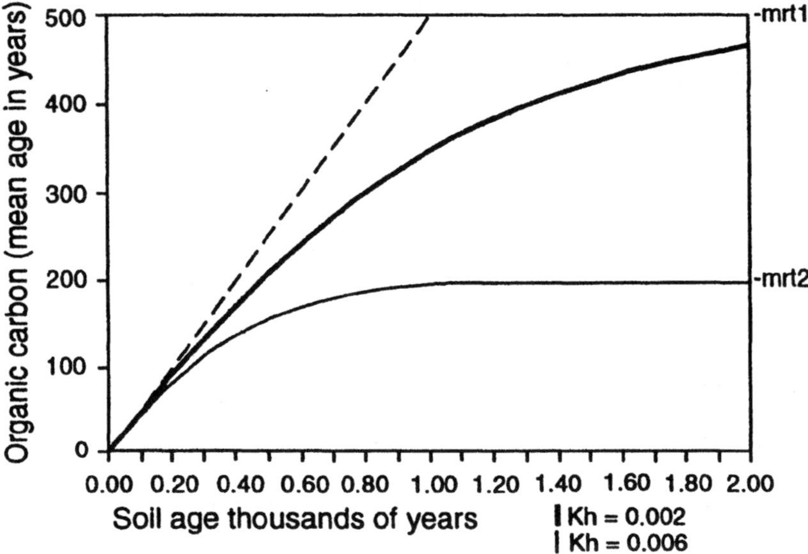

Figure 6.5. Mean age of organic matter accumulated in a virgin soil from time = 0, with two different values for k_h. The asymptotic value of each curve is the Mean Residence Time.

MEAN RESIDENCE TIME

Figure 6.5 shows how the mean age of the soil organic carbon changes with soil age for two values of k_h. Initially, the organic matter age is close to half the soil age (indicated by the broken line). However, as the soil gets older, the increase in organic matter age levels off to a steady state value of $1/k_h$ year, which is the Mean Residence Time (MRT) of the organic matter (fraction).

Figure 6.5 shows that, in soils with a large pool of low-MRT organic matter, the soil can be much older than the ^{14}C age of the organic matter. The model illustrates a few important aspects of humus accumulation but is of course, a gross oversimplification. The time necessary to reach a steady state differs for various fractions of humic material.

Question 6.5. *Figure 6.5 shows that the mean age of the organic matter fraction is lower than the soil age. a) Is the discrepancy stronger for a system with low k_h (slow turnover) or for one with high k_h? b) What is the soil age when the ^{14}C age of organic matter is 300 years at a humus decomposition rate of 1) 0.002 yr^{-1}, or 2) 0.006 yr^{-1}?*

Question 6.6. *The models illustrated in Figures 6.4 and 6.5 contain a number of unrealistic assumptions. List three of these assumptions and comment on the errors that they introduce in the calculation.*

CARBON ISOTOPES AND ORGANIC MATTER DYNAMICS

Fresh plant material contains a specific amount of (radioactive) ^{14}C. In dead organic matter, the content of this isotope decreases with time by radioactive decay. To measure the decrease of ^{14}C activity, it is necessary to know its original content in the fresh plant material.

Plants favour the light carbon isotope (^{12}C) and discriminate against the heavier isotopes ^{13}C and ^{14}C. This discrimination is called *isotope fractionation.* The fractionation by plants of both heavy isotopes is equal, and therefore the original ^{14}C content can be obtained by measuring the content of the stable isotope ^{13}C. Carbon isotope fractionation occurs with respect to carbon in the atmosphere, but because the ratio in the atmosphere is too variable, the deviation of the $^{13}C/^{12}C$ ratio is expressed relative to the ratio in a standard marine carbonate. This standard carbonate is a belemnite from the PeeDee Formation, or PDB, but nowadays a proxy, the Vienna PDB standard (VPDB) is commonly used. The relative deviation of the $^{13}C/^{12}C$ ratio with respect to that of the PDB is called the $\delta^{13}C$.

$$\delta^{13}C = [(^{13}C/^{12}C)_{sample} - (^{13}C/^{12}C)_{PDB}]*1000 / (^{13}C/^{12}C)_{PDB} \qquad (6\text{-}3)$$

The atmosphere has a $\delta^{13}C$ of -8 per mil. The exclusion by plants depends on their photosynthetic pathway. C-3 plants (most plants from temperate climates, tropical trees) have a fairly strong isotope fractionation ($\delta^{13}C$ is around -25 to -30 per mil), while C-4 plants (most tropical grasses, maize) have a much lower fractionation ($\delta^{13}C$ is around -14 per mil). These values are reflected in soil organic matter. When a C-3 vegetation is replaced by a C-4 vegetation, as happens when a forest is cleared and maize is planted, the ^{13}C of the soil humus adapts to the new situation. The speed with which the $\delta^{13}C$ adapts to that of the new plants gives insight in humus dynamics.

MODELS AND POOLS

Various simulation models describe the change of soil organic matter contents in relation to time, land use, etc. Examples of such models are CENTURY (Parton et al., 1987) and the Rothamsted model (Jenkinson, 1990). Models use hypothetical fractions of organic matter, called 'pools'. Pools are attributed a size and a turnover time (MRT) to fit the models. The number of assumed pools varies from one to more than five. Most model pools have little to do with humus fractions that can be isolated chemically or physically from soils (see Chapter 4). This is especially the case for 'slow' ($1/k_h$ = 100-500 years) or 'stable' ($1/k_h$ = >1000 years) fractions. Too many factors determine the MRT of such pools. Only 'rapid' (short MRT) pools can be identified with some ease: microbial organic matter, free sugars, low-molecular weight acids, and such.

HUMUS FORMS IN DIAGNOSTIC SURFACE HORIZONS

The relations between the four diagnostic surface horizons that are used in Soil Taxonomy are given in Figure 6.6. The general properties of these horizons indicate that the division is indirectly linked to soil organic matter dynamics:

mollic: Physically stable, usually Ca-saturated humic material, dominated by 'humic acids'. Dominant humus form is mull. Mull is common on calcareous parent materials. Vegetation usually herbaceous; but also under agricultural land. MRT of non-bound fractions is low (tens to hundreds of years), of clay bound fractions is high (thousands of years). Horizons can be very deep in unconsolidated calcareous sediments, but C contents are usually below 10%.

umbric: Less physically stable, under-saturated or Al/Fe saturated humic material, dominated by fulvic acids. Dominant humus form is moder, but at low pH and in cold climates, mor-type humus forms develop. Common on felsic rocks in all climates. Vegetation: variable, from tundra to tropical forest and poor range land. MRT of metal-bound organic matter can be high, of other fractions low. Horizons are usually shallower than mollic horizons because biological mixing is less intense, but may have higher carbon contents.

histic: High organic matter accumulation, usually due to wetness. Plant remains are partly decomposed; moder and mor humic material dominate in the decomposed part. Little stabilisation of humic material, except in calcareous peats. Fulvic acids dominate. MRTs are high (thousands of years). Vegetation: marsh, fen and bog communities. Histic horizons can be several meters thick and may have up to 60% organic carbon in the dry matter fraction.

ochric: Horizons either too low in organic matter, too thin, or too light in colour to qualify for any of the other horizons. Consequently, ochric horizons are not genetically determined, occur in virtually all climates and their properties vary accordingly. Ochric horizons are common in eroded soils or soils that have been depleted in organic matter by intensive cultivation.

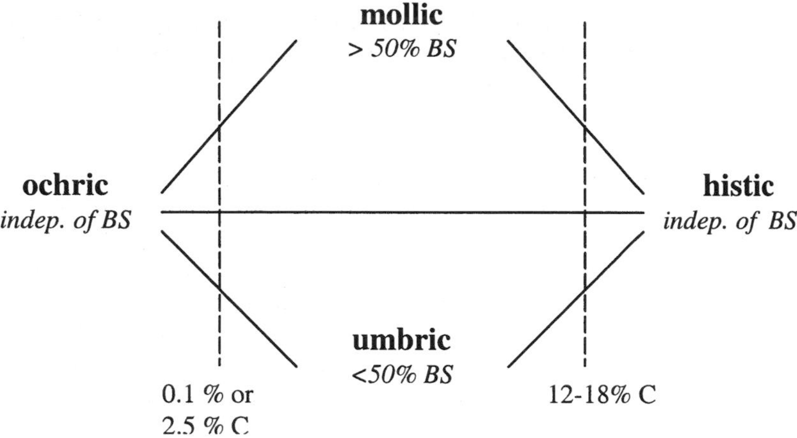

Figure 6.6. Relations between USDA diagnostic surface horizons. BS = base saturation (pH7); C = organic carbon.

6.4. Problems

Problem 6.1

Reclamation of forest and grassland for arable agriculture greatly affects soil organic matter. Under a given land use, the organic matter content strongly varies with temperature and precipitation, but also with clay content. This illustrates that also in agricultural land the quantity (as well as MRT) of the organic matter strongly depends on factors that are largely beyond man's control.

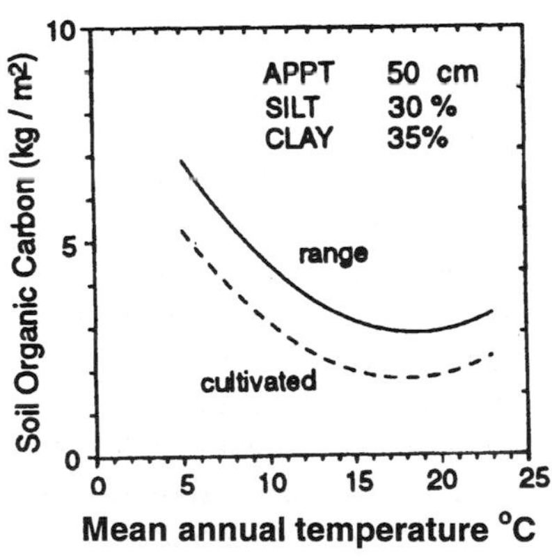

Figure 6A. Organic matter content in dependence of temperature and land use in Chernozems of the US Central Plains. APPT = annual precipitation. From Burke et al., 1989.

Explain the effects of climate, soil texture, and land use on organic matter contents in Figures 6A (preceding page) and 6B. 'Range' means grazing land; cultivated lands are under maize.

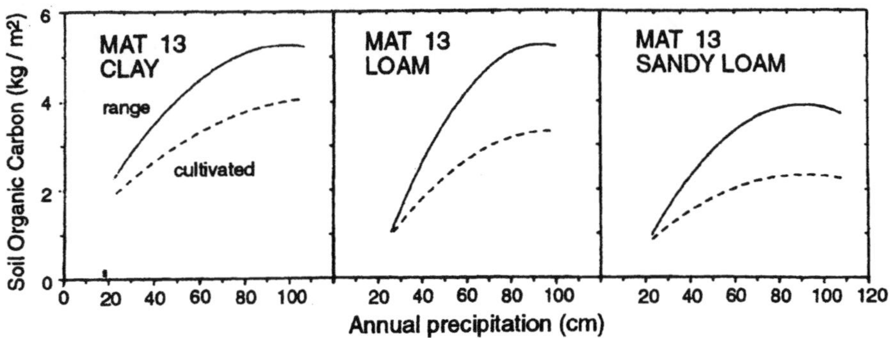

Figure 6B. Relation between organic matter content, land use, and precipitation for Chernozems of different texture in the US Central Plains. MAT = mean annual temperature. From Burke et al., 1989.

Problem 6.2

a. Show that at steady state (addition equals removal) the organic carbon pool is equal to $(1-k_f)/k_h$ times the annual litter input, and that MRT (= mean age of organic matter at steady state) equals $1/k_h$.

b. Argue that the MRT in young soils is close to half the soil age.

Problem 6.3

Assuming that the bulk density of an ectorganic litter layer is 0.1 g/cm^3, calculate the thickness of that layer under each of the vegetation types presented in Table 6C. Discuss the results (see also Chapter 4, Table 4.1).

Problem 6.4

Calculate the mean rate of decomposition of plant litter from data on litter input and litter layer, and the MRT of the organic matter in that litter layer under the vegetation types of Table 6C. Discuss some of the assumptions involved in making those calculations. Discuss how realistic those assumptions might be.

Problem 6.5

Does the discrepancy between real soil age and MRT of organic matter affect the accuracy of dating of a sediment-layer by ^{14}C analysis of an underlying peat layer?

Table 6C. Biomass standing crop (s.c.), root mass, and dead litter (litter layer) of various vegetation types in 10^3 kg/ha, and annual production (a.p.) of (mainly leaf) litter in 10^3 kg ha^{-1} yr^{-1}. Mass refers to organic matter (about 50% C) (after Kononova, 1975).
1 = arctic; 2 = brushwood; 3 = northern taiga; 4 = central taiga; 5 = southern taiga; for. = forest. The litter layer is equivalent to the organic O-horizon.

	tundra		spruce forests			oak for.	steppes		semi desert	subtr decid for.	savanna		humid trop. for.
	1	2	3	4	5		mod dry	dry			dry	moist	
biomass (s.c.)	5	28	100	260	330	400	25	10	4	410	27	67	500
roots (s.c)	4	23	22	60	73	96	21	9	4	82	11	4	90
roots (% of biomass)	70	83	22	23	22	24	82	85	87	20	4	<1	18
litter fall (a.p.)	1	2	4	5	6	7	11	4	1	21	7	12	25
litter layer (s.c.)	4	83	30	45	36	15	6	2	-	10	-	1	2

Problem 6.6

Organic matter in the deeper horizons of a coastal sediment dated at 800 year old on the basis of historic evidence has a ^{14}C age of 2000 years. This is at variance with the model of Fig. 6.5, where the real age is always greater than the radiocarbon age. Can you give an explanation? Remember that sediments may contain organic matter upon deposition.

Problem 6.7

In virtually all soils, the ^{14}C age of the organic matter increases with depth. Figure 6D (next page) gives examples of changes with depth for populations of andosols, podzols, and chernozems. Name one or more possible reasons for this increase.

Problem 6.8

Redraw Figure 6.5 for a situation where a soil with a vegetation producing a mean humus MRT of 200 years is invaded after 1000 years by a vegetation with a mean humus MRT of 500 years. How does this affect the relation between ^{14}C age of the organic matter and depth?

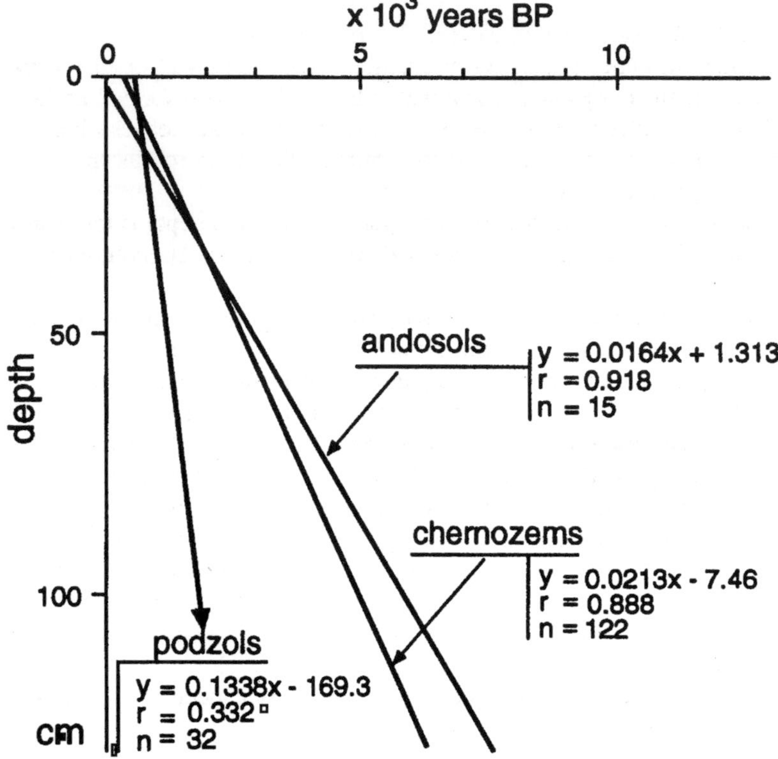

Figure 6D. Gradient of radiometric age with depth for three groups of soils. From Guillet, 1987.

6.5. Answers

Question 6.1

Endorganic horizons are mineral horizons that contain organic matter: Ah, Bh, (Ap). Arable soils do not have ectorganic horizons because topsoils are artificially mixed by ploughing.

Question 6.2

a. In clayey soils, occlusion in aggregates and binding to clay minerals are important. In calcareous clayey soils, there are no Fe- and Al-humus complexes.
b. In sandy soils, there is more possibility of hydrophobic arrangement and of protection by Fe and Al, but no occlusion.

Question 6.3.

The following information can be extracted from Figure 6.3:

- In cold climates, either under low vegetation (Gelic Leptosol) or under forest (Haplic Podzol), an ectorganic horizon with a high C/N ratio is formed at the soil surface. Organic matter contents decrease sharply with depth because biological homogenisation is restricted. In the podzol, humus illuviation results in a second maximum of organic C in the subsoil.
- In the Dystric Cambisol, the decrease of organic matter with depth is less marked. There is no significant ectorganic horizon, C/N ratios are lower, but homogenisation is not very deep.
- In soils with very strong biological activity and predominant vegetation of deep-rooting grasses, such as the Chernozem, the Kastanozem, and the soils of warmer climates, homogenisation is very deep. Organic carbon contents at a depth of 1 meter may still exceed 1 percent, while organic C contents in the surface horizon are only slightly higher. The organic carbon *content* depends on climate and vegetation, it may be high both under grass (Chernozem) and under primary forest (Humic Ferralsol). Biological activity is mainly due to earthworms in chernozems and to termites and/or ants in tropical soils. In the Vertisol, mixing to great depth is partly due to swelling and shrinking (see chapter 10), and organic matter is sometimes of sedimentary origin.
- The Ferralsol (upper right) is an exception in that it contains significant amounts of organic C only in the upper few dm of the soil profile. This is probably due to ironstone at shallow depth (see Chapter 13), which restricts deep rooting and deep biological homogenisation.

Question 6.4

The equilibrium values are 20x and 50x the annual input (see Problem 6.2). The equilibrium is reached after 500 and >2000 years, respectively.

Question 6.5

a. The discrepancy is stronger when k_h is larger.

b. From Figure 6.5 you can see that at $k_h = 0.002$, the MRT is larger than 300 years, so that, if you measure a ^{14}C age of 300 years, equilibrium has not been reached and the real soil age is at least 2x the ^{14}C age. At $k_h = 0.006$, the MRT is about 200 years, so a ^{14}C age of 300 years is unrealistic.

Question 6.6

Unrealistic assumptions are, e.g.:

- Litter production is constant with time. Normally, a bare soil is colonised gradually and the vegetation goes through various development stages.
- Decomposition rate is the same for all organic matter fractions. Humus is not homogeneous, and has both readily decomposable and very refractory fractions.
- The climate was constant throughout the development sequence.
- There is no influence of biological mixing.

Problem 6.1

Figure 6A. Soil organic matter first decreases with increasing temperature. This effect can be explained by a shift in the equilibrium towards faster decomposition. The upward shift towards 25°C may be due to increased production of litter. Although the mean rainfall is the same for the whole curve, it is unlikely that the seasonal distribution is also the same. A strong dry spell at higher temperatures may decrease decomposition in the topsoil considerably. The difference between rangeland and cultivated land will be due mainly to different tillage (ploughing stimulates decomposition), and amounts of litter deposition (removal of crop residues from cultivated land).

Figure 6B. The increase in organic matter content with precipitation is probably due to higher litter (a.o. roots) production at higher rainfall. Be aware that the range spans a rainfall of 500 to 1200 mm. In both rangelands and cultivated lands the organic matter content decreases on coarser-textured soils. This will be due to less protection in the latter: less occlusion in aggregates, less opportunity to bind to the clay fraction.

Problem 6.2

a. In steady state, addition equals removal. The input equals the litter input (F_l) times the litter humification fraction, or:

$$\text{Input} = (1-k_f) * F_l$$

The decomposition equals [pool size] * k_h. Input = output gives:

$$(1-k_f) * F_l = [\text{pool size}] * (k_h), \text{ so the pool size equals:}$$
$$[\text{pool size}] = \{(1-k_f)*F_l\}/k_h$$

In steady state, the decomposition rate equals k_h, and the final pool equals 100%. This means that the turnover time of the total pools equals $1/k_h$. This is the MRT.

b. In young soils, where the influence of the decay of the humus fraction can still be neglected, the mean age of the organic matter pool (at constant input) equals (1+2+3+..+n)/n. The solution of this progression is ½ * n * (first term + last term) /n; which is very close to half the last term (age).

Problem 6.3

Assuming that the bulk density of the litter layer is 0.1 $kg.dm^{-3}$, we find that:

In tundra 1: 4 tons of litter per ha equals 4000 kg, or a volume of 4000/0.1 = 40.000 dm^3. The surface of 1 ha equals 10^6 dm^2, so the thickness of the litter layer is 0.04 dm.

In tundra 2: The thickness is 0.83 dm, etc.

There are large differences in aboveground and belowground biomass in the various systems. In forests, the aboveground mass dominates, in grasslands, the belowground biomass. Most forests also have a considerable litter layer, with the exception of the humid tropical forest, where decomposition is fast due to high temperature.

Problem 6.4

If each system is at steady state, the MRT equals [pool size]/[litter fall]
This would give MRT's of 4, resp. 42 years in Tundra 1 and Tundra 2.
The decay rate k_h equals 1/MRT, so the decay rates for Tundra 1 and Tundra 2 are 0.25 and 0.2 yr^{-1}, respectively.

	tundra		spruce forests			oak for.	steppes		semi desert	subtr decid for.	savanna		humid trop. for.
	1	2	3	4	5		mod dry	dry			dry	moist	
litter layer (s.c.)	4	83	30	45	36	15	6	2	-	10	-´	1	2
litter fall (a.p.)	1	2	4	5	6	7	11	4	1	21	7	12	25
MRT (yrs)	4	41.5	7.5	9	6	2.1	0.5	0.5	--	0.5	--	0.1	0.1

Problem 6.5

Because peat grows upwards without mixing, the top layer always represents the most recent growth. Decay speed is very low, so the ^{14}C age of the unhumified top of the peat layer probably reflects its real age and not its MRT. The dating would be correct.

Problem 6.6

This is only possible if the organic matter was not formed *in situ*, but is reworked material of an older deposit (e.g. old peat). Fresh alluvial and marine sediments usually have a varying organic matter fraction that was sedimented.

Problem 6.7

a. There are various processes that cause an increase of organic matter age (or MRT) with depth:

- breakdown is slower at greater depth (e.g. because of lower temperatures, less oxygen and available nutrients).
- refractory components with a high MRT are more common in the deeper parts of the soil (there is more addition of fresh litter in the topsoil).
- mixing of fresh components at greater depth lags behind in time.

The combined effect of the second and third causes is reflected in Figure 6E (next page).
The correlation for the podzol samples is very bad, because the mechanism for accumulation with depth is very different, and podzols may have very different dynamics (see Chapter 11).

Problem 6.8

From the moment of invasion by the new vegetation, the curve will become steeper. The new asymptotic value will be the MRT of 500 years (see Figure 6.5). Because mixing of fresh material will lag behind, deeper horizons may exhibit shorter MRT's for some time.

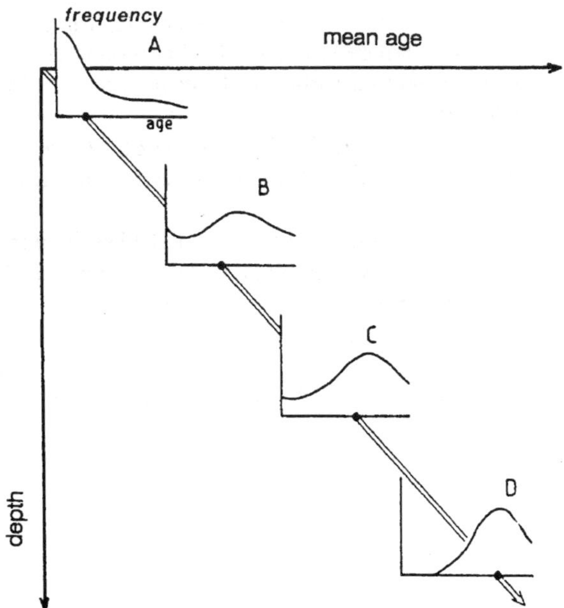

Figure 6E. Model of changing age distribution of organic matter with depth, causing an increase of mean age. From Guillet, 1987.

6.6. References

Babel, U., 1975. Micromorphology of soil organic matter. In: J.E. Gieseking (ed): *Soil Components. I. Organic Components*, p 369-474. Springer, Berlin.

Bal, L., 1973. *Micromorphological analysis of soils*. PhD. Thesis, University of Utrecht, 175 pp.

Burke, I.C., C.M. Yonker, W.J. Parton, C.V. Cole, K. Flach & D.S. Schimel, 1989. Texture, climate, and cultivation effects on soil organic matter content in U.S. grassland soils. Soil Science Society of America Journal, 53:800-805.

Emmer, I.M., 1995. Humus form and soil development during a primary succession of monoculture *Pinus sylvestris* forests on poor sandy substrates. PhD Thesis, University of Amsterdam, 135 pp.

FAO-Unesco, 1990. FAO-UNESCO Soil map of the world, Revised Legend. FAO, Rome.

Flaig, W., H. Beutelspacher, and E. Rietz, 1975. Chemical composition and physical properties of humic substances. In: J.E. Gieseking (ed): *Soil Components I. Organic Components*, p 1-211. Springer, Berlin.

Gadgil, R.L., and P.D. Gadgil, 1975. Suppression of litter decomposition by mycorrhizal roots of *Pinus radiata*. New Zealand J. of Forest Science, 5:35-41.

Guillet, B., 1987. L'age des podzols. In: D. Righi and A. Chauvel, *Podzols et Podzolisation*, p131-144. Institut National de la Recherche Agronomique.

Jenkinson, D.S., 1990. The turnover of organic carbon and nitrogen in soil. Phil. Trans. R. Soc. London, B329:361-368.

Klinka, K., R.N. Greene, R.L. Towbridge, and L.E. Lowe, 1981. *Taxonomic classification of humus forms in ecosystems of British Colombia*. Province of B.C., Ministry of Forests, 54 pp.

Kononova, M.M., 1975. Humus in virgin and cultivated soils. In: J.E. Gieseking (ed): *Soil Components. Vol. 1. Organic Components*. Springer, New York, pp 475-526.

Parsons, J.W., and J. Tinsley, 1975. Nitrogenous substances. In: J.E. Gieseking (ed): *Soil Components. Vol. 1. Organic Components*. Springer, New York, pp 263-304.

Parton, W.J., D.S. Schimel, C.V. Cole, and D.S. Ojima, 1987. Analysis of factors controlling soil organic matter levels in Great Plains grasslands. Soil Science Society of America Journal, 51:1173-1179.

Soil Survey Staff, 1992. Keys to Soil Taxonomy. Soil Management Support Services Technical Monograph 19. USDA

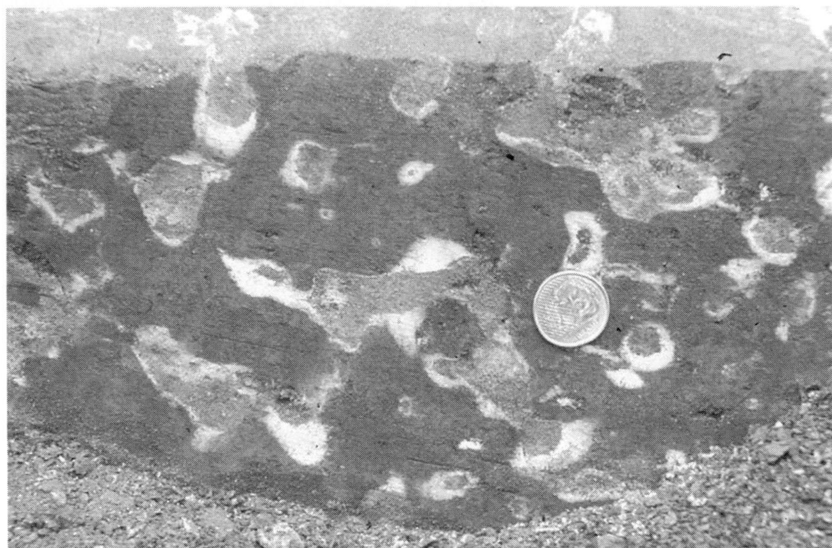

Plate K. Two phases of animal burrows in the deep saprolite of an Oxisol, Minas Gerais, Brazil. Note depletion of iron in the older burrows due to preferential water movement. Coin diameter is 3 cm. Photograph P. Buurman.

Plate L. Extremely developed Planosol (ferrolysis) on an old river Allier terrace in Central France. Note the extremely bleached topsoil (E). At the head of the hammer, the former top of the argillic horizon is strongly accentuated by iron and manganese accumulation. In the lower part, bleaching is found along cracks. Photograph P. Buurman.

CHAPTER 7

HYDROMORPHIC SOILS

7.1. Introduction

Prolonged water saturation and seasonal alternation between water logging and drainage has profound effects on soil chemical and morphological properties. Changes in the degree of water saturation affect the supply of O_2 to the soil, which in turn affects the oxidation state of important elements. The oxidation state of iron, manganese and sulphur strongly influences their solubility and colour, explaining the brown, grey, blue, black and yellow mottles often seen in periodically wet, so-called hydromorphic, soils. Redox processes often involve production or consumption of H^+ and so have an important effect on soil pH too.

STAGNIC AND GLEYIC PROPERTIES; ACID SULPHATE SOILS
In the FAO-Unesco classification, 'gleyic properties' refer to soil materials with signs of chemical reduction arising from saturation with groundwater: temporarily low redox potentials, the presence of marked concentrations of dissolved Fe(II), and the presence of a grey-coloured soil matrix with or without black, or brownish to reddish mottles due to Mn- and Fe oxides. 'Stagnic properties' come from water stagnation at shallow depth, resulting in seasonal water saturation of the surface soil (a 'perched' water table). 'Gleyic properties' are characteristic for Gleysols, 'stagnic properties' for Planosols and for 'anthraquic phases' of other soils, which have a seasonally perched water table from, e.g., irrigation for wetland rice. Some hydromorphic soils are strongly influenced by the pH effects of redox processes. Examples are acid sulphate soils (Thionic Fluvisols, FAO-Unesco, or Sulfaquepts, Soil Taxonomy), with characteristic, very acid, yellow-mottled B-horizons), and ferrolysed soils. Planosols have a pale-coloured surface horizon low in clay and organic matter due to surface water stagnation and ferrolysis. Podzoluvisols have tongues of bleached E horizon material penetrating in a more clayey B, due to preferential water movement (and saturation) in such tongues.

7.2. Gley soils

MORPHOLOGY OF GLEY SOILS

As has been described in Chapter 3.3, gas diffusion becomes very slow when pores are filled with water. Therefore, in waterlogged soils supply of O_2 to heterotrophic micro-organisms is much slower than their consumption of O_2. This leads to the disappearance of free O_2. In the absence of O_2, the oxidation of organic matter can continue in the presence of oxidised soil components, including Fe(III), Mn(IV) and Mn(III). These are reduced to Fe(II) and Mn(II), which are far more soluble than their oxidised counterparts. As a result, Fe and Mn can be transported more easily in waterlogged than in well-drained soils. The Fe(II) and Mn(II) can be reoxidised and precipitated, elsewhere or at some other time, in the presence of O_2.

Question 7.1. *List forms of Fe and Mn that can be important in oxidised and reduced soils.*

Permanent or periodic saturation of the soil by water, which is the starting point of redox processes causing the gley phenomena, can occur either from below (groundwater) or from above (rain or irrigation water).

Soils with 'gleyic properties' are so-called *groundwater gley soils*, i.e. they have a permanently waterlogged (non-mottled, greyish) subsoil and a seasonally wet (brown and grey mottled) shallower horizon. 'Stagnic properties' are typical for *surface water gley soils*. Formerly these were called 'pseüdogley' soils. They have a seasonally water-saturated, brown and grey mottled surface horizon on a predominantly aerated, unmottled subsoil. The water table can be temporarily perched on a subsoil layer that is slowly permeable to water. An argillic horizon may act as a slowly permeable layer. In wetland rice soils, tillage of the water-saturated soil ('puddling') turns the surface layer into a (temporary) watery, finely porous, very slowly permeable layer, and the subsurface layer into a (more permanent) physically compacted ploughing or 'traffic' pan ('traffic' of water buffaloes or people in muddy rice fields).

Figure 7.1 shows typical landscape positions of ground water and surface water gley soils. Figure 7.2 illustrates soil profiles of a chronosequence of surface water gley soils and of a ground water gley. In ground water gleys, the interior of soil structural elements tends to be greyish, with the brown to black Fe(III) and Mn(IV) oxides concentrated around the larger pores and cracks through which O_2 penetrates from above (Plate M, p.191; Plate O, p.192). In surface water gley soils, greyish colours predominate in the most porous parts, which fill up periodically with water. Here, Fe(III) and Mn(IV) oxides accumulate in the transition zone to the less permeable, predominantly aerated and brownish soil below and adjacent to seasonally water-saturated pockets of grey soil. If vertical water movement is strongly impeded, lateral translocation of iron predominates. Along water ducts, Fe and Mn are completely removed (Plate N, page 192).

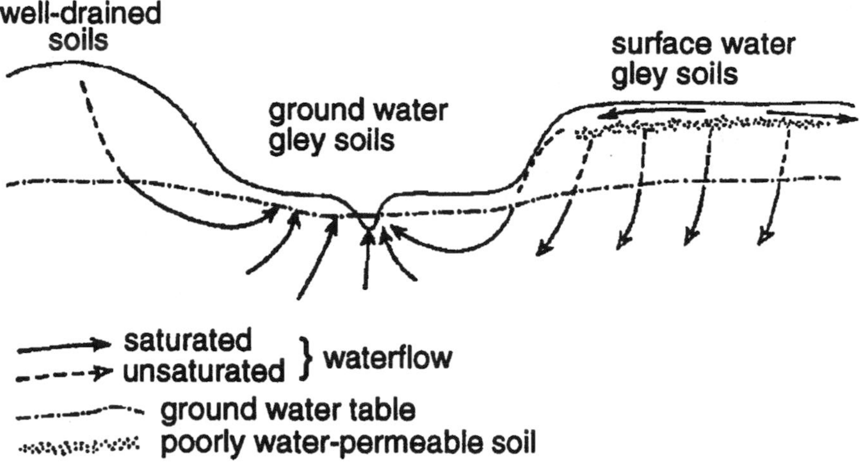

Figure 7.1. Position of groundwater and surface water gley soils in a hypothetical landscape.

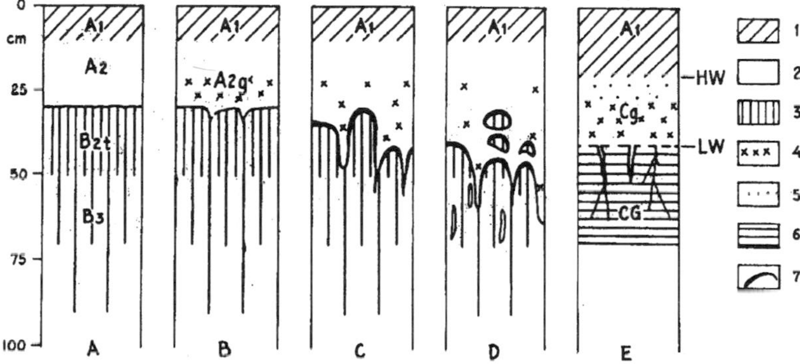

Figure 7.2. Surface water and groundwater gley profiles. A-D: soils with textural-B horizons. A: well-drained, B-D with increasingly expressed stagnic properties. E: ground water gley soil. 'A2' stands for E horizon.
From Buurman, 1980; reproduced with permission of Blackwell Science Ltd., Oxford.

Question 7.2. *Write a legend that explains the numbers 1 to 7 in Figure 7.2.*

Aerobic soils that contain free iron oxides have warm colours, ranging from red via brown to yellow. Colours in reduced soils are cool: neutral grey to greenish or bluish grey, depending on the concentration of divalent iron compounds, or black (rapidly turning grey when exposed to the air) where finely divided FeS (mackinawite) has stained the soil. Pyrite (FeS_2) can only persist in reduced (greyish coloured) soil, but it has no specific effect on soil colour. Red, brown and yellow soils that become permanently water-saturated will turn grey only if at least a little metabolizable organic matter is present as an energy source, e.g., a few tenths of a percent.

Question 7.3. *A soil contains 1 % Fe(III) as FeOOH. How much organic matter (CH_2O, mass fraction of the soil in % C) would be needed to reduce all Fe(III) to Fe(II)?*

Without organic matter there is no activity of the anaerobic, heterotrophic micro-organisms, whose respiration drives the reduction process, and brown Fe(III) mottles can persist for centuries to millions of years.
The time needed to form a well-developed gley horizon is probably in the order of years to decades. Thousands of years or more may be needed to form a horizon with a strong absolute accumulation of iron and or manganese by supply of dissolved Fe^{2+} and Mn^{2+} from elsewhere. However, Fe(III) oxide mottles can form in just a few days by aeration of a soil with dissolved or adsorbed Fe^{2+}.

GLEY PROCESSES

The following sequence of individual processes explains the segregation of iron into Fe(III)oxide-rich mottles, concretions and accumulation zones in hydromorphic soils:
i) Solid Fe(III) oxide reduces to dissolved Fe^{2+} when (1) organic matter is present as a substrate for micro-organisms, (2) the soil is saturated with water so the diffusion of O_2 is slowed down, (3) micro-organisms capable of reducing Fe(III) are present (see Chapter 3.3).
ii) Dissolved Fe^{2+} moves by diffusion or by mass flow. Diffusion takes place along a concentration gradient that may be caused by spatial variation in the rates of production or removal of Fe^{2+} by various processes (e.g. oxidation). Mass flow may be driven by gravity or by capillary action.
iii) Dissolved Fe^{2+} is removed from solution by (1) precipitation of solid Fe(II) compounds, (2) adsorption on clay or Fe(III) oxides, (3) by oxidation of dissolved, adsorbed or solid Fe(II) by free O_2 to Fe(III) oxide, and (4) Fe(III) oxides are formed when oxygen enters the reduced soil and meets with dissolved, adsorbed, or solid Fe(II). Gaseous O_2 can be conveyed into the soil by diffusion or, if a waterlogged soil is drained rapidly, by convection in open channels.

Question 7.4. Sketch by means of arrows the direction of movement *of O_2 and of Fe^{2+} in the right hand panel of Fig. 7.3, and indicate as a thick line the zone where FeOOH precipitates*

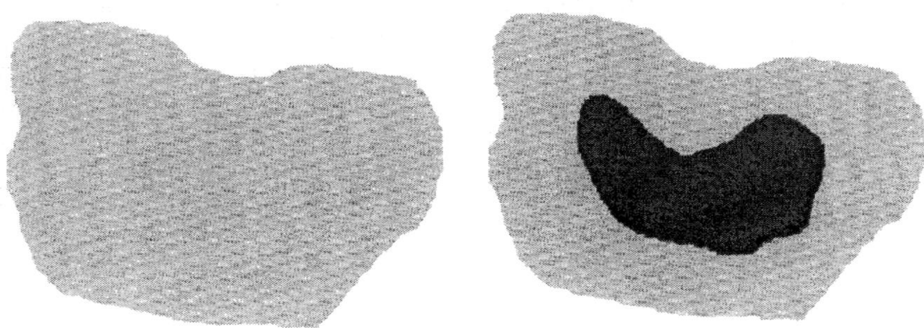

Figure 7.3. Soil aggregate before (left) and after (right) wetting an originally well-drained soil, resulting in anoxic (black) interior of otherwise oxygenated ped.

Transport of Fe^{2+} can be over distances varying from mm to km. In groundwater gleys in valleys, absolute accumulations of iron, conveyed from surrounding areas may occur. The processes listed above can act in various combinations and influenced by vertical and horizontal variation in porosity, to produce a bewildering variety of visible gley phenomena. In addition to physical-chemical processes, gley phenomena can be strongly influenced by the activity of soil fauna. Both vesicular and nodular patterns in iron concentrations in hardened plinthite (laterite, see Chapter 13), for instance, have been attributed to termite activity.

Similar processes are involved in segregation of Mn. The processes involving gley by Fe and Mn are basically similar, but the oxidation of Fe^{2+} to Fe(III) oxide occurs at much lower O_2 concentrations (and lower redox potentials) than the transformation of Mn^{2+} to Mn (III, IV) oxide. Similarly, MnO_2 is reduced to Mn^{2+} at higher redox potentials than FeOOH is reduced to Fe^{2+} (see problem 7.4). As a result, black Mn oxides generally accumulate closer to the source of oxygen in the soil than FeOOH, which is formed closer to the reduced zone (Plate M, p.190). Furthermore, Fe is quite abundant in many soils, while Mn occurs in relatively low concentrations. So, whereas brown Fe mottles are normally present in gley soils, black Mn mottles are often rare or absent.

Question 7.5. *Sketch in the right hand part of the Fig 7.3 the zone where Mn(IV)-oxide would precipitate, relative to that of FeOOH.*

PLACIC HORIZON

The placic horizon (Plate P, p.192) is a thin (several mm to a cm thick) cemented band of iron (+ organic matter and manganese) accumulation that is frequently found in podzols, andosols, gley soils, and under peat, and has nothing to do with the podzolisation process itself. The placic horizon may form when, e.g., a spodic (Al-Fe-humus illuviation) horizon becomes dense enough to cause stagnation of rainwater. This stagnation causes temporary waterlogging of the topsoil. Remaining iron in the topsoil may be reduced and may accumulate at the boundary with the (still aerated) subsoil.

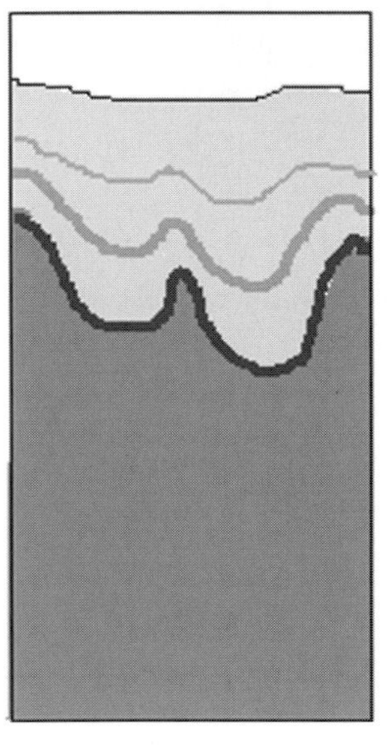

Figure 7.4. Three hypothetical stages in the development of a placic horizon.
1: first stage
2 and 3: later stages
Light grey: temporary anoxic surface horizon; dark grey: permanently oxidized subsoil.

Question 7.6. *The placic horizon may move downward with time (stages 1, 2 and 3 in Fig. 7.4). We believe deeper parts have the tendency to become still deeper, so that the horizon tends to become more irregular with age. Explain the rationale of this hypothesis.*

7.3. Ferrolysis

Oxidation and reduction processes involving Fe not only influence the distribution of Fe oxides in the soil profile, but lie at the root of a soil forming processes called *ferrolysis* (= 'dissolution by iron'). Ferrolysis involves acidification and clay destruction under the influence of alternating reduction and oxidation of iron, as will be detailed below. It explains the formation of gleyed, acidic, pale-coloured surface horizons low in clay and organic matter on nearly level land of marine or riverine terraces that are seasonally flooded by rainwater, and the light-coloured tongues of E material in the B-horizons of podzoluvisols. The light-coloured, light textured surface horizons usually satisfy the requirements for an albic E-horizon, as exemplified by Albaqualfs or Planosols. Albic E-horizons, however, are usually associated with the very different process of podzolisation (chapter 11). This has caused much confusion in soil science literature. Brinkman (1970), who recognised ferrolysis as a separate soil forming process, solved the problem.

A TWO-STAGE PROCESS

Ferrolysis is a surface-water gley process that proceeds in two stages. Because reduction causes fairly high concentrations of dissolved Fe^{2+} compared to other cations, part of the Fe^{2+} formed during soil reduction always attaches to the exchange complex and forces formerly adsorbed other cations into the soil solution. In case of a near neutral (not-yet ferrolysed) soil, the displaced cations are base cations: Ca^{2+}, Mg^{2+}, K^{+} and Na^{+}. These displaced cations are highly mobile and can easily be lost from the soil profile, either by percolation, or by diffusion into surface water standing on the field, followed by lateral drainage.

Question 7.7. *Loss of those cations represents loss of alkalinity, because the cations are normally balanced by the anion HCO_3^-. Why would the anion be HCO_3^-?*

If supply of cations from other sources (weathering, flood or irrigation water) is low, the adsorption complex may eventually become depleted of bases. The loss of bases should lead to a decrease in soil pH. Base cation depletion is not immediately apparent in the reduced stage when the pH remains between 6 and 7 (Chapter 3.3). Upon aeration of the soil, however, exchangeable Fe(II) is oxidised to insoluble Fe(III) oxide. During that oxidation H^{+} is formed, which competes with remaining basic cations for exchange sites.

Question 7.8. *Balance the reaction equation: adsorbed Fe^{2+} plus O_2 gives FeOOH*

CLAY DESTRUCTION

The formation of H^{+} results in a marked decrease in pH on the clay surface. Within a few days, the unstable H^{+}-clay converts to clay with exchangeable Al^{3+}. The Al^{3+} is dissolved from octahedral positions in clay minerals. This partly destroys the clay minerals, leaving plate edges of essentially amorphous silica. During the following water logging period, Fe(III) is reduced to Fe^{2+}, which in turn can displace adsorbed Al^{3+}. The pH increases under the influence of reduction of Fe(III) to Fe^{2+} so that dissolved or exchangeable Al^{3+} can hydrolyse and precipitate as Al hydroxide (see Chapter 3).

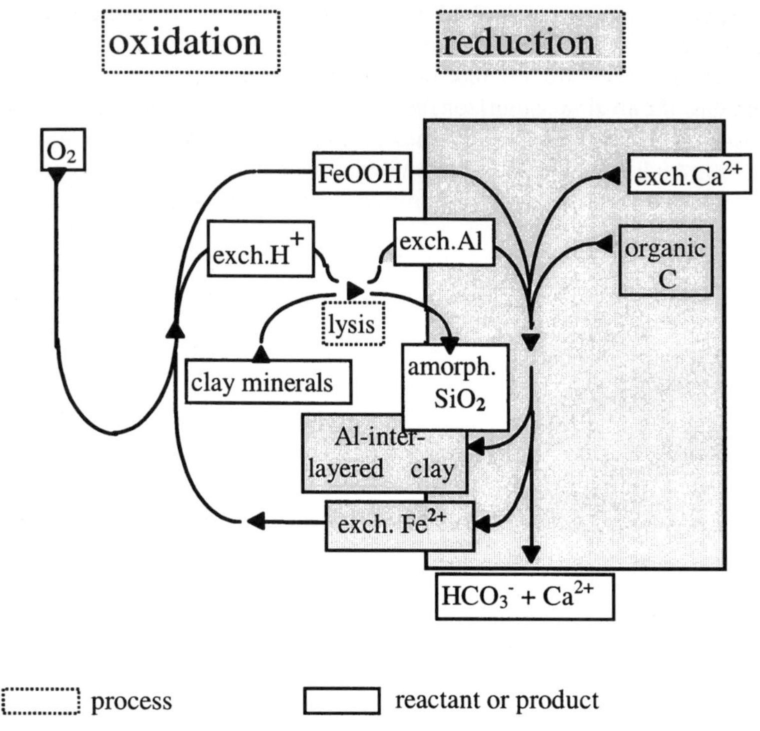

Figure 7.5. Pathways of the processes involved in ferrolysis. Modified after Brinkman, 1979.

Al-HYDROXYDE INTERLAYER

Intermediate products of Al-hydrolysis such as $[Al(OH)^{0.5+}{}_{2.5}]_n$, where n may be 6 or more, can form Al-hydroxy interlayers between plates of 2:1 clay minerals. Adsorbed Fe^{2+} and other cations, trapped between the plates that are 'glued' together by Al-hydroxy interlayers. The resulting Al-interlayered 2:1 clays ('soil chlorites') have a lower capacity to swell and shrink upon wetting and drying, and have a lower cation exchange capacity. Figure 7.5 summarises the individual processes involved in ferrolysis.

Question 7.9. *Less Fe^{2+} is formed and attached to the adsorption complex in near-neutral soils than in acidic soils. a) Why is that? b) What does this imply for the intensity of ferrolysis and of ferrolysis-induced chloritisation in near neutral and in acidic soils?*

Question 7.10. *Indicate which reactants and products in Figure 7.5 are cycled, and which are consumed, formed or lost.*

Ferrolysis is likely to occur in flat, relatively high land surfaces, such as riverine or marine terraces of early Holocene-late Pleistocene age. In these soils, decalcification and desilication have removed the most easily weatherable minerals. Moreover, clay illuviation has often led to conditions favouring seasonal ponding with (rain) water. So, ferrolysis has often been preceded by clay translocation down the soil profile (compare Figure 7.2). An example of a strongly ferrolysed soil is given in Plate L (p.158). In extreme cases, ferrolysis may be succeeded by podzolisation (see Chapter 11).

Question 7.11. *In a soil with an argillic horizon, but without ferrolysis, the clay balance may be closed: the amount of clay lost from the surface soil equals the extra clay found in the argillic horizon.*

a) Explain how a study of the clay balance could help you to distinguish between a soil with only clay illuviation, and a soil that has undergone ferrolysis after clay illuviation.

b) Give three possible reasons for the low CEC of the A- and E- horizons of a ferrolysed soil.

c) What other soil characteristics would support the hypothesis that a given soil has undergone ferrolysis?

Question 7.12. *The interlayered 2:1 clay minerals formed during ferrolysis seem to be quite stable. Why would those clay minerals not be destroyed easily by further ferrolysis?*

DISTINGUISHING BETWEEN FERROLYSIS, DESILICATION, AND PODZOLIZATION

In a soil subject to ferrolysis, part of the silica liberated by weathering of clay minerals stays behind as silt- to sand-sized amorphous silica aggregates, especially in the surface horizon. In thin sections, the ferrolysed clay cutans appear less birefringent and are characteristically grainy.

In soils undergoing 'normal' desilication, such amorphous silica is not normally formed. The clay cutans and the ground mass appear bleached, by the loss of part of the free iron oxides. The iron oxide has been concentrated elsewhere in nodules or mottles, partly also in surface horizons but mainly in the deeper horizons.

In well drained acidic (pH 4.5-5.5) soils subject to desilication but without ferrolysis, 2:1 clay minerals such as smectites and vermiculite tend to become interlayered with Al, or to disappear completely. The interlayers do not appear to be stable over long periods, and the interlayered 2:1 clay minerals are eventually transformed (by desilication) into the 1:1 clay mineral kaolinite. Under extreme leaching in well-drained soils, desilication may go so far that only gibbsite ($Al(OH)_3$) stays behind as a weathering product. In contrast, the interlayered 2:1 clay minerals formed during ferrolysis seem to be stable with respect to kaolinite, possibly because desilication is not strong enough.

During cheluviation in podzols, clay is decomposed too. Moreover, a bleached eluvial horizon may be formed by loss of iron oxides, resembling the eluvial horizon formed during ferrolysis. During cheluviation, however, Al as well as Fe are removed as organic complexes, as will be discussed in Chapter 11. The (small amounts of) clay remaining in the E horizon of a podzol usually consist of a beidellite-like smectite, a swelling clay mineral without aluminous interlayers and with a high CEC. Interlayered 2:1 clay minerals (soil chlorites) are absent from albic E-horizons in podzols, but may be present in the podzol B-horizon. The major clay mineralogical differences between soils subject to cheluviation and ferrolysis therefore occur in the A-, E- horizons and B-horizons The differences between the effects of desilication, cheluviation and ferrolysis on clay minerals in the upper soil horizons are summarised in Table 7.1.

Table 7.1. Clay mineralogical changes by different soil processes.

Process	*Horizon affected*	*Nature of changes*
desilication	whole profile	Effect decreases with depth. 2:1 minerals → kaolinite → gibbsite (intermediate steps lacking with strong leaching).
cheluviation	A and E	Decomposition of most clay minerals: removal of Al-interlayers → Al-smectite (see Chapter 3.2).
	B	2:1 minerals become Al-interlayered
ferrolysis	A and E	2:1 minerals → soil chlorite (with some Fe trapped in interlayers) and secondary silica.
	B	little change

7.4. Acid sulphate soils

The strong acidity of a particular group of hydromorphic soils, acid sulphate soils, has caused, and still causes, much harm to farmers in many coastal areas. Acid sulphate soils release dissolved Al^{3+} from soil minerals in concentrations that are often toxic to plants, and that lead to strong phosphorus deficiency. The problem is particularly important because most coastal areas are otherwise very suitable for agriculture and planners and engineers often overlook the occurrence of potential acid sulphate soils.

JAROSITE AND SULFURIC HORIZONS

The root of the problem lies in the strongly acid character of sulfuric acid, the stable form of S in the system S-H_2O under oxic conditions. Typical acid sulphate soils have a 'sulfuric horizon', with a pH below 3.5 and conspicuous, pale yellow mottles of jarosite $KFe_3(SO_4)_2(OH)_6$, formed by aeration of reduced sediments rich in iron sulphides, mainly pyrite (cubic FeS_2).

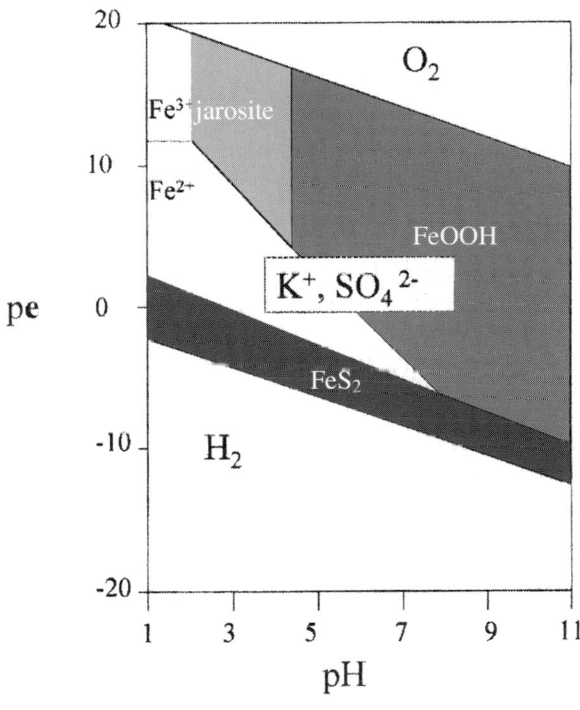

Figure7.6. Approximate pe-pH diagram of fine-grained goethite, jarosite and pyrite. Solid phases are indicated by hatching. The boundaries between solids and dissolved species are for solute concentrations of about 10^{-4}M for Fe^{2+}, 10^{-2}M for SO_4^{2-}, and 10^{-3}M for K^+. The SO_4^{2-} field encompasses the whole area of the diagram above the FeS_2 field. Increased concentrations would cause an expansion of jarosite and goethite fields at the expense of the dissolved Fe fields; decreased concentrations would cause shrinkage of the stability area of those minerals. After Van Breemen, 1988. Compare with Figure 3.11.

Soils with a sulfuric horizon are classified as Thionic Fluvisols (FAO-Unesco) or Sulfaquepts (Soil Taxonomy). Some soils that have a pH<3.5 from sulfuric acid lack jarosite mottles, and some do contain jarosite, but have a pH between 3.5 and 4. Neither of these satisfies the criteria for Thionic Fluvisols or Sulfaquepts. Because the processes in all these cases are similar, we define acid sulphate soils somewhat broader: soils that have acidified to pH <4 by oxidation of iron sulphides.

Question 7.13. *Illustrate that pyrite oxidation leads to acidification by writing out the reaction equation:* $FeS_2 + O_2 + H_2O$ *produces* SO_4^{2-}.

SULFIDIC MATERIAL

The unoxidized, potentially acid, sulfidic parent material is called 'sulfidic material'. Acid sulphate soils occur frequently in recent coastal plains, where pyritic tidal marsh sediments have been exposed to the air following drainage, either naturally or by man. In the absence of sufficient neutralising substances such as calcium carbonate, the sulfuric acid liberated during oxidation of pyrite may cause a drop in soil pH to 3-4 or sometimes lower. Acid sulphate weathering takes place also in pyritic sedimentary rocks outside recent coastal plains, and in pyritic mine spoils. The formation of potential acid sulphate soils, of acid sulphate soils and their eventual transformation to less acid soils without jarosite can be illustrated well with the pe-pH diagram shown in Figure 7.6.

Question 7.14. *a) Describe in words the conditions where jarosite, goethite, pyrite and dissolved* Fe^{2+} *are stable ('acid', 'near neutral', 'oxidised', 'reduced').*
b) What is the stable form of sulphur outside the pyrite field?
c) According to the diagram, which of the following pairs cannot exist at the specified concentrations and other conditions: Fe^{3+}-Fe^{2+}, Fe^{3+}-*FeOOH*, Fe^{3+}-*jarosite*, Fe^{2+}-*jarosite*, Fe^{3+}-FeS_2, Fe^{2+}-FeS_2, *FeOOH-jarosite*, *FeOOH*-FeS_2, *jarosite*-FeS_2, Fe^{2+}-SO_4^{2-}, Fe^{3+}-SO_4^{2-}, *FeOOH*- SO_4^{2-}, *jarosite*-SO_4^{2-}, *and* FeS_2- SO_4^{2-}, O_2-*FeOOH*, O_2-Fe^{3+}, O_2- Fe^{2+}, *and* O_2- FeS_2*?*

PYRITE FORMATION

In most potential acid sulphate soils, pyrite is the dominant Fe(II) sulphide with concentrations in the order of 1-5% by mass. Formation of pyrite is initiated by sulphate reduction. During sulphate reduction, bacteria oxidise organic matter by means of sulphate, which in turn is reduced to S(-II), the form of sulphur in H_2S. Black ferrous monosulphide (FeS) may be formed first, but is unstable and is eventually transformed to pyrite. Pyrite is a disulphide in which S has a higher oxidation state (S(-I)) than in H_2S or FeS (S(-II)), so an **oxidant** is needed to form pyrite from S(-II).

Question 7.15. *Write reaction equations for: a) the formation of an equimolar mixture of FeS and FeS_2 plus water from FeOOH and H_2S; b) the formation of FeS_2 plus water from FeOOH, H_2S and O_2. What are the reductants and oxidants in each of these reactions? Use these equations to show that O_2 (or some other oxidant) would be necessary for complete pyritisation of H_2S plus sedimentary iron.*

So, formation of pyrite involves (1) reduction of sulphate to sulphide by **dissimilatory** sulphate reducing bacteria, (2) oxidation of sulphide to disulphide, and reaction of sulphide and disulphide with iron minerals. If Fe_2O_3 stands for the iron source, the formation of pyrite can be described by the reaction equation:

$$Fe_2O_3 + 4SO_4^{2-} + 8CH_2O + 1/2O_2 \rightarrow 2FeS_2 + 8HCO_3^- + 4H_2O \quad (7\text{-}1)$$

Wherein CH_2O represents organic matter. Thus, essential ingredients for pyrite formation are sulphate, iron-containing minerals, metabolizable organic matter, sulphate reducing bacteria, and alternating anaeroby, in time or space, with limited aeration. In fresh-water sedimentary environments, sulphate concentrations are too low for appreciable pyrite formation. In tidal marshes, particularly those with mangrove vegetation, supply of both sulphate (from seawater) and organic matter (from the mangroves) is abundant. Supply of sulphate and oxygen are aided by tidal movement through the high permeability of mangrove soils, which is due to numerous biopores from crabs, roots, and decaying organic matter. In such conditions, most of the iron available for pyritisation (fine-grained Fe(III) oxide) is eventually transformed into pyrite.

Pyrite usually occurs in raspberry-shaped aggregates, so-called framboids, of 2 to 40 μm diameter, or as single, fine-grained crystals of 0.1 to 5 μm diameter. It is formed relatively rapidly (mass fractions of several % of FeS_2 can be formed in years to decades) in mangrove and brackish reed marshes with intensive tidal influence due to the presence of many tidal creeks. Pyrite formation is much slower in sediments with stagnant water, e.g. at depth below which tidal flushing ceases, or in areas with few tidal creeks. The rate of pyritisation is increased by tidal flushing which 1) increases the supply of O_2, necessary for complete pyritisation of Fe(III) oxides, and 2) facilitates removal of bicarbonate formed during sulphate reduction (see eq. 7.1). Removal of HCO_3^- lowers the pH (from 8 to about 6) which greatly speeds up pyrite formation. Pyrite formation in fresh water systems is restricted by the availability of sulphate (e.g. from gypsum deposits, or oxidised marine clays).

OXIDATION OF PYRITE

As can be read from Figure 7.6, pyrite is unstable in the presence of O_2. It can be transformed to dissolved Fe^{2+} and SO_4^{2-}, to jarosite or to goethite, but regardless of the end product, pyrite oxidation produces H^+. The pH will drop only, however, if the H^+ formed is not completely neutralised by a neutralising substance, e.g. calcium carbonate. Pyritic material without sufficient acid buffering capacity is called 'potentially acid'.

Question 7.16 *a) Under what conditions of pe and pH is pyrite oxidised to each of the phases 1) dissolved Fe^{2+} and SO_4^{2-}, 2) jarosite, or 3) goethite?*
b) Illustrate the formation of H^+ during oxidation of pyrite to dissolved Fe(II) sulphate by means of an appropriate reaction equation.

In the absence of acid buffering, the pHs of pyritic soil material often drop from near neutral to values below 3.5 within several weeks after the start of aeration. Although pyrite oxidation does proceed without the help of micro-organisms, chemoautotrophic sulphur bacteria (*Thiobacillus* sp.) greatly speed up the reaction rate.

Question 7.17. *What is the meaning of 'chemoautotrophic'?*

These bacteria seem to be invariably active wherever pyrite oxidation takes place, and some can stand pH values below 1. In their presence the oxidation rate of pyrite is usually limited by the rate of diffusion of O_2.

Question 7.18. *Spoils of pyritic material along canals and ditches, which are exposed directly to atmospheric O_2, acidify much more rapidly than pyritic material present at some depth in the soil. Why?*

Temporary oxidation of some pyrite probably occurs in many environments where pyrite actually forms, such as tidal marshes at low tide. This does not produce much acid, and has little or no effect on soil pH. But when potentially acid soil is aerated for weeks or months, dramatic acidification may occur. After the initial appearance of dissolved sulphate and brown spots of Fe(III), the pH may drops to values below 3, and pale yellow jarosite mottles may form.
Prolonged aeration requires sustained lowering of the groundwater table. In a mangrove marsh this is only possible if the tidal action stops, e.g. by empoldering land, or more slowly, by decreasing tidal influence as a result of coastal aggradation.

Question 7.19. *Why must tidal action stop before mangrove marshes can turn into acid sulphate soils?*

While the pale yellow jarosite mottles show the presence of acid sulphate soils, not all acid sulphate soils do have jarosite. Jarosite does not form at pH below 4, and when pe values remain relatively low as in highly organic, very wet acid sulphate soils of equatorial regions, e.g. in Indonesia. In those conditions, iron from oxidised pyrite remains as dissolved Fe^{2+}. This Fe^{2+} can be leached from the soil, ultimately into ditches and creeks, where it is oxidised to brown iron III oxide precipitates. As a result of this process, such acid sulphate soils are low in iron, which has important consequences for their practical use.

Question 7.20. *Explain the absence of jarosite from many acid sulphate soils in the perhumid zone, with reference to Fig. 7.6.*

BUFFERING OF ACIDITY - ACID NEUTRALIZING CAPACITY

Acid sulphate soils form if the amount of potential acidity in the form of pyrite exceeds the amount of neutralising substances. Neutralising substances that make up the so-called acid neutralising capacity (ANC) are calcium carbonate, exchangeable cations, and weatherable silicate minerals. Any calcium carbonate present can rapidly neutralise sulphuric acid, giving CO_2 (which largely disappears into the air) and dissolved calcium sulphate. $CaCO_3$ is dissolved by H_2SO_4 faster than H_2SO_4 can be formed by oxidation of pyrite. Neutralisation by weathering of silicate minerals, however, is usually much slower than pyrite oxidation. These differences in relative rate of acid formation and buffering influence the acidification process.

Question 7.21. *A pyritic soil sample with an acid neutralising capacity (ANC) that exceeds the potential acidity is exposed to the air in a beaker in the lab (no leaching). What do you expect to happen with the pH of that sample, a) if the ANC is made up of $CaCO_3$ only, b) if ANC is $CaCO_3$ plus exchangeable base cations, and c) if ANC consist only of weatherable silicates.*

EXCHANGEABLE H^+ and Al^{3+}

If the amount of H_2SO_4 formed exceeds the ANC provided by any $CaCO_3$ present, any exchangeable cations present will be exchanged for H^+, and the pH will drop below 6. If enough acid is formed to lower the soil pH below about 5, clay minerals are attacked by the acid, and H^+ is neutralised by removal of cations (mainly Al^{3+}, Mg^{2+} and K^+) from the clay structure. Al^{3+} largely replaces exchangeable H^+, while Mg^{2+}, K^+, and silica dissolved from the clay appear in the soil solution. If the pH drops below 4, appreciable amounts of Al^{3+} come into solution too. In a period with excess rainfall or during inundation, base cations such as Ca^{2+} or K^+ are removed from the soil as sulphates, together with a part of the dissolved Al^{3+}, either downward with the percolating water or upward by diffusion into the inundation water. In very dry periods, efflorescences of soluble sulphate salts of Fe, Al and base cations can be formed on ped faces and on the soil surface

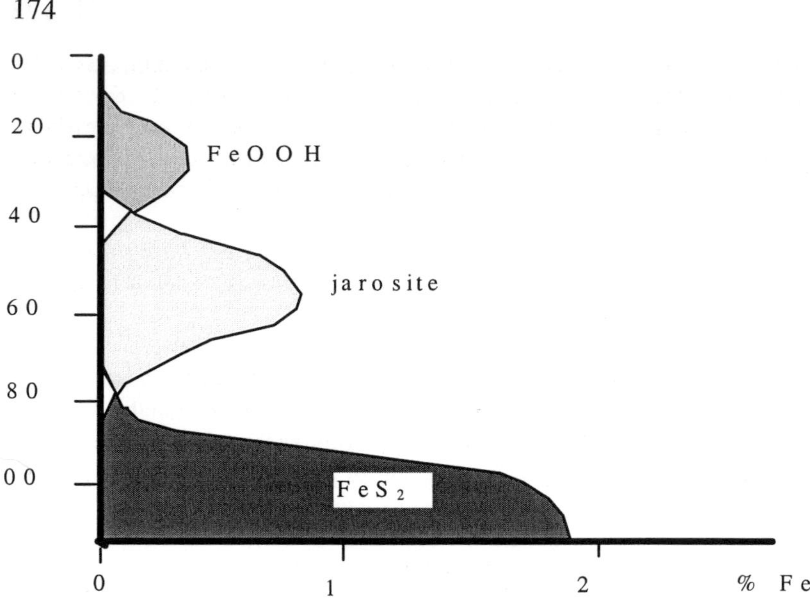

Figure 7.7. Distribution with soil depth (cm) of pyrite, jarosite and goethite in an acid sulphate soil in Thailand. The more deeply developed the soil (= the more deeply drained for a considerable time of the year), the deeper the pyritic substratum and the two accumulation horizons.

If all pyrite in the surface horizons is oxidised, the pH of acid sulphate soils tends to increase in the course of decades. This is due to leaching of free acidity slow acid neutralisation during weathering of clay minerals and other silicate minerals. As the pH rises, jarosite becomes unstable and is hydrolysed, producing goethite and some sulphuric acid. During hydrolysis, the yellow jarosite mottles gradually turn brown. Such mottles with mixed colours, in contrast to the pure yellow ones, indicate that soil acidity has decreased.

Question 7.22*. Write a reaction equation describing the transformation of jarosite plus water to FeOOH,* SO_4^{2-} *and* K^+.

SOIL PROFILE DEVELOPMENT

Jarosite tends to accumulate in a distinct horizon above the pyritic horizon (Fig. 7.7) The transition zone between the jarosite horizon and the pyritic substratum is acidic but not sufficiently oxidised for jarosite formation. With increasing soil age, the pyritic substratum and the jarosite horizon are found at progressively greater depth, due to (1) oxidation of pyrite at the upper boundary of the pyritic substratum, (2) downward extension of the lower boundary of the jarosite horizon and (3) hydrolysis of jarosite to goethite at the upper part of the jarosite horizons. These processes usually occur to a greater depth along vertical cracks and channels, so that the upper horizons appear to protrude into lower adjacent horizons along such zones.

Question 7.23. *How could jarosite and pyrite occur at the same depth (Fig. 7.7) although theoretically they cannot co-exist (Fig 7.6)?*

Upon continued drainage, all pyrite is eventually oxidised and all jarosite is hydrolysed. Complete removal of jarosite takes several years to centuries, depending on the intensity of leaching. Acid sulphate soils are therefore always young soils, and eventually lose most of their acidity.

ACID SULPHATE SOILS FOR WETLAND RICE

Coastal lowlands are important rice growing areas, and they commonly have acid sulphate soils. How do rice and acid sulphate soils get along? Not bad. The pH of most wetland rice soils rises to near neutral values within a few weeks to months of flooding, which counteracts Al toxicity and other problems associated with low pH.

This rise in pH upon flooding is one of the reasons why wetland rice cultivation is a good way to use acid sulphate soils. In many very young acid sulphate soils, the pH indeed increases quickly from 3-3.5 to 5.5-6, while dissolved Fe^{2+} also rises steeply. In most older acid sulphate soils, however, the pH increases very slowly after water logging, and sometimes does not rise above 5.5.

> A slow rise in pH can be attributed to (i) slow reduction perhaps because of lack of easily decomposable organic matter, or (ii) appreciable reduction (fermentation) in the absence of inorganic reducible substances, e.g Fe(III) oxides. In case (i), neither pe nor pH will change much after flooding. In case (ii), the pe will drop without concomitant increase in pH. Slow reduction has often been observed, and has been attributed to a low content of metabolizable organic matter, and/or to adverse effects of low pH, high dissolved Al, and poor nutrient status on the activity of microbes. The strong reduction and associated increase in pH upon flooding of very young acid sulphate soils suggests that the presence of relatively undecomposed organic matter from the original (mangrove) vegetation, and high chemical fertility may be more important for strong reduction than a low pH by itself.
>
> Acid sulphate soils in Kalimantan (Indonesia) show little or no increase in pH after reduction following flooding. This has been attributed to low contents of

Fe(III) oxides relative to the base neutralising capacity of the soil (mainly exchangeable acidity associated with organic matter) (Konsten et al., 1994). Apparently, the combination of highly organic parent material and perhumid climate limits oxidation of Fe(II) and accumulation of appreciable quantities of Fe(III) oxides and jarosite, and causes leaching of most of the iron that was once present in pyrite.

RECLAIMING ACID SULPHATE SOILS

Reclamation of potential or actual acid sulphate soils can be seen as applied soil genesis. A good understanding of the soil forming processes involved can be very helpful to improve such soils for agriculture, or to limit environmentally adverse effects (e.g. on surface water quality) of reclamation.

The only sure way to curtail pyrite oxidation is cutting the supply of O_2, e.g. by keeping the soil waterlogged. Once formed, soil acidity can be removed by leaching or by neutralisation. Neutralisation can be done by adding acid-neutralising substances, normally lime ($CaCO_3$), or by so-called 'self-liming', utilising the acid neutralising effect of reduction of Fe(III) oxides upon flooding.

Soluble and exchangeable acidity should be removed as much as possible by leaching before applying amendments or fertilisers. In principle, leaching with fresh water is efficient in removing free H_2SO_4 and soluble Fe and Al salts. By leaching with salt or brackish water exchangeable Al can be replaced by Na, Ca and Mg. The possibilities for leaching depend, among other things, on the structure of the soil. Water may percolate easily through a highly permeable soil, and bypass the interior of soil aggregates, which tends to decrease leaching efficiency. The contact between leaching water and soil could be increased by ponding and puddling, followed by surface drainage. While leaching can be a relatively cheap and effective measure to improve soils, it also removes nutrients from the soil, and it pollutes surface waters.

The amount of base required for the neutralisation of acidity produced by oxidation of one per percent of oxidizable sulphur is about 30 tons of $CaCO_3$ per 10 cm of soil of bulk density 1 kg/dm^3. Exchangeable cations provide a neutralising capacity equivalent to 3 to 30 tons of $CaCO_3$ on the same basis, depending on clay content and mineralogy. Most of the oxidizable sulphur in excess of that must be removed by leaching or liming. Considering that contents of oxidizable sulphur are commonly between 1 and 5 %, this illustrates the huge amount of lime that would be needed if reclamation of acid sulphate soils depended on it. In practice, liming is efficient only after most of the water-soluble acidity has been leached. In case of lowland rice, applications of 2 to 10 ton of $CaCO_3$/ha on leached acid sulphate soils often have a distinct beneficial effect, while large doses are rarely economical.

7.5. Problems

Problem 7.1

Figure 7A shows the profiles of free iron oxide in a chronosequence of polder soils, in Japan. The land is drained, and the water table is generally kept at about 0.5 to 1 m below the surface throughout the year. In summer, wetland rice is grown, and the fields are flooded, resulting in a perched water table.

a. Describe and explain the change in the Fe oxide concentration with depth as the soils become older (note that the (presumed) Fe_2O_3 content of the parent material varies somewhat, between 5 and 6%).
b. Where do you expect Mn accumulation to occur?

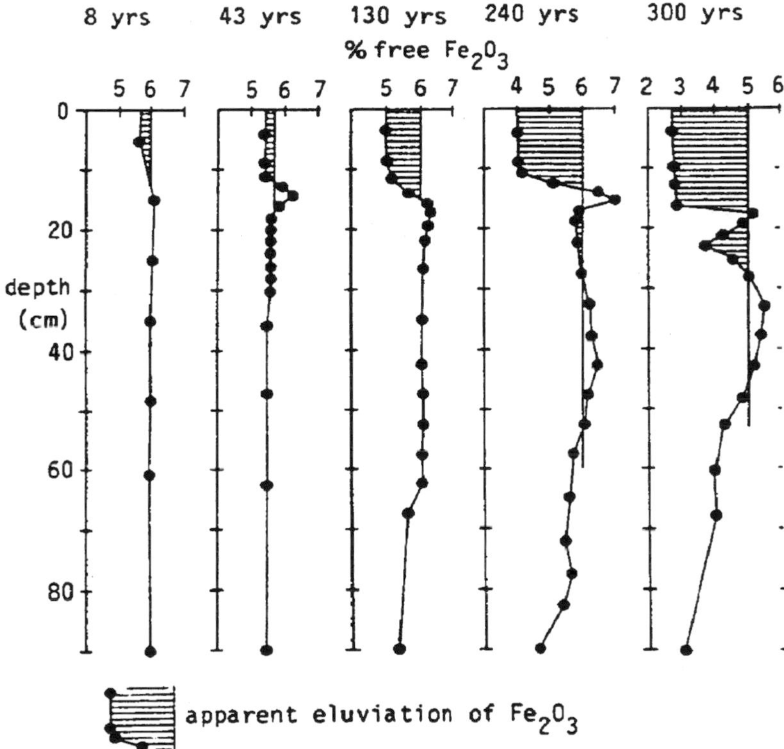

Figure 7A. Profiles of free Fe(III)oxide contents with depth in a chronosequence of rice soils from Japan (From Kawaguchi and Matsuo, 1975).

Problem 7.2

Artificially stagnic soils under rice cultivation in Japan, as shown in Figure 7A, sometimes have lost practically all iron oxides from the seasonally flooded surface soil. They tend to produce low rice yields. These "degraded padi soils" are usually rather acidic, and tend to remain so after flooding, in contrast to non-degraded soils where the pH usually rises to 6.5 - 7 after flooding. Low yields are attributed to toxicity of organic acids and H_2S that may reach high concentrations at low pH. Traditionally, farmers improve such soils by adding "red earth" from upland areas. Explain why pH values of "degraded padi soils" remain low upon flooding. What could be the reason why these soils are improved by adding red earth?

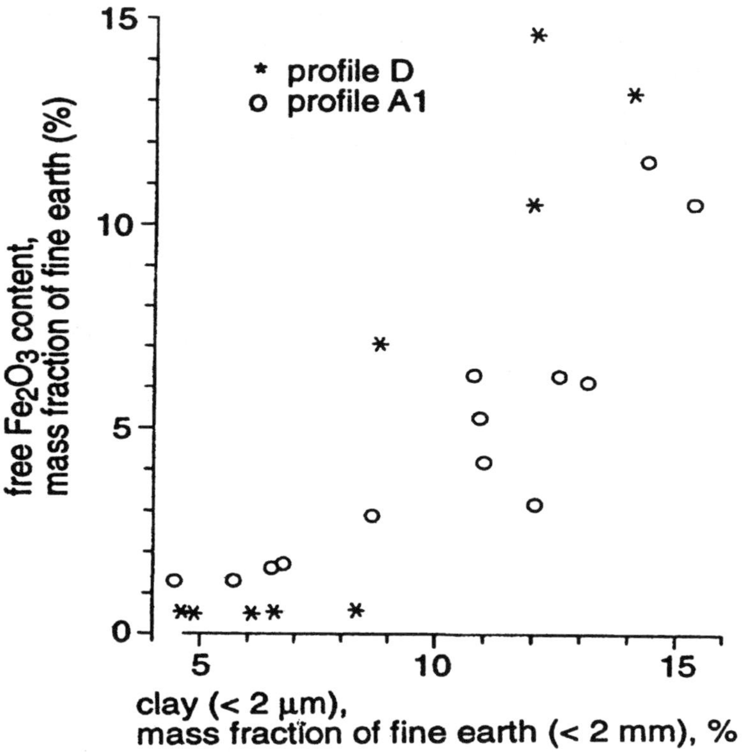

Figure 7B. Relationship between the clay and free Fe(III) contents in the mottled horizon of two ground water gley soils (o, *), influenced by upwelling ground water near Warnsveld, The Netherlands. From Van Breemen et al., 1987.

Problem 7.3

Figure 7B is a plot of the clay content against the content of free Fe_2O_3 in soil samples from various depth in the mottled horizon of two low-lying ground water gley soils (*, o) in the Achterhoek, the Netherlands. The parent material contains a few % of total Fe.

a. What process can account for the high Fe_2O_3 contents?
b. Explain the positive correlation between the clay content and the Fe_2O_3 content of individual samples.

Problem 7.4

Figure 7C illustrates the variation in redox conditions with time and depth in a very thin surface layer of a waterlogged soil used for wetland rice.

a. Discuss the differences between Mn, Fe and S in i) speed of reduction, and ii) the depth at which reduced Mn, Fe and S appear.
b. Thin, "mini horizons" of Mn oxide and Fe(III) oxide develop within weeks to month at the soil surface. Sketch the positions of these mini horizons.
c. Could these mini-horizons be used diagnostically to classify seasonally waterlogged soils?

Problem 7.5

The data in table 7D refer to a surface water gley soil from Bangladesh. Assume that the soil bulk density is 1 kg/dm^3 throughout the soil profile.

a. Can you explain the profile of free Fe_2O_3 on the basis of oxidation-reduction processes?
b. Calculate the clay balance of the profile, using ε based on (sand plus silt) as index (so, calculate τ_{clay} for each horizon). Explain the excess or deficit of clay.

Table 7.D. Clay contents and chemical data for a profile of Chhiata series, Bangladesh. (Brinkman 1977). Reproduced with permission of Elsevier Science, Amsterdam.

horizon	depth	clay	pH water	org. C %	Ca	Mg	K	Al	CEC	free Fe_2O_3
	cm	%		%		mmol(+) kg^{-1}				%
Apg1	0- 8	12.5	4.9	0.65	22	0	3	13	40	0.9
Apg2	8- 13	25.1	5.0	0.42	38	4	3	3	63	1.7
Eg1	13- 18	27.1	5.0	0.46	55	6	1	4	79	1.5
Eg2	18- 30	29.7	4.9	0.24	51	5	2	4	66	2.5
Eg3	30- 41	33.3	4.9	0.12	81	9	3	9	87	2.3
ECg	41- 58	41.6	5.1	0.02	101	19	3	6	114	1.7
Cg1	58- 97	42.1	5.3	0.02	112	25	2	1	138	1.8
Cg2	97-127	42.9	5.6	n.d.	130	30	3	1	156	1.5
Cg3	127-152	44.2	5.8	0.18	141	31	3	0	173	1.6

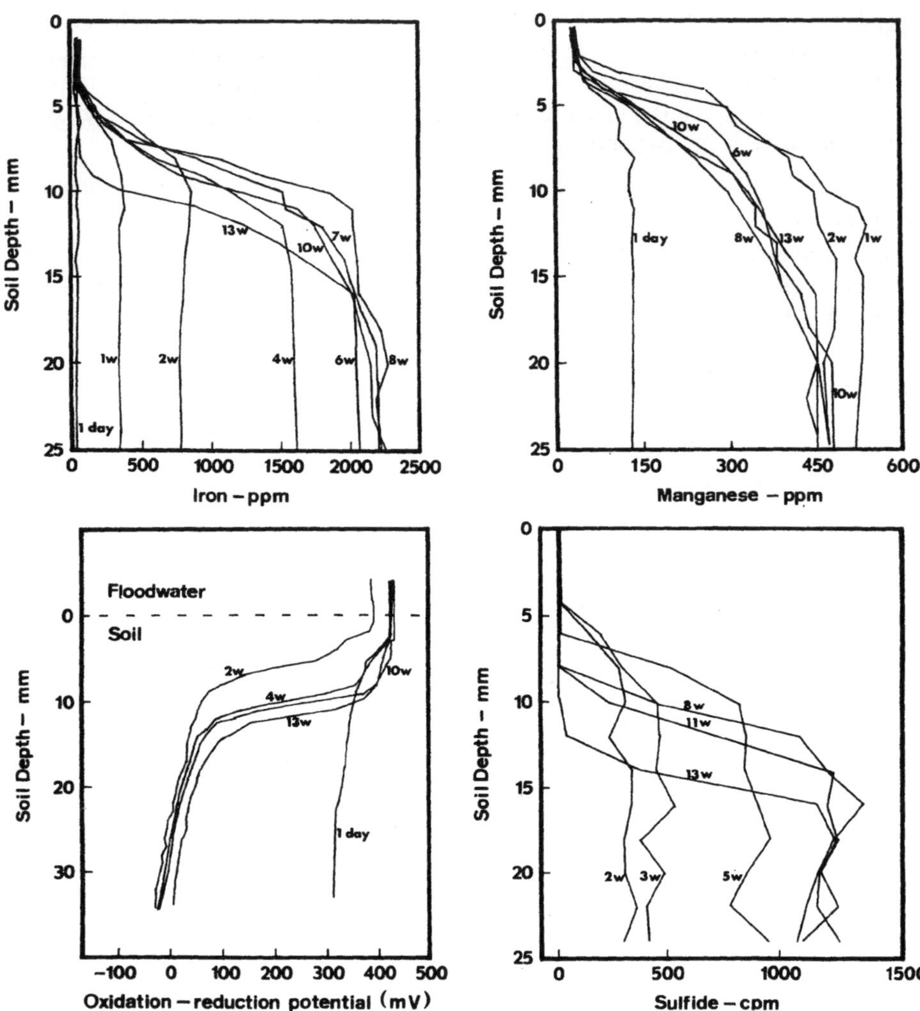

Figure 7C. Distribution with depth of the redox potential and Na-acetate extractable Mn(II), Fe(II), and total S(-II) in the upper 25 mm of a reduced (water logged) soil at various times after the onset of flooding. Fe and Mn extractable by Na-acetate (pH=2.8) include solid plus dissolved Mn(II) and Fe(II), formed after reduction.

From Patrick and Delaune, 1972.

c. Calculate the specific CEC of the clay fraction ($mmol_c$/kg clay) and of the organic C fraction ($mmol_c$/kg org. C) in the Apg1 and the Apg2 horizons, assuming that they do not differ between these two horizons. Next calculate the specific CEC of the clay fraction of all horizons, assuming that the specific CEC of the organic fraction does not vary with depth. Explain the change in CEC_{clay} with depth.

Problem 7.6

Table 7E contains analytical data on two acid sulphate soils of different age from Thailand. Both soils have developed from similar parent materials containing a few percent of pyrite.

Table 7E. pH and sulphur fractions in acid sulphate soil profiles in Thailand (Van Breemen, 1976).

depth (cm)	pH		sulphur fractions (%S)		bulk density (kg/dm^3)
	in the field	after 3 weeks of aeration	FeS_2	jarosite	
young soil, deepest groundwater table 40 cm					
0 - 10	4.5	4.8	0.0	0.1	1.1
10 - 18	3.3	3.2	0.1	0.1	0.9
18 - 30	3.6	2.1	0.6	0.0	0.7
30 - 40	4.5	2.5	2.0	0.0	0.6
40 - 60	6.1	3.1	2.3	0.0	0.6
old soil, deepest groundwater table 150 cm					
0 - 25	4.2	4.2	0.0	0.0	1.4
25 - 38	3.9	4.0	0.0	0.0	1.3
38 - 58	3.6	3.6	0.0	0.1	1.2
58 -130	3.6	3.6	0.0	0.4	1.0
130 -140	3.7	3.6	0.1	0.2	0.9
150 -200	4.5	2.9	1.0	0.0	0.9

a. Explain the pH profiles before and after aeration.
b. How much acid would have been formed ($kmol.ha^{-1}$) in the old soil if the original pyrite content were 1.0 % (S) throughout?
c. If none of this acid would have been neutralised and all of it would have remained in the soil, what would be the soil pH now (assume that the water content in the old soil is 50% by volume). Explain the difference between actual and calculated pH.
d. If all acid produced by oxidation of 1% of pyrite is completely neutralised by weathering either of kaolinite or of montmorillonite (with, among others, Al^{3+} and H_4SiO_4 as end products), how much of these clay minerals (g clay/kg of soil) will be consumed? Would this greatly affect the clay mineral composition or the clay content of the soil if the original clay content were 50 %, with equal amounts of kaolinite and montmorillonite?

Problem 7.7
The soils in problem 7.6 have a CEC of about 300 mmol/kg of soil, and have a base saturation of 100 % at pH > 6. How much pyrite (in % S in the soil) must at least be oxidised to cause a decrease in base saturation to 40%.

7.6. Answers

Question 7.1.
In oxidised soils: FeOOH (goethite), Fe_2O_3 (hematite), $Fe_2O_3.nH_2O$ (ferrihydrite), MnO_2 (birnesite), MnOOH (manganite).
In reduced soils: soluble and exchangeable Fe^{2+} and Mn^{2+}, green rust (mixed FeIIFeIII hydroxides), $FeCO_3$ (siderite), FeS (mackinawite) and pyrite (FeS_2).

Question 7.2
1=Ah horizon, 2=E horizon, lowered in clay, 3=Bt horizon, with illuviated clay, 4=Fe(III)oxide mottles, 5=Mn oxide mottles, 6=permanently reduced C horizon, below lowest ground water table, 7=Fe(III) oxide accumulation at the interface between oxic zone (here: Bt horizon) and seasonally anoxic zone (here E horizon)

Question 7.3
According to Question 3.32, 0.25 mole of metabolizable C (presented as 'CH_2O') are needed to reduce one mole of FeOOH. So 0.25 * 12 g C could reduce 1* 56 g Fe(III). To reduce 1 % Fe(III), 0.25*12/56 = 0.054 % C would suffice.

Question 7.4
See sketch. Of course, Fe^{2+} and O_2 move radially in all directions. The thick line at the boundary between the oxic and anoxic zones represents the FeOOH precipitate.

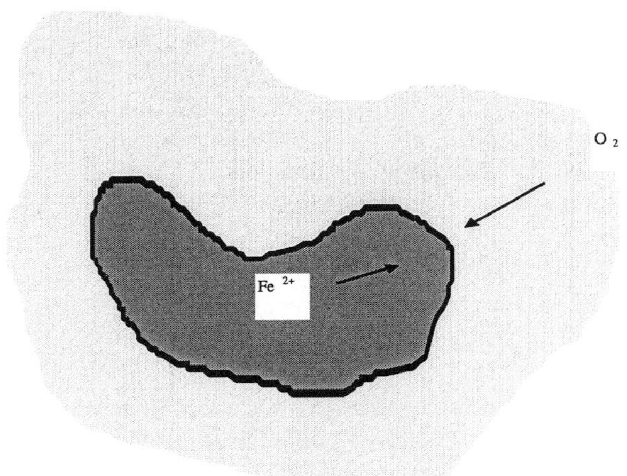

Question 7.5
The Mn(IV) oxide accumulation would lie outside the FeOOH accumulation zone, so in a zone of higher **pe** or O_2 partial pressure.

Question 7.6
The incipient placic horizon further inhibits drainage, and so stimulates water logging of the surface horizon. Fe(III) at the upper part of the placic horizon will reduce to Fe^{2+}, which in part diffuses through pores and cracks to the bottom of the placic horizon where it will reprecipitate after oxidation to Fe(III). The dissolution occurs most frequently in small depressions in the placic horizon, which become saturated more often and therefore move down more quickly than higher parts.

Question 7.7
HCO_3^- is the anion normally formed along with dissolved Fe^{2+} (see, e.g., the answer to Question 3.3), and remains the counter ion of the cations exchanged for Fe^{2+} at the exchange complex.

Question 7.8
$adsFe^{2+} + 1/4O_2 + 3/2H_2O \rightarrow 2adsH^+ + FeOOH$

Question 7.9
a) Soils that are near-neutral under oxic conditions have little exchangeable Fe^{2+} after reduction, 1) because they are well supplied with base cations which compete with Fe^{2+} for exchange sites, and 2) because the solubility of FeII compounds is less at higher pH. In acidic soils that become reduced, exchangeable Al undergoes forced exchange because Al precipitates as hydroxide due to the increase in pH.

b. Because the clay exchange complex in initially acidic reduced soils has a fairly high saturation with Fe^{2+}, the intensity of ferrolysis can be expected to increase initially, as the pH decreases under the influence of ferrolysis.

Question 7.10

Cycled: FeOOH and Fe^{2+}; *consumed*: O_2, org C, exchangeable Ca^{2+} and 2:1 clay minerals; *formed*: exchangeable Al, Al-interlayered clay, amorphous SiO_2, and Ca^{2+} plus HCO_3^-. Only the last two are lost from the soil, according to the Figure. Actually, part of the FeOOH and SiO_2 is lost too, at least from the surface soil, the former by transfer of Fe^{2+} to the oxidised B-horizons and perhaps to the drainage water.

Question 7.11

a) A ferrolysed soil will have a negative clay balance: more clay is lost from the surface horizon (due to eluviation plus destruction of clay) than is illuviated into the B-horizon (See e.g. the soil in problem 5.1). However, Chapter 8 will tell you that many other processes may be involved in surface loss of clay.
b) The CEC of the A and E horizons is low because of 1) clay eluviation, 2) clay destruction, and 3) blocking of part of the negative charge by Al hydroxy interlayers.
c) A surface water gley morphology with a bleached horizon (see section 7.2).

Question 7.12

The interlayered 2:1 clay minerals are more slowly destroyed by further ferrolysis because they have a very low CEC, and therefore cannot adsorb much Fe^{2+} during the reduction stage or much H^+ during the oxidation stage.

Question 7.13.

$FeS_2 + 15/4\ O_2 + 5/2\ H_2O \rightarrow FeOOH + 4H^+ + 2SO_4^{2-}$

Question 7.14.

a) From Figure 7.6 you can see that jarosite occurs under acid and oxic conditions; goethite under slightly acid to alkaline, anoxic to oxic conditions; pyrite under acid to strongly alkaline, anoxic conditions.
b) Sulphate.
c) Unstable pairs under the conditions specified for the pe-pH diagram are those with non-adjacent stability fields: Fe^{3+}-FeOOH, Fe^{3+}-FeS_2, jarosite-FeS_2, O_2-Fe^{2+}, and O_2-FeS_2.

Question 7.15

-----oxidants------- --reductants---

a) 2FeOOH + 3 $H_2S \rightarrow FeS + FeS_2 + 4H_2O$

b) FeOOH + ¼ O_2 + 2 $H_2S \rightarrow FeS_2 + 2½ H_2O$

So in the absence of O_2, only half of all iron sulphide can be transformed into pyrite, the other half remains as (black) FeS. The fact that most sulfidic tidal sediments are pyritic and not black indicates that partial oxidation by O_2 plays a role in pyrite formation.

Question 7.16

a) From Figure 7.6 you can see that pyrite is oxidised:
 * to dissolved Fe(II) sulphate at very low pH and high pe, or at moderate pH and moderate pe,
 * to jarosite (via Fe^{2+} and SO_4^{2-}, at high pe and pH <4), and
 * to goethite, at pH>4.

b) $FeS_2 + {}^7/_2\, O_2 + H_2O \rightarrow Fe^{2+} + 2H^+ + 2\, SO_4^{2-}$

Question 7.17

Chemoautotrophic organisms use chemical energy to transform CO_2 into cell material. *Thiobacilli* utilise the energy released during oxidation of reduced S compounds and (in case of *Thiobacillus ferrooxidans*) Fe(II).

Question 7.18

Slow diffusion of O_2 through (usually wet) soil hampers its supply to pyrite particles in soil in-situ, and thereby depresses the rate of (bacterial) pyrite oxidation. Pyritic spoils are more directly exposed to O_2 in the air.

Question 7.19

A soil or sediment subjected to tidal action remains very wet, and largely anoxic, because of daily (or twice-daily) high tides.

Question 7.20

In very wet conditions or in the presence of large quantities of organic matter (peaty soils) the soil may be sufficiently oxic to transform pyrite to Fe^{2+} and SO_4^{2-} (see Question 7.16), but not enough for further oxidation of dissolved $FeSO_4$ to jarosite (or goethite). The pe-pH conditions are those of the Fe^{2+} field in Figure 7.6.

Question 7.21

a) The pH will remain near neutral because there is excess $CaCO_3$ over H_2SO_4,

b) The pH will remain near neutral because ion exchange is also fast enough to keep up with the rate of acid formation,

c) The soil pH will initially decrease, because acid is formed quicker than it can be neutralised; if incubated sufficiently long (for months of years) however, all the acid will be neutralised by weathering of the silicate minerals, and the pH should rise again to near neutral.

Question 7.22

$KFe_3(SO_4)_2(OH)_6 \rightarrow 3FeOOH + K^+ + 2SO_4^{2-} + 3H^+$ No electron transfer!

Question 7.23

Where jarosite and pyrite occur at the same depth, they are spatially separated: jarosite occurs along vertical cracks where **pe** can be much higher than inside adjacent finely porous, wet soil peds, where pyrite can persist longer.

Problem 7.1

a. In the artificially flooded and puddled surface soil (0-15 cm depth), water stagnates in the puddled layer and onto the developing traffic pan: perched water table. Fe(III) oxide is reduced seasonally to Fe^{2+}, which moves by mass flow and diffusion to the more oxic sub surface soil, where it is oxidised and precipitated as FeOOH. The surface layer becomes progressively depleted of Fe(III) oxide, and the Fe accumulation horizon becomes more strongly expressed with soil age. Note, however, that the older soils seem to have lost Fe from the whole profile, probably because of removal of Fe^{2+} with drainage water. The lowering of the Fe(III) content below 60 cm depth may reflect effects of reduction below the winter water table, while the slight accumulation of Fe in the oldest soil may reflect accumulation from ground water.
b. Because this is a surface water gley system, Mn will accumulate below Fe.

Problem 7.2

The increase in pH after flooding and soil reduction is mainly due to consumption of H^+ during reduction of Fe(III) oxide ("self-liming" by reduction). Soils depleted of Fe(III) oxide do not "self-lime" anymore when flooded. Red earth is rich in Fe(III) oxide.

Problem 7.3

a) Higher free Fe_2O_3-Fe contents in the soil than total Fe in the parent material implies an absolute accumulation of Fe by supply from elsewhere. In these low-lying gley soils, the Fe probably came with ground water draining from higher areas, that welled up in the low area (Dutch: "kwel").
b) The positive correlation between the contents of Fe_2O_3 and clay suggests that Fe^{2+} was adsorbed on clay when ground water tables were shallow during strong upwelling (e.g. in winter), followed by oxidation of exchangeable Fe^{2+} to Fe_2O_3 during lower ground water tables in summer. The exchangeable H^+ formed at the same time must have been neutralised by the alkalinity (Ca and Mg -HCO_3^-) that

was supplied with dissolved Fe^{2+} every winter.

Problem 7.4

a) Reduced Mn reaches its maximum within one week, faster than Fe(II) and S(II-) (8 weeks). Reduced Mn appears highest in the profile, then Fe, then S. This reflects the progressively lower values of **pe** or P_{O2} at which Mn(II), Fe(II) and S(II-) can exist. The thickening of the oxidised surface layer (by O_2 continuously diffusing into the soil from the atmosphere through the floodwater) reflects the decrease of reductive capacity (Mn(II), Fe(II), S(II-), and metabolizable organic matter) which is consumed by anaerobic organisms.
b) The accumulation zone of Mn(IV) oxide is probably in the upper 5 mm, that of Fe(III) oxide in the 5-10 mm zone. These zones grow thicker (deeper) with time.
c) The very thin accumulation zones at the surface are too easily disturbed to be of any use for classification purposes. They do form every year again, however, and do indicate earlier waterlogging if observed.

Problem 7.5

a) See answer to problem 7.1.
b) For the clay balance see table on next page. Clearly, clay is lost from the whole profile. Possible causes are lateral removal of suspended clay following flooding and puddling (wet soil tillage) of the (rice) field for many centuries, and ferrolysis. c) The total CEC of the Apg1 horizon is made up of contributions by the clay fraction (a), and by soil organic C (b):
   0.125 (kg clay/kg soil) * a ($mmol_c$/kg/clay) + 0.0065 (kg org.C/kg soil) * b ($mmol_c$/kg org.C) = 40 $mmol_c$/kg soil)
   With a similar equation for the Apg2 horizon, you have two equations with two unknowns. These yield a = 217 $mmol_c$/kg clay, and b = 1975 $mmol_c$/kg org. C. Substituting the specific CEC value for the organic C fraction in similar equations for each horizon gives the following specific CECs for the clay fractions ($mmol_c$/kg/clay):

Horizon	Apg1	Apg2	Eg1	Eg2	Eg3	Ecg	Cg1	Cg2	Cg3
CEC_{clay}	217	217	258	206	254	274	328	362	391

The specific CEC of the clay is considerably lower in the surface horizons, which may be explained by hydroxy Al interlayering as a result of ferrolysis.

						clay kg/m²		
	thickness	clay%	s+s, %	ε	τ	now	original	lost
Apg1	8.00	12.50	87.50	-0.36	-0.82	10.00	55.45	45.45
Apg2	5.00	25.00	75.00	-0.26	-0.58	12.50	29.70	17.20
Eg1	5.00	27.10	72.90	-0.23	-0.53	13.55	28.87	15.32
Eg2	12.00	29.70	70.30	-0.21	-0.47	35.64	66.82	31.18
Eg3	11.00	33.30	66.70	-0.16	-0.37	36.63	58.12	21.49
ECg	17.00	41.60	58.40	-0.04	-0.10	70.72	78.64	7.92
Cg1	3.90	42.10	57.90	-0.04	-0.08	164.19	178.87	14.67
Cg2	30.00	42.90	57.10	-0.02	-0.05	128.70	135.69	6.99
Cg3	25.00	44.20	55.80	0.00	0.00	110.50	110.50	0.00
							sum	160.23

Problem 7.6

a) The pH dropped by pyrite oxidation upon aeration of the pyritic subsoil (below 18 cm in the young soil, below 140 cm in the old soil), while there was little effect of aeration in the shallower layers, which contain no pyrite (anymore).

b) From the bulk density data it follows that the 0-140 cm layer contains 15.7 * 10^6 kg soil per ha, or 0.01* 1.57 * 10^6= 1.57 * 10^4 kg S. One kmol S (= 32 kg S) produces a maximum of 2 kmol H^+, so the maximum acid production is 2* 1.57 * 10^4 kg * 1/32 kmol/kg = 9.8*10^3 kmol H^+.

c) If all that acid remained in solution (total volume 0.5* 10^4*1.4= 7*10^3 m^3) and was not neutralised and/or leached the pH would be - log (9.8/7) = - 0.15. The actual pH is in the order of 4 indicating that practically all H^+ that was formed has been buffered and/or leached.

d) Congruent dissolution of 1 mol of kaolinite or montmorillonite would neutralise 6 mol of H^+. So 9.8*10^3 kmol of H^+ can be neutralised by 1.63*10^3 kmol of kaolinite or montmorillonite. The mass of 1 kmol of kaolinite is 258 kg and of 1 kmol of montmorillonite 367 kg, so 4.2* 10^5 kg of kaolinite or 6*10^5 kg montmorillonite could be dissolved by the acid released. The total amount of montmorillonite or kaolinite in 1 ha* 140 cm depth is 0.5* 0.5* 15.7 * 10^6 kg per ha = 3.92 * 10^6 kg/ha. So a maximum of about 10 to 15% of the clay originally present could be dissolved.

Problem 7.7

The decrease in base saturation equals (1-0.4)*300 mmol(+)/kg of soil, which is equivalent to 240 mmol of H^+ or 120 mmol of S. The mass of 120 mmol of S is 32 mg/mmol * 120 = 3.84 g of S. So the oxidation of 0.38 % of pyrite-S can bring down the base saturation from 100 to 40 %. This also implies that most clay soils with a fairly high CEC (normal for many tropical marine sediments), can contain up to 0.4 % of pyrite-S without much risk of strong acidification.

7.7. References

Brinkman, R., 1970. Ferrolysis, a hydromorphic soil forming process. Geoderma, 3:199-206.

Brinkman, R., 1977. Surface-water gley soils in Bangladesh: genesis. Geoderma, 17:111-144.

Brinkman, R., 1979. *Ferrolysis, a soil forming process in hydromorphic conditions.* Agricultural Research Reports 887: vi + 106 pp., PUDOC, Wageningen.

Buurman, P., 1980. Palaeosols in the Reading Beds (Paleocene) of Alum Bay, Isle of Wight, U.K. Sedimentology, 27:593-606.

Kawaguchi, K., and Y. Matsuo, 1957. Reinvestigation of active and inactive oxides along soil profiles in time series of dry rice fields in polder lands of Kojima basin, Okayama prefecture, Japan. Soil Plant Food, 3: 29-35.

Konsten, C.J.M., N. van Breemen, Supardo Suping, and J.E. Groenenberg, 1994. Effects of flooding on pH of rice-producing acid sulphate soils in Indonesia Soil Science Society of America Journal, 58,871-883.

Patrick Jr., W.H. and R.D. Delaune, 1972. Characterization of the oxidized and reduced zone in flooded soil. Soil Science Society of America Proceedings, 36: 573-576.

Van Breemen, N., 1976. *Genesis and solution chemistry of acid sulfate soils in Thailand.* Agricultural Research Reports 848, 263 pp. PUDOC, Wageningen.

Van Breemen, N., 1988. Redox processes of iron and sulfur involved in the formation of acid sulfate soils. p. 825-841 in J.W.Stucki et al. (eds): *Iron in Soils and Clay Minerals.* D. Reidel Publ. Co., Dordrecht. The Netherlands, 893 pp.

Van Breemen, N., W.F.J. Visser, and Th. Pape, 1987. *Biogeochemistry of an oak-woodland ecosystem in the Netherlands, affected by acid atmospheric deposition.* Agricultural Research Reports 930, 197 pp. PUDOC, Wageningen.

Van Mensvoort, M.E.F. and Le Quang Tri, 1988. Morphology and genesis of actual acid sulphate soils without jarosite in the Ha Tien Plain, Mekong Delta, Viet Nam. p. 11-15 in: Dost, H. (ed.): *Selected Papers of the Dakar Symposium on Acid Sulphate Soils*, Dakar, Senegal, January 1986, ILRI Publication 44.

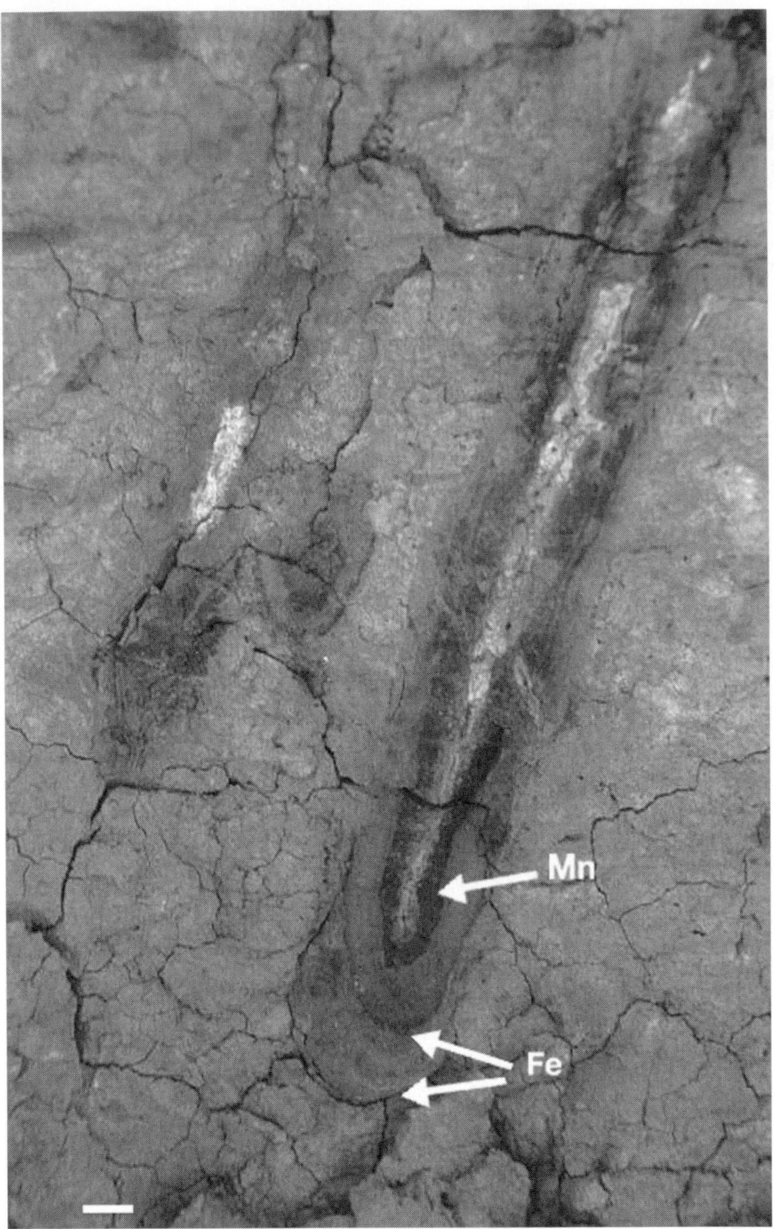

Plate M. Accumulation of iron (Fe) and manganese (Mn) oxides along a former root channel in a clayey gley soil. Scale is 1 cm. Pleistocene clay, southern Spain. Photograph P. Buurman.

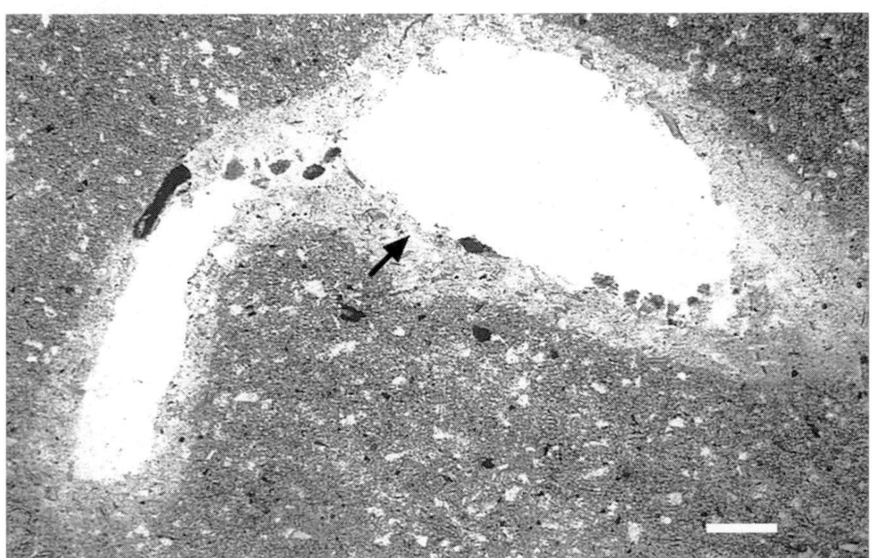

Plate N. Removal of iron along a pore (arrow: depleted zone). Surface-water gley. Plain light. Scale bar is 215µm. Photograph A.G. Jongmans

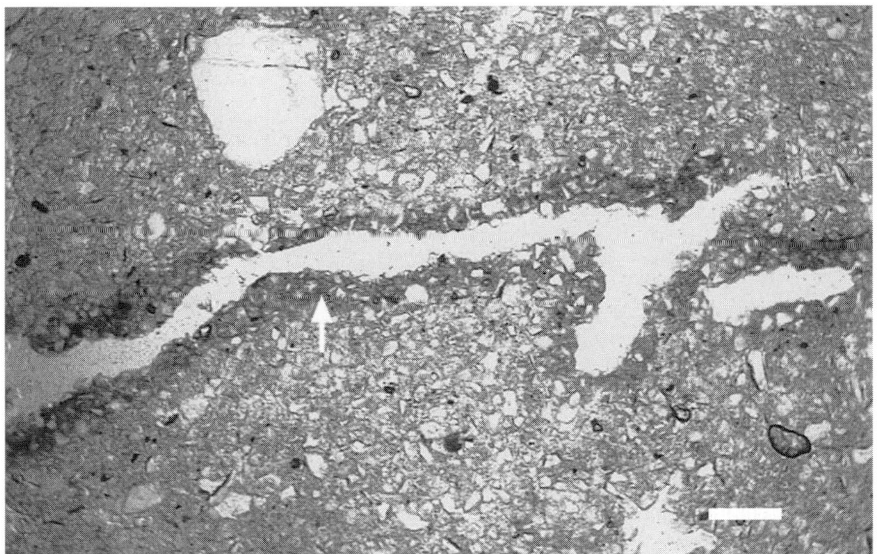

Plate O. Accumulation of iron (arrow) along a pore. Groundwater gley. Plain light. Scale bar is 215µm. Photograph A.G. Jongmans

Plate P. Thin iron pan (arrows) at the contact between two sandy deposits, the Netherlands. Above, general view; below: detail. Left-hand side has humic podzol-B horizon. Note thin humus bands (B). Photographs P. Buurman.

CHAPTER 8

TEXTURAL DIFFERENTIATION

8.1. Introduction

Textural differences between parent material, topsoil, and subsoil are common in soils of virtually all climates. Sometimes, such differences are inherited from the substrate (e.g., textural variations in sediments). Frequently, however, the differences are due to soil forming processes. Textural differentiation leads to topsoils and subsoils that are either finer, or coarser than the parent material. At least eight processes can result in textural differentiation:

1. Physical and chemical weathering of parent material,
2. Upward vertical transport of fine fractions by biological activity,
3. Downward transport of clay suspended in percolating soil water,
4. Superficial removal of clay by erosion without illuviation,
5. Superficial removal of clay due to tillage in wetland rice agriculture ('puddling'),
6. Clay formation in the subsoil by precipitation from solution,
7. Weathering/dissolution of clay,
8. Vertical movement of soil (matrix) material.

In soil classification systems, process (3) is emphasised, but other causes of texture differentiation may be more important in specific cases. The following discussion of the various processes leading to textural differentiation is adapted from Buurman (1990) and Soil Survey Staff (1975).

8.2. Processes of textural differentiation and their characteristics

PHYSICAL AND CHEMICAL WEATHERING

As discussed in Chapter 2, physical weathering caused by temperature differences, and ice formation reduces the size of rock fragments and mineral grains. Such physical weathering is strongest in the topsoil, where temperature differences are most pronounced. Physical weathering reduces the size of gravel, sand and silt fractions, but hardly produces any clay-size material. Mild chemical weathering results in size reduction through, e.g., exfoliation of micas. Strong chemical weathering processes related to different soil forming processes are discussed elsewhere (Chapters 7, 10-13).

BIOLOGICAL ACTIVITY

Burrowing animals cannot directly move soil particles that much exceed their own size. By bringing fine material to the surface, however, they will indirectly cause a downward transport of coarser fragments. Slowly, the coarser fragments sink down to the lower boundary of animal activity. This is especially visible in soils with a conspicuous coarse fraction, such as gravel, in a fine matrix. Examples of this process are plentiful:

- Ground squirrels in North America (pocket gophers) bury relatively large stones, while homogenising coarse-textured, gravelly soils.
- Earthworms deposit part of their fine-textured ingested material as casts at the soil surface. Coarse fragments are not ingested and sink away. This is perfectly illustrated in the ruins of Roman villas in England. The wall foundations of such structures may still be in place, while the tile floors between the foundations are found 20-50 cm below the original surface, completely covered with a dark surface soil. After removing this soil, earthworms again deposited their excrements on top of the tiles (Darwin, 1881).
- In tropical areas, termites build nests in the soil, or large mounds on top of the soil. Both nests and mounds are predominantly built of clay and silt. The nests decay again with time, causing finer textures in the topsoil. Coarse material is not brought up and may sink to a depth of more than 2 metres below the surface, forming distinct *stonelines*. In addition, mound-building termites carry fine material from the surroundings to the mound site, causing not only vertical differentiation, but also horizontal variability in fine material contents (Wielemaker, 1984).

DOWNWARD TRANSPORT OF SUSPENDED CLAY (Eluviation and Illuviation)

Eluviation (removal) and illuviation (addition) of clay in a soil depend on a number of conditions. In a dry soil, all clay is flocculated and bound in aggregates or on grain surfaces. Abrupt wetting of a dry soil may disrupt the aggregates and mobilise clay by *air explosion*. Air explosion is the sudden disintegration of a dry structural element when it is wetted: water enters the pores by capillary force and the entrapped air builds up sufficient pressure to blow the aggregate apart. For air explosion to be effective, the soil must be dry. This implies, that clay illuviation is favoured by climates with distinct dry periods (during which the soil dries out), followed by rainfall of high intensity.

Air explosion occurs in all soils that are periodically dry, but it does not automatically bring clay particles into suspension. As outlined in Chapter 3, clay particles will disaggregate and remain in suspension when the electrical double layer expands. This expansion is favoured by 1) low electrolyte concentrations, and 2) a high concentration of monovalent ions (Na^+). Expansion is hampered by di- and trivalent ions, such as Ca^{2+},

Mg^{2+}, and Al^{3+} on the exchange complex. Fe^{3+} concentrations in solution are usually too low to influence the composition of the cation exchange complex, but finely divided iron oxyhydrates, because of their positive surface charge at soil pH (see Chapter 13), tend to keep clays flocculated. The influence of organic molecules is poorly understood: some increase dispersion while others form very stable complexes with clay that are difficult to disperse (see Problem 8.1). In soils, conditions for flocculation are related to pH and pH-related properties, as is illustrated in Figure 8.1.

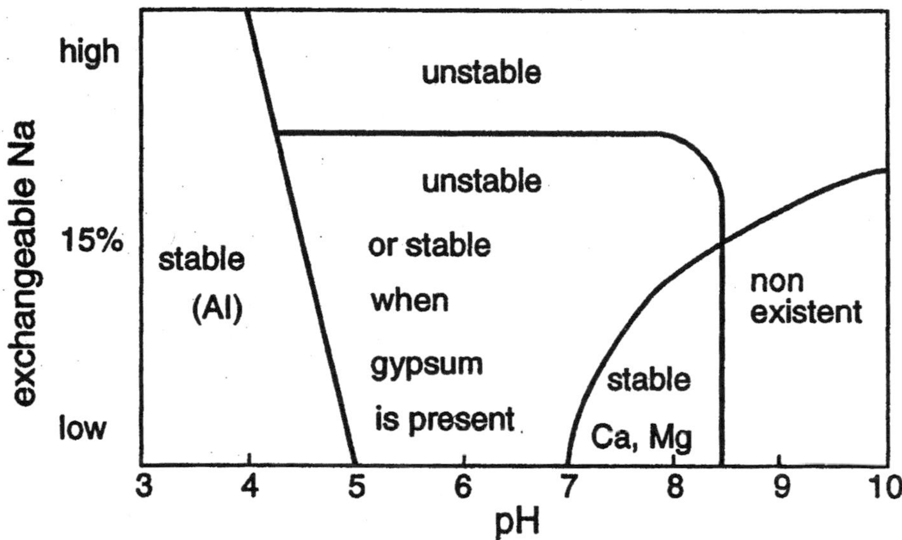

Figure 8.1. Stability of clay aggregates in dependence of pH and adsorbed Na.

Question 8.1. a) Explain the stability of clay aggregates as a function of pH as shown in Figure 8.1. b) Why does the presence of gypsum (which has no influence on pH) increase the stability in the near-neutral pH range?

Once in suspension, the clay may move with the percolating water. Clay movement usually involves the finest clay particles (<0.2 μm). Suspended clay mainly moves in non-capillary voids, where water fluxes are sufficiently high. There are three main mechanisms by which clay movement stops.

- First: in fine-textured, moist soils, water percolating in large voids is drained away laterally by capillary withdrawal into the matrix. Suspended clay particles are filtered out against the walls of the non-capillary pores and fine clay orients itself with its plates parallel to the wall of the void. Thick accumulations of such oriented clay have a specific birefringence under the polarising microscope (see Figure 8.2).

Such accumulations are called *argillans* when low in iron, and *ferri-argillans* when they are coloured brown or red by iron oxides.

- Second: when water movement stops, clay movement stops too. This is the main cause of clay illuviation in sandy soils which have a limited capacity to retain water. Water stagnates also where fine pores widen suddenly to coarser ones, e.g. when fine-textured layers overly coarser layers (see Chapter 2). Upon evaporation of stagnating water, all suspended matter is left behind just above such a contact. The clay platelets tend to form parallel arrangements around sand grains and in pores, which may result in birefringent coatings (see next paragraph and 8.3).
- Finally, suspended clay may flocculate when electrolyte concentration of the soil solution increases as, e.g., in contact with carbonate-containing subsoils. Because such flocculation may not cause parallel orientation of clay plates, the resulting accumulations may not have the birefringence that is typical of the first two mechanisms.

Illuviated clay is frequently found in carbonate-rich soils of arid regions. This can be explained by two subsequent phases of soil formation: a phase in which clay was moved (in a non-calcareous or a saline soil), followed by a phase of carbonate accumulation.

Question 8.2. *Why do we think that clay movement preceded accumulation of carbonate?*

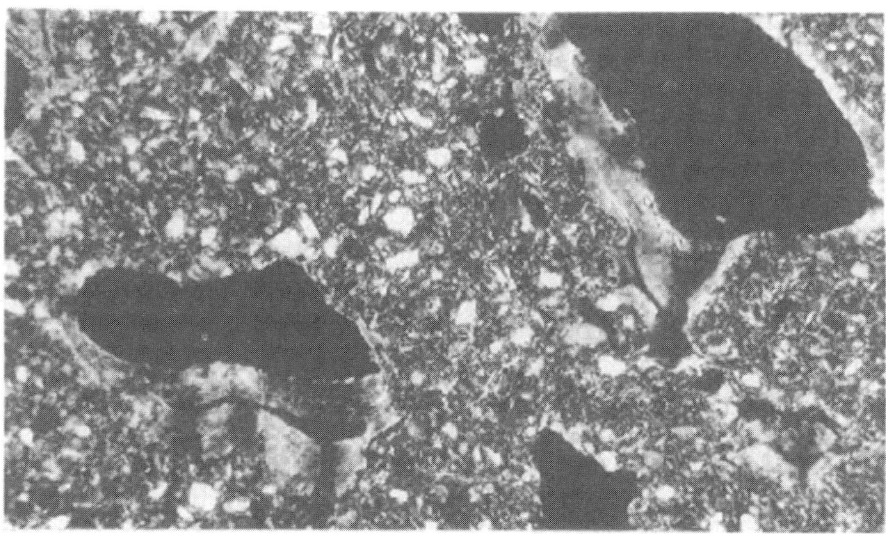

Figure 8.2. Laminated clay coatings in pores of a Podzoluvisol from Russia. Crossed polarisers. Width of picture is 1.15 mm. Pores are black. The coatings are visible in crossed polarised light because the parallel-oriented clay plates cause refraction of the polarised light. Photograph A.G. Jongmans.

RECOGNITION OF CLAY ILLUVIATION

In soils with clay eluviation and illuviation, the topsoil is depleted in fine clay. The subsoil with illuviated clay has the following properties:

- It contains argillans along pores and on structural elements. In the field, argillans can only be identified with a hand lens in a soil that is not too wet. Argillans usually have surfaces that are smooth. They superficially resemble 'slickensides', shiny pressure surfaces that are formed by sliding structural elements in smectite clay soils that swell and shrink (see Chapter 10), but lack the striations (linear patterns) typical of slickensides. Many argillans are darker than the matrix and they do not contain coarse grains. Argillans are best-recognised micromorphologically, in thin section (Figure 8.2). Some argillans, however, have formed by chemical precipitation of clay minerals from solution (see 'clay formation', below), but these are usually not confined to linings along pores and peds.
- It has higher clay content and higher fine clay to clay ratio than the overlying and underlying horizons. The fine clay fraction (<0.5 or <0.2 μm) is transported preferentially, so that the illuvial horizon has a relatively high fine clay content. If the parent material is stratified, the clay content in the illuvial horizon is not necessarily higher than that of the horizon below it.
- It has a higher CEC-clay than the overlying and underlying horizons. Because fine clay has a larger specific surface than coarse clay, it also has a higher CEC per unit mass of clay. This is reflected in a relatively high CEC-clay of the illuvial horizon in soils of mixed mineralogy.
- Its clay fraction contains more smectite than that of the overlying and underlying horizons, because the finest clay fraction is usually dominated by smectite.

Whereas illuviated clay is initially deposited as strongly parallel-oriented material in cutans, these argillans do not persist forever. Part of the cutans will pass through intestines of soil-ingesting fauna, and end up as small separate lumps of oriented clay, called *'papules'*. Given sufficient time, the orientation may disappear altogether.

SUPERFICIAL REMOVAL OF CLAY BY EROSION: ELUTRIATION

Overland flow of water is a common process in many landscapes, e.g., under tropical rainforest. Such overland flow may not result in features we commonly associate with erosion, such as gullies or sheet erosion, because it covers only short distances at a time, and the water is again taken up by the soil. However, in the long run, overland flow may cause significant removal of fine fraction from the topsoil; a process called *elutriation* (Eng.) or *appauvrissement* (Fr.).

Canopy drip (concentration of rainwater by the leaf form or tree structure, resulting in localised drip of water instead of even distribution) usually results in increased dispersion of clay in the drip area, partly as a result of larger drops. During overland flow, the suspended clay moves with the runoff over the surface of the soil until the water infiltrates into the soil. Eventually, this process leads to removal of fine material from the surface horizon into the drainage system. Biological homogenisation spreads the effect of surface removal over the depth of the homogenised layer (usually the Ah horizon). The contrast in clay content between the Ah horizon and the underlying horizon may be quite sharp.

Soil fauna further stimulates the preferential removal of fine fractions from the surface soil because they bring relatively fine material to the surface (e.g. earthworm casts, termite hills). On old, stable land surfaces this may lead to a thick so-called *faunal mantle* (Johnson, 1990), consisting of a one to two meter thick medium-textured soil, deprived of both the very fine and very coarse soil particles. Within this faunal mantle, the Ah horizon may be markedly lighter-textured again.

Question 8.3. *a) Characteristics of surface removal of clay are different from those of clay illuviation. How can you recognise surface removal? (Consider argillans, fine/coarse clay ratios, clay and silt movement, clay balance, and CEC_{clay}). b) Why do 'faunal mantles' lack very coarse soil particles?*

TEXTURAL DIFFERENTIATION AS A RESULT OF PUDDLING

Wet rice cultivation is widespread throughout the humid tropics. The practice calls for ploughing of waterlogged fields, to decrease permeability of the subsoil and to soften the soil before planting of the rice seedlings. This so-called puddling results in loss of clay from the surface soil if, as in many irrigation systems (e.g. in Indonesia, the Philippines), the water constantly flows downhill from one field to the next. The flowing water carries off all suspended matter, mainly clay. The long-term effect is a distinct removal of clay from the puddled layer.

Question 8.4. *How can you, by investigating the soil profile, distinguish such losses through puddling from losses through eluviation?*

CLAY FORMATION

Clay formation accompanies weathering in practically all soils that have weatherable minerals. Although weathering is usually strongest at shallow depth, clay formation is usually stronger in B-horizons than in A-horizons. Reasons probably are that 1) clay formation in A horizons may be inhibited by organic matter (see Chapter 3.2), and 2) solutes are transported from the A to the B horizon, where their concentration may be increased by evaporation, causing stronger super-saturation with secondary (clay) minerals. This leads to B-horizons that have more clay than the overlying A-horizons.

The amount of clay formed depends on the parent material and on the time and intensity of the weathering process. Strongly weathered soils, such as Ferralsols (Chapter 13) may have up to 90% clay, mainly due to new formation.

Question 8.5. *What is the effect of a) low contents of weatherable minerals in the parent material, and b) young soil age, on the amount of clay formed by weathering?*

Clay that is formed by crystallisation from the soil solution *may* be strongly oriented. In that case, it is difficult to distinguish from illuviation coatings. However, illuviation coatings tend to have a zoned structure (Figure 8.2) with thin bands of slightly coarser particles, which is absent in clay precipitates (Figure 8.3).

Recrystallisation of allophane and imogolite in Andisols may lead to clay coatings that are very similar to the (ferri)argillans that are due to translocation of clay. Because Andosols usually have a high Al-activity and because phyllosilicates are present in only minor amounts, 'argillans' observed in such soils should usually be ascribed to recrystallisation, or 'clay formation' (Buurman and Jongmans, 1987; Jongmans et al., 1994). See also Chapter 12.

Differences in clay formation in stratified parent materials (unequal grain size, unequal amounts of weatherable minerals) are a common source of differences - but not of differentiation - in texture. Layers with higher contents of weatherable minerals will usually develop higher clay contents. This is a very common feature in layered volcanic deposits. Such stratification is also reflected in composition of the sand and silt fractions and is easy to spot when detailed grain-size analyses or sand mineralogy are available. If the coarser-textured layer is encountered in the topsoil, the difference with the subsoil is often (erroneously) attributed to clay movement.

TEXTURAL DIFFERENTIATION THROUGH BREAKDOWN OF CLAY

The weathering of clay that accompanies surface-water gleying in acid soils has been discussed in detail in Chapter 7. This breakdown usually enhances already existing differences in texture that were either formed by illuviation, or were due to stratification of the parent material.

MOVEMENT OF MATRIX MATERIAL

Downward movement of 'whole soil' material (usually fine silt, organic matter, and clay) is common in cultivated soils that have topsoils with low structure stability. In such soils, coatings of unsorted fine material can be found on the vertical walls of pores or, in sandy materials, as caps on the top of sand grains. Transport of matrix material occurs preferentially during heavy showers on freshly ploughed soil. This process is purely a suspension transport which ends when the water stops flowing in the larger pores.

8.3. Intensity and expression of textural differentiation by clay illuviation

Clay movement is a purely mechanical process, which can be relatively rapid if the conditions are optimal. Clay illuviation has been demonstrated in deposits exposed by glaciers less than 200 years ago. In general, however, more than a thousand years will be necessary to develop a distinct illuvial horizon.

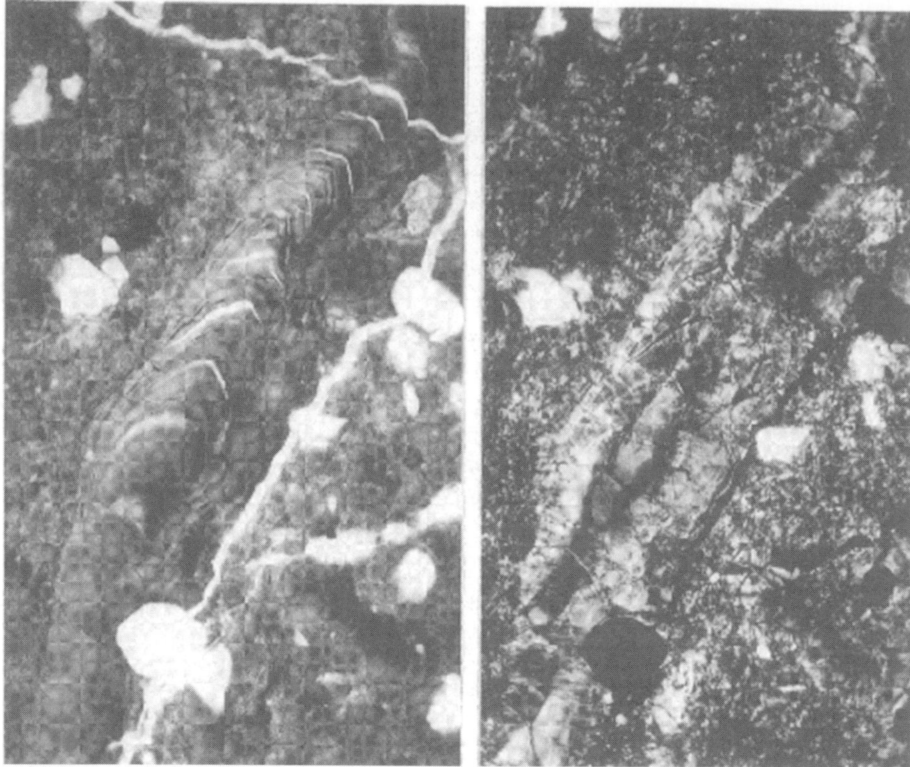

Figure 8.3. Oriented clay due to chemical precipitation. Tertiary soil on limestone; the Netherlands. Left: plain polarized light; right: crossed polarisers. Width of each picture is 1.8 mm. Pore is completely filled with clay. Note difference in birefringence with coating of Figure 8.2. Photographs by A.G. Jongmans.

Clay illuviation in strongly weathered soils may be older than 10,000 years. In loess soils of southern Germany, clay movement appears to have started earlier than 5500 years BP (before present) and ended before 2800 years BP, which would indicate that in this area it is a fossil process (Slager & Van de Wetering, 1977).

Comparison between present environmental conditions and conditions that lead to clay illuviation frequently indicates that this illuviation is a fossil process. In Western Europe, most soils with clay movement have been under deciduous forest during this process; a smaller part has been under coniferous forest or grasslands. In the tropics, they are more common under savannah than under rain forest.

Question 8.6. *The following table lists the eight processes that cause textural differentiation. Indicate for each process, what its influence is on topsoil and subsoil texture. Use the terms 'finer', 'coarser' and 'unchanged'. Add additional characteristics in the fourth column.*

Process	Topsoil	Subsoil	Remarks
1. Physical weathering			
2. Biological activity			
3. Clay illuviation			
4. Clay erosion			
5. Puddling			
6. Clay formation			
7. Clay breakdown			
8. Matrix transport			

Question 8.7. *Which of the (sub)processes involved in clay illuviation (paragraph 8.2 and Question 8.6) is unlikely under rainforest?*

EFFECT OF PARENT MATERIAL TEXTURE

The expression of clay movement strongly varies with soil texture. In sandy soils, where the clay fraction amounts to a few percent only, illuviated clay usually concentrates in a series of disconnected, more or less horizontal, wavy bands or lamellae. In the Netherlands, such clay illuviation is common in older coversands, where the clay forms bridges between sand grains. The lamellae grow upwards with time (Van Reeuwijk & De Villiers, 1985).

Eluvial and illuvial horizons are most clearly expressed in loamy soils, e.g. loess or fluvial loams. The illuvial horizon is usually continuous and several decimetres thick. It is commonly darker than the overlying and underlying horizons. Clay coatings on ped faces and in biopores are clearly recognisable with a hand lens. The illuvial horizon usually has a strong blocky structure, while the eluvial horizon is structureless or platy. The illuvial horizon tends to grow upwards with time, while its clay content increases and its upper boundary becomes sharper (Figure 8.4).

In residual soils on carbonate-rich parent materials, the amount of oriented clay tends to be very high. This is probably due to the fact that the clay is liberated from the carbonate rock in small amounts, and that clay movement close to the carbonate parent rock is extremely restricted.

In soils with very high clay contents, the textural differentiation becomes less expressed and oriented clay is less abundant. In kaolinitic soils of the humid tropics, coatings of oriented clay may be virtually absent, but it is not clear whether this is due to absence of illuviation or to strong biological homogenisation. Profiles tend to be deeper in

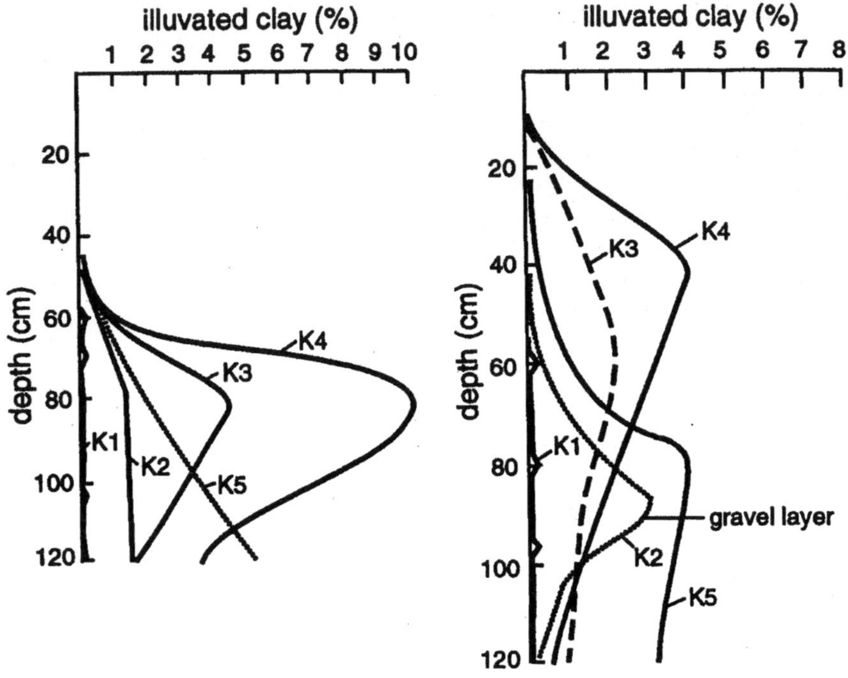

Figure 8.4. Stages of clay illuviation in river terraces. Left: Macley River; right: Gooroomon Ponds, Australia. K1 to K5 are successive stages of development. From Brewer, 1972. Reproduced with permission of the Geological Society of Australia.

tropical climates than in temperate climates, which may be due to both higher rainfall (deeper percolation) and usually longer time of soil formation in the tropics.

In soils with swelling and shrinking clays, physical movements may be strong enough to obliterate effects of clay movement. Moreover, oriented clay is not visible in a matrix that is dominated by pressure structures.

Many profiles with clay movement have younger overprints of other processes, such as podzolisation, surface water gley, or homogenisation.

8.4. Clay minerals in soils with clay eluviation and illuviation

Fine-grained clay minerals, such as smectite, vermiculite, and illite, appear to be more mobile than kaolinite, which is relatively coarse-grained. This may lead to a mineralogical differentiation of eluvial and illuvial horizons.

Weathering and clay mineral assemblage in soils with clay movement are variable. In temperate regions, clay mineral assemblages in the illuvial horizon usually resemble those of the parent material, but with higher contents of smectite minerals. In the tropics, profiles with clay movement can be strongly weathered and very similar to Oxisols. In these soils, kaolinite is the dominant - and sometimes the only - clay mineral. Many soils with a textural contrast in the tropics are acid, with high contents of exchangeable adsorbed aluminium. Few have been investigated micromorphologically, and it is not sure whether illuviation or elutriation has played a dominant role in causing vertical texture differentiation in such soils.

Question 8.8. *The presence of illuviated clay in soils with much exchangeable* Al^{3+} *suggests that clay movement is not an active process in these soils. Why?*

In temperate regions, weathering is much more restricted, e.g. vermiculite formation from illite in the eluvial horizon, and interlayering of illites to form soil chlorites in the illuvial horizons (see Chapter 3.2).

8.5. Texture-based diagnostic horizons in soil classification

Changes in texture in a soil profile may dominate soil-forming processes and affect agronomically important soil properties, such as rootability, water availability, and water percolation. For these reasons, changes in texture are used in soil classification. Differences in texture due to movement and/or depletion of clay are quantified in the *arg(ill)ic and natric* (B) of USDA and FAO and in the *kandic* horizon of USDA (SSS, 1987 and later). The *agric* horizon of the USDA is due to transport of matrix material.

Both the argillic and the natric horizon should have a certain amount of illuviated clay, either indicated by grain-size analysis, or recognised as oriented clay in thin section. The concept of these diagnostic horizons is based on clay eluviation and illuviation, but their definitions do not necessarily exclude textural differences that were formed in a different way.

Many tropical soils have a relatively light-textured surface soil, but do not show clear clay illuviation cutans in the finer B-horizon. Such soils would not have an argillic horizon. The absence of cutans is generally explained by the theory that kaolinitic clays do not form birefringent cutans. To accommodate such soils, the *kandic horizon* was created (from *kandites,* a general term for 1:1 clays). The kandic horizon is defined by its upper boundary rather than by properties of the horizon itself: it should have a sharp increase in clay content at its top, but need not have evidence of illuviated clay. Chemically and mineralogically, the kandic horizon is similar to the oxic horizon (see Chapter 13: Ferralitisation).

Although there is only one argillic horizon for temperate and tropical climates, classification does take account of the weathering state (base saturation and CEC of the clay fraction) of the argillic horizon. In temperate regions, the illuvial horizon is usually mildly under-saturated and pH values are between 5 and 7, except if they have been affected by ferrolysis (see Chapter 7). Ferrolysed B horizons can be strongly depleted in exchangeable bases and strongly weathered, and have lower pH values.

In tropical regions, argillic horizons tend to be low in exchangeable bases, and strongly weathered. The clay mineral assemblage is dominated by kaolinite, but some w-eatherable minerals may remain in the silt and sand fractions. Colours are usually redder than in temperate regions (see also Chapter 3). In the wet tropics, where the soil does not dry out seasonally, soils rarely show illuviated clay, and argillic horizons are scarce. If the textural contrast is sufficient, horizons are classified as kandic.

The natric horizon is an argillic horizon high in exchangeable sodium. It is characterised by prisms showing typical rounded tops (see Chapter 9). These horizons probably form upon leaching of saline soils. When NaCl-rich soils are leached, dilution of the electrolyte results in peptisation of sodium clay. The dispersed clay is transported downward, where it may fill up the pore system and reduce permeability (see Chapter 7). Because adsorption of Na also disrupts the conformational structure of soil organic matter (see Chapter 4), clay illuviation is frequently accompanied by illuviation of organic matter.

On stable land surfaces, soils with eluviation and illuviation of clay have an eluvial (E) and an illuvial (Bt; argillic) horizon. The E horizon is usually characterised by lighter texture and colour than the overlying and underlying horizons, but it is not white enough to qualify for an albic horizon. The albic horizon is clearly expressed in soils with clay movement that have also periodically stagnating water on the B-horizon, causing ferrolysis (see Chapter 7). In eroded soils, the E horizon has disappeared, and an Ah horizon may directly overlie the Bt.

8.6. Problems

Problem 8.1

Figures 8A and 8B illustrate some of the possible interactions between clay and organic acids in dependence of electrolyte concentration ($CaCl_2$). Fulvic acid (see Chapter 4.5), parahydroxy-benzoic acid and salicylic are used as proxies of dissolved of organic matter. The turbidity is a measure of the amount of material in suspension.

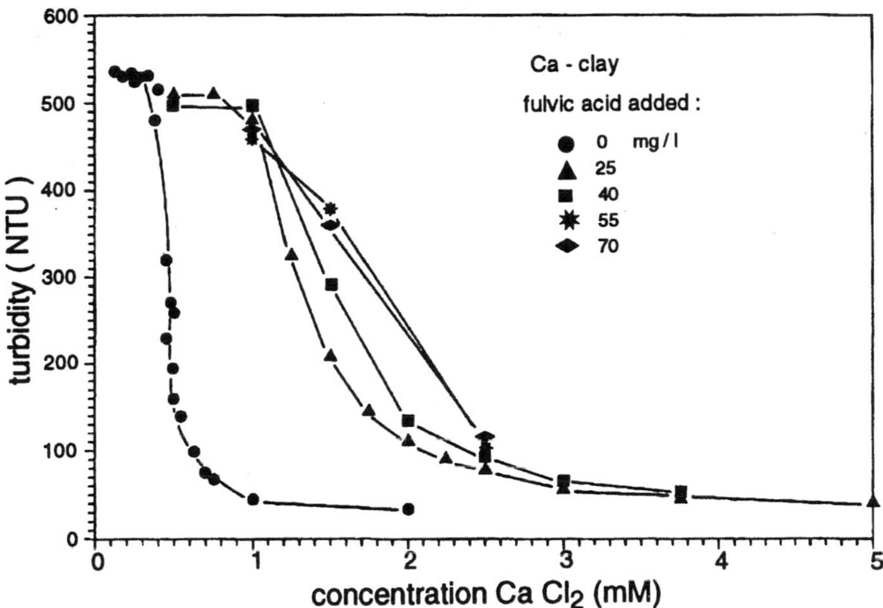

Figure 8A. Effect of addition of fulvic acid to a Ca-clay suspension as a function of $CaCl_2$ concentration. From Van den Broek, 1989.

a. Discuss the effects of electrolyte concentration on the flocculation as a function of the amount of fulvic acid added (Fig. 8A).
b. Plot the initial situation of Figure 8B (next page) in Figure 8A.
c. Discuss the effect of fulvic acid, citric acid, and parahydroxy-benzoic acid (phb)/ salicylic acid on the flocculation of the suspension (Figure 8B).

Problem 8.2

Which of the following situations is most favourable to the suspension and transport of clay in a soil profile: a) heavy rainfall on a moist soil; b) moderate rainfall on a dry soil;

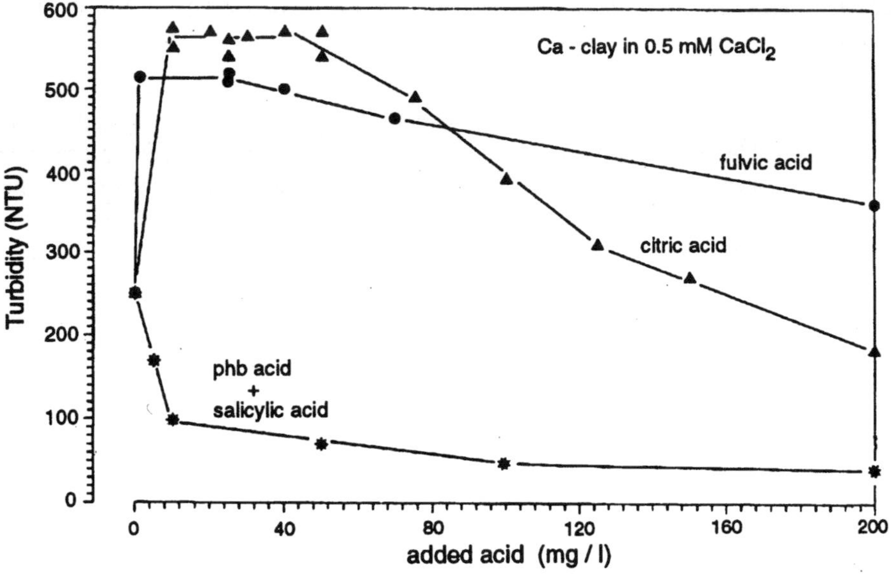

Figure 8B. Effect of the addition of organic acids on the flocculation of a Ca-clay suspension at constant electrolyte concentration. From Van den Broek, 1989.

c) a water table on top of a wet soil; d) continuous low rainfall on a moist soil. Consider which of the situations if favourable for suspension of clay particles and which for transport in macropores.

Problem 8.3

In Figure 8.4, the clay illuviation maximum seems to move up in the soil profile first (K1 to K4), but is found at greater depth in the oldest (most expressed) stage. What sequence of soil forming processes could explain these observations?

Problem 8.4

Table 8C presents data on a soil with an argillic horizon in loess from the Southern Netherlands.

a. Calculate the volume of oriented clay in a pedon of $1 m^2$ surface and 3.27 m depth.
b. Calculate the amount of clay (kg/m^2) removed from the A and E horizons and that accumulated in the lower horizons, assuming that: (i) there is no clay formation in the profile; (ii) sand+silt fractions have remained unchanged; (iii) specific weight of clay and non-clay are 2500 kg/m^3, and bulk density of all horizons is 1.4 $g.cm^{-1}$; (iv) the original clay content was 15% throughout the soil.
c. Explain the differences between a and b, disregarding changes in bulk density.

Table 8C. Clay contents and oriented clay in the soil profile Heerlerheide (Typic Hapludalf, Orthic Luvisol) developed in loess. From Van Schuylenborgh et al., 1970.

horizon	depth cm	Clay (mass%)	oriented clay (volume %)		
			total	in situ ferri-arg	papules
Ah	0- 13	13.2	n.d.	n.d.	n.d.
E	13- 21	13.6	0.3	0.0	0.3
Bt1	21- 33	15.5	2.2	1.1	1.1
Bt2	33- 92	21.5	4.1	2.3	1.8
Bt3	92-123	22.2	2.8	2.3	0.5
BC	123-307	19.2	1.2	1.0	0.2
C	307-327	15.0	0.3	0.3	0.0

Problem 8.5

Figures 8D and 8E (next page) give textural profiles and fine clay to clay ratios for a number of soil profiles from West Kalimantan, Indonesia. Which processes could be responsible for the texture differentiation? Assume that the in each case the lower horizon represents the C horizon.

8.7. Answers

Question 8.1

a) The stable field below pH 5 is caused by Al^{3+} in solution; the stable field around pH 8 by the presence of calcium (and magnesium) carbonates, which cause both the presence of divalent cations in solution and a sufficiently high concentration of these cations to keep clays flocculated.

b) The presence of gypsum ($CaSO_4.2H_2O$) has no influence on pH. It is more soluble than calcite and will therefore stabilise clays with exchangeable Ca^{2+} ions.

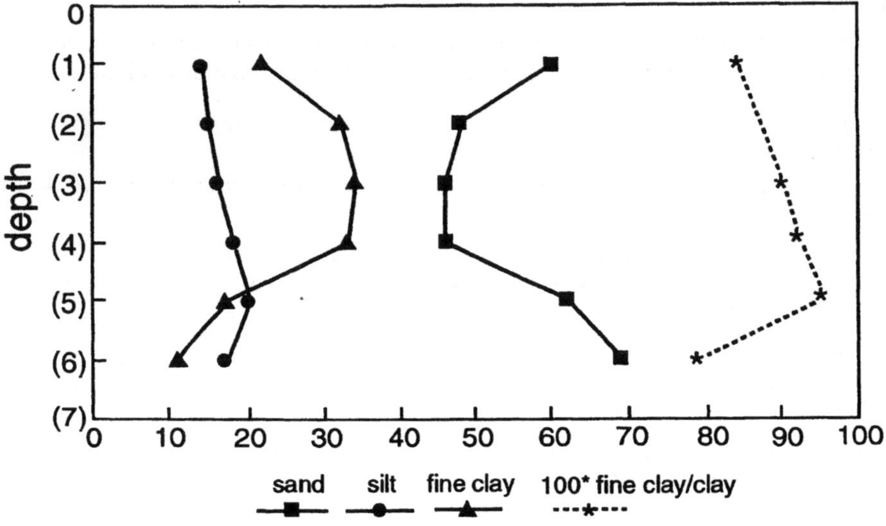

Figure 8D. Texture profile of profile Kalimantan 2. From Buurman and Subagjo, 1980.

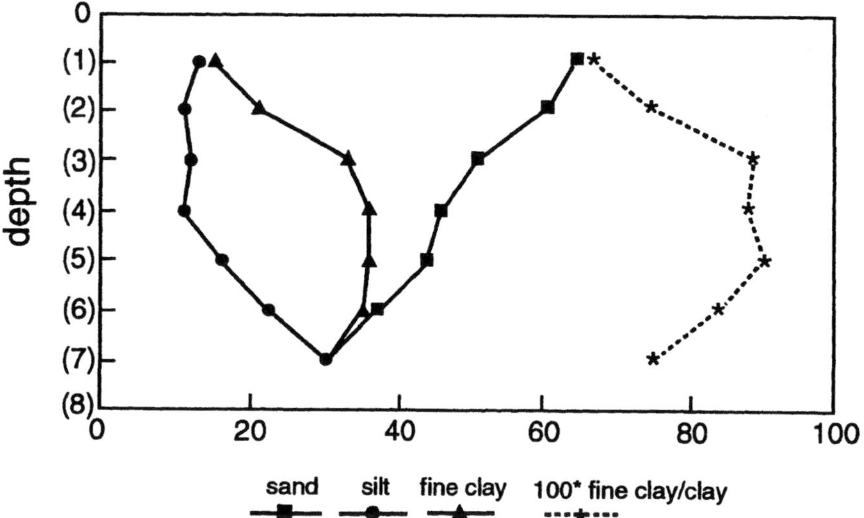

Figure 8E. Texture profile of profile Kalimantan 3. From Buurman and Subagjo, 1980.

Question 8.2

The main cause of major changes in soil forming processes is a change in climate. In the present case it should have been a change from a fairly moist to a drier climate.

Question 8.3

a). Superficial removal of clay without illuviation can be spotted by the following features:

i) Absence of argillans. Illuviated clay cutans are absent from the B-horizon or deeper in the profile. ii) Lower clay contents and fine clay contents in the topsoil without higher fine clay/clay ratios and fine clay contents in the subsoil. Fine clay to clay ratios should be more or less constant with depth. iii) Slight loss of fine silt from the topsoil. When detailed grain-size analyses are available, this effect can be measured. The patterns obtained for clay illuviation and clay removal are very different. iv) A negative clay balance: removal of clay from the surface soil exceeds the clay increase in the B-horizon. v) No increase in CEC_{clay} in the subsoil. If there is no accumulation of fine clay, the CEC of the clay fraction is not affected.

b) Bringing only relatively fine soil material to the soil surface causes a downward movement of all coarser material. This tends to accumulate below the zone of faunal activity (stoneline).

Question 8.4

The characteristics by which such a removal can be distinguished from clay loss and gain by illuviation are similar to those of elutriation. The process is essentially the same.

Question 8.5

Low amounts of weatherable minerals result in a smaller potential to form clay. In young soils, clay formation has not reached its maximum development.

Question 8.6

Process	*Topsoil*	*Subsoil*	*Remarks*
1. Physical weathering	finer	unchanged	little clay formation
2. Biological activity	finer	(coarser)	stoneline formed
3. Clay illuviation	coarser	finer	fine clay, CEC, coatings
4. Clay erosion	coarser	unchanged	also silt loss
5. Puddling	coarser	unchanged	also silt loss
6. Clay formation	unchanged	finer	
7. Clay breakdown	coarser	unchanged	ferrolysis
8. Matrix transport	coarser	finer	cultivation

Question 8.7

Air explosion requires the presence of dry aggregates. This is unlikely in a permanently humid environment.

Question 8.8

High amounts of exchangeable Al^{3+} would stabilise the clay and prevent peptisation. This suggests that clay movement is not an active process, but occurred before acidification of the profile.

Problem 8.1

a) In all cases, the addition of $CaCl_2$ stimulates flocculation of the clay/organic matter suspension, as expected from increasing concentrations of exchangeable divalent cations and increasing electrolyte concentrations in solution. The addition of fulvic acid decreases flocculation. When more fulvic acid is added, flocculation occurs at higher $CaCl_2$ concentrations. This may be explained by binding of part of the calcium to fulvic acid.

b) The starting situation of the second graph is on the zero fulvic acid point of graph A, where the graph is at 0.5M $CaCl_2$ concentration.

c) PHBA and salicylic acid strongly increase flocculation even at low additions, while very high additions of fulvic acid and citric acid are necessary to cause flocculation. We have no explanation for these differences, but the lower molecular size of PHBA and salicylic acid, in comparison to fulvic acid, may play a role.

Problem 8.2

The situation where moderate rainfall falls on a dry soil is most favourable to clay transport, because is causes both air explosion and transport. If the soil is already moist, air explosion does not occur. If precipitation is too low, water transport through macropores is absent.

Problem 8.3

Clay illuviation starts at various depths of rainfall penetration and/or textural changes. The fact that clay is accumulated in a certain layer increases the textural contrast and the water stagnation on this layer. This would explain upward growth of the Bt horizon. In the last stage, the top of the Bt horizon appears to move downwards. The sharp upper boundary of the B-horizon in that stage suggests breakdown of clay through ferrolysis, caused by water stagnation on the Bt.

Problem 8.4

a) The total volume of oriented clay in a pedon of 1 m^2 surface area and a depth of 327 cm is calculated by adding up the values of the separate horizons. For each horizon, the total volume is calculated. This volume is multiplied by the fraction of oriented clay. For the Bt2 horizon, the calculation is as follows:

Volume of horizon: (0.92-0.33) m x 1 m^2

Fraction of oriented clay: 0.041

Total volume oriented clay: 0.59 x 0.041 m^3 = 0.024 m^3

In this way, the total volume of oriented clay for the whole profile is calculated as: 0.058m^3.

Depth (cm)		bulk density (kg/m^3)	clay (mass %)	oriented clay (vol %)	Volume of horizon (m^3)	oriented clay in horizon (m^3)
top	bottom					
0	13	1400	13.2		0.13	0
13	21	1400	13.6	0.3	0.08	0.000
21	33	1400	15.5	2.2	0.12	0.003
33	92	1400	21.5	4.1	0.59	0.024
92	123	1400	22.2	2.8	0.31	0.009
123	307	1400	19.2	1.2	1.84	0.022
307	327	1400	15	0.3	0.2	0.001
					Total volume (m^3)	**0.058**

b) To calculate the amount of clay that is lost or gained, we need a reference material. Because sand and silt fractions are supposed to be unchanged, we can use the sum of these fractions as reference.

In the C-horizon, there is 15% clay and 85% (sand+silt). See the table below for the calculation of the strain of the sand and silt fraction, and the τ of the clay fraction.

depth (cm)		bulk density (kg/m3)	clay (mass %)	sand + silt (fract.)	strain sand + silt	clay mass now (kg)	τ	clay original (kg/m^2)	clay lost (kg/m^2)
top	bottom								
							Eq. 5.4	Eq. 5.5	
0	13	1400	13.2	0.868	-0.021	24.0	-0.14	27.9	3.9
13	21	1400	13.6	0.864	-0.016	15.2	-0.11	17.1	1.8
21	33	1400	15.5	0.845	0.006	26.0	0.04	25.1	-1.0
33	92	1400	21.5	0.785	0.083	177.6	0.55	114.4	-63.2
92	123	1400	22.2	0.778	0.093	96.3	0.62	59.6	-36.8
123	307	1400	19.2	0.808	0.052	494.6	0.35	367.3	-127.3
307	327	1400	15	0.85	0.000	42	0	42	0
								TOTAL	222.50

c) The volume of oriented clay is less than one third of the clay gain in the B-horizons. This can be explained by destruction of domains of oriented clay by soil-ingesting fauna.

Problem 8.5

a) Profile 8D has a distinct increase in clay in horizons 3-5, which is offset mainly by changes in sand content. There is no loss of silt from the topsoil. Clay illuviation in these horizons should also have depressed silt contents. The fine clay/clay ratio increases with depth, but its maximum is below the clay maximum. The combination suggests a formation of clay from silt-size minerals, perhaps combined with some clay illuviation in the subsoil. There does not seem to be surface erosion.

b) Profile 8E shows, towards the top, an increase in sand fraction and a loss in fine clay and silt fraction. Together these suggest a loss of fine fractions by superficial erosion. The fine clay/clay ratio seem to suggest an illuviation of clay in horizons 3-5, and it is possible that surface erosion and clay illuviation have acted together.

8.8. References

Brewer, R., 1972. Use of macro- and micromorphological data in soil stratigraphy to elucidate surficial geology and soil genesis. Journal of the Geological Society of Australia, 19(3):331-344.

Buurman, P., and A.G. Jongmans, 1987. Amorphous clay coatings in a lowland Oxisol and other andesitic soils of West Java, Indonesia. Pemberitaan Penelitian Tanah dan Pupuk, No. 7:31-40.

Buurman, P., and Subagjo, 1980. Soil formation on granodiorites near Pontianak (West Kalimantan). In: P. Buurman (ed): *Red soils in Indonesia*, 106-118. Agricultural Research Reports 889, Pudoc, Wageningen.

Buurman, P., 1990. Soil catenas of Sumatran landscapes. Soil Data Base Management Project, Miscellaneous Papers No. 13:97-109.

Darwin, C.H., 1881. The formation of vegetable mould through the action of worms, with observations on their habits. John Murray, London, 298 pp.

Johnson, D.L., 1990. Biomantle evolution and the redistribution of earth materials and artifacts. Soil Science, 149:84-102.

Jongmans, A.G., F. Van Oort, P. Buurman, and A.M. Jaunet, 1994. Micromorphology and submicroscopy of isotropic and anisotropic Al/Si coatings in a Quaternary Allier terrace, France. In: A.J. Ringrose and G.S. Humphreys (eds.): *Soil Micromorphology: studies in management and genetics*, pp. 285-291. Developments in Soil Science 22, Elsevier, Amsterdam.

Slager, S., and H.T.J. van de Wetering, 1977. Soil formation in archeological pits and adjacent loess soils in Southern Germany. Journal of Archeological Science 4:259-267.

Soil Survey Staff, 1975. *Soil Taxonomy*. Agriculture Handbook No. 436. Soil Conservation Service, USDA, Washington

Soil Survey Staff, 1990. *Keys to Soil Taxonomy*. Soil Management Support Services Technical Monograph 19. Blacksburg, Virginia.

Van den Broek, T.M.W., 1989. *Clay dispersion and pedogenesis of soils with an abrupt contrast in texture - a hydrochemical approach on subcatchment scale*. PhD Thesis, University of Amsterdam, 1-109.

Van Reeuwijk, L.P., and J.M. de Villiers, 1985. The origin of textural lamellae in Quaternary coast sands of Natal. South African Journal of Plant and Soil, 2:38-44.

Van Schuylenborgh, J., S. Slager and A.G. Jongmans, 1970. On soil genesis in temperate humid climate.VIII. The formation of a 'Udalfic' Eutrochrept. Netherlands Journal of Agricultural Science, 18:207-214.

Wielemaker, W.G., 1984. *Soil formation by termites - a study in the Kisii area, Kenya*. PhD Thesis, University of Wageningen; 132 pp.

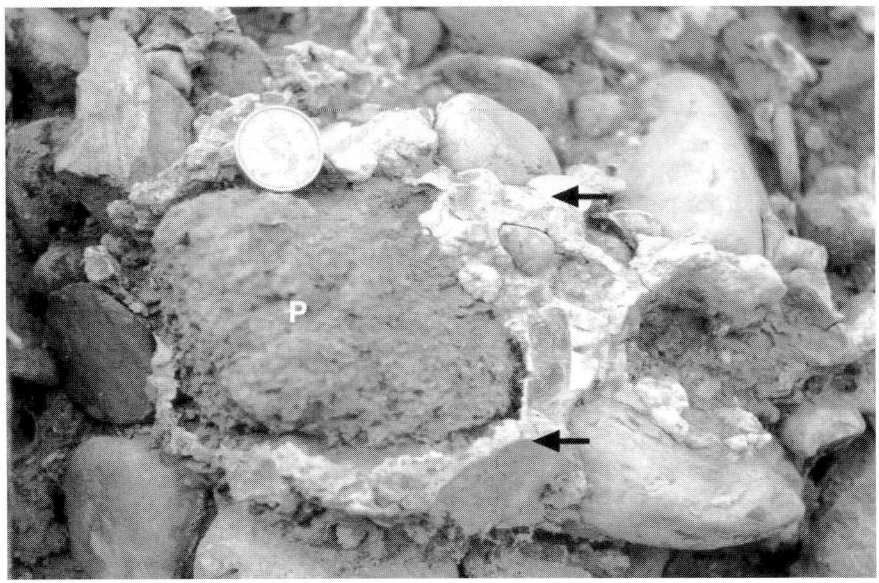

Plate Q. Massive calcite cement (arrow) in a river terrace, southern Spain. The weathered pebble (P) consist of serpentinite. Diameter of coin is 3 cm. Photograph P. Buurman

CHAPTER 9

FORMATION OF CALCIC, GYPSIFEROUS, AND SALINE SOILS

9.1. Introduction

Many soils in areas with low rainfall have accumulations of relatively soluble minerals such as carbonates, sulphates, and chlorides of Ca, Mg, and Na. These strongly affect their properties and the plants and crops that can grow there. The presence of carbonates maintains the pH of the soil at values above 7-8. High concentrations of soluble salts increase the osmotic pressure of the soil solution, thereby decreasing the availability of water to plants. Calcite, gypsum, and more soluble salts such as chlorides, and sulphates of Mg, K and Na, and carbonates of Na and K are common in certain sedimentary rocks and soil, but not in primary (igneous) rocks. The constituent cations are found in common rock-forming silicates. Sulphur is common in magmatic ore deposits (sulphides). Sulphur and chlorine are common in volcanic gases, while the atmosphere usually contributes carbon dioxide.

Calcium and magnesium carbonates ($CaCO_3$, calcite and aragonite; $CaMg(CO_3)_2$, dolomite), gypsum and anhydrite ($CaSO_4.2H_2O$ and $CaSO_4$), halite (NaCl) and sylvite (KCl) are common constituents of sediments from marine and arid environments. Such salts form, e.g., when seawater evaporates. The composition of 'standard' sea water is given in Table 9.1.

Table 9.1. Average composition of seawater (From Garrels and Christ, 1965).

Cations	$mmol_c/l$	*Anions*	$mmol_c/l$
Na^+	475	Cl^-	555
Mg^{2+}	108	SO_4^{2-}	57
Ca^{2+}	21	HCO_3^-	2.7
K^+	10		
Total	614		615

When seawater evaporates, part of the Ca first precipitates as $CaCO_3$ and the rest as gypsum. The other cations form mainly chlorides and sulphates. Carbonates of monovalent ions are not formed, because insufficient bicarbonate is available.

In silicate rocks, the counter-ion for Ca, Mg, Na, and K is usually a silicate. At pH <10, silica in solution occurs as uncharged $H_4SiO_4^o$, and therefore a different counter-ion must be supplied upon weathering of silicates. This counter-ion is usually HCO_3^-, derived from atmospheric CO_2 - dissolved in water - which acts as a weathering agent (see 3.1.). Calcium carbonate is the least soluble of the salts that usually precipitate if such a solution loses water by evaporation. This is the reason that virtually all soils contain calcium carbonate when weathering products are not removed by percolation. Where HCl or H_2SO_4 act as weathering agents dissolved chloride and sulphates form that may precipitate under dry conditions.

Question 9.1. *Show by means of reaction equations how a calcium bicarbonate soil solution is formed during weathering of a plagioclase ($CaAl_2Si_2O_8$), and how $CaCO_3$ can be formed by evaporation of that soil solution.*

SOLUBILITY

Most carbonates, sulphates, and halides are much more soluble in water than silicate minerals. This means that they are easily dissolved in percolating rainwater and transported and redistributed with flowing groundwater, both within a profile and a landscape. The occurrence in soils of such soluble compounds therefore depends on the presence of a source, the difference between annual precipitation and evapotranspiration, porosity of the soil, and groundwater movement. We can distinguish three main groups of soluble compounds (see also Table 9.2):

1. *Calcium and magnesium carbonates.*
   These have a relatively low solubility. Their occurrence is influenced by concentrations of Ca^{2+}, Mg^{2+}, HCO_3^-, CO_2 and pH.
2. *Gypsum ($CaSO_4.2H_2O$) and anhydrite ($CaSO_4$).*
   These are compounds of medium solubility. Anhydrite only forms under extremely dry conditions. The presence of gypsum is governed by concentrations of Ca^{2+} and SO_4^{2-}.
3. *Very soluble salts.*
   These include carbonates of Na and K, chlorides and sulphates of Ca, Mg, Na, and K, and mixed salts of these ions.

Question 9.2. *Show, through reaction equations, that the solubility of calcite depends on pH, and that the solubility of gypsum does not.*

The differences in solubility between the three groups cause a strong relation between accumulated soluble compounds and excess of evapo-transpiration. In humid areas, where rainfall far exceeds evapo-transpiration, calcite and more soluble components are eventually removed. In somewhat dry climates, with a relatively small excess evapo-transpiration, calcium carbonate will usually persist, or even accumulate, but more soluble compounds are removed. Very soluble salts can only persist in soils of arid climates, with a strong excess evapo-transpiration. Only soils with such salts are called 'saline'. Saline soils virtually always have calcium carbonate and/or gypsum accumulation in addition to more soluble salts.
In the following, we will discuss the processes of calcite and gypsum accumulation separate from those of 'salt' accumulation.

9.2. Calcium carbonate and gypsum

CALCIUM CARBONATE
Calcium carbonate easily dissolves in humid surface soils, under the influence of relatively high CO_2 pressure provided by root respiration and decomposing organic

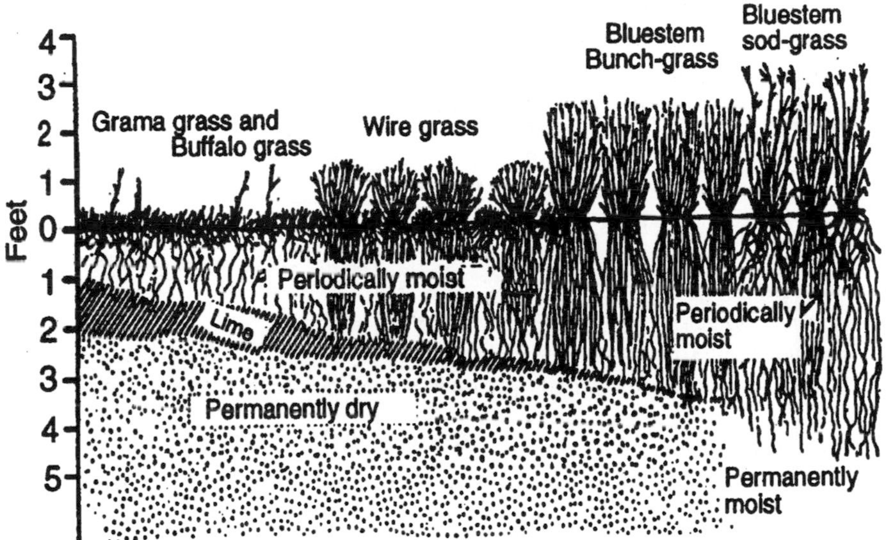

Figure 9.1. Depth of calcium carbonate (lime) accumulation in relation to rainfall (left: low; right: high) in a savannah sequence. 1 Foot = 30 cm. From Jenny, 1941. Reprinted by permission of McGraw Hill Publishing Company, Maidenhead.

matter (bacterial respiration). Calcium carbonate precipitates again when the calcite solubility product is exceeded (see Chapter 3) by a decrease in pCO_2 and/or an increase in solute concentrations.

Supersaturation or undersaturation with respect to calcite may vary on a scale of mm to cm, depending on variations in P_{CO2} caused by the presence of roots and pores. This may lead to intricate patterns of precipitation and dissolution. In moderately dry climates, surface soils are slowly decalcified and secondary $CaCO_3$ is accumulated at some depth. Decreasing CO_2 production below the rooting zone and increasing solute concentration by uptake of percolating water by roots may cause $CaCO_3$ to

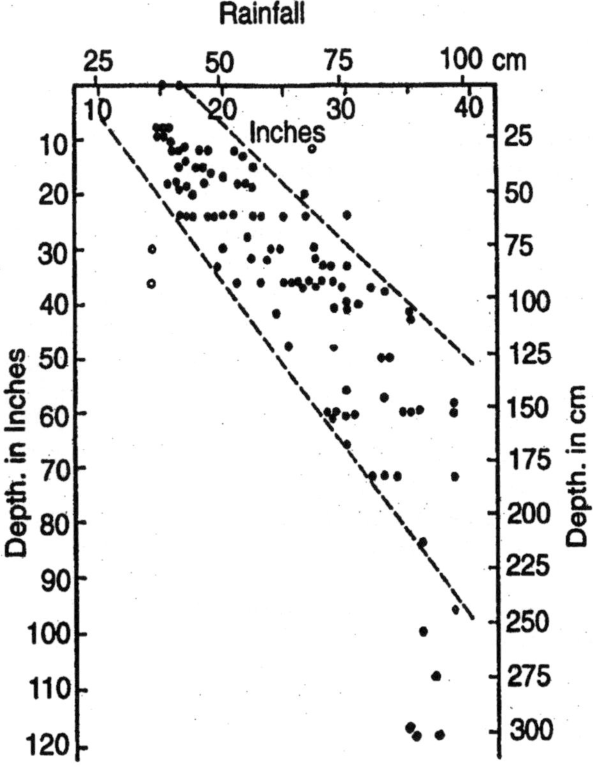

Figure 9.2. Relation between depth of occurrence of calcite and mean annual rainfall in loess soils of the USA. From Jenny, 1941; reprinted by permission of McGraw Hill Publishing Company, Maidenhead.

precipitate at a certain depth. The actual depth of calcium carbonate accumulation increases with increasing precipitation (Figure 9.1, 9.2). The depth of accumulation depends on the mean penetration depth of individual precipitation events. In a dry climate, penetration is shallow, because the dry soil absorbs most rainstorms at shallow depth. With increasing precipitation, also the mean penetration depth increases, because subsoils become less dry. If rainfall increases beyond a certain limit (about 900 mm in Fig. 9.2), the subsoil rarely dries out and solutes are removed through groundwater flow rather than by precipitation when the soil dries out.

Question 9.3. *In Figure 9.2, the amount of calcite (suggested by the thickness of the accumulation horizon) decreases with increasing rainfall. Why would this be the case? Assume that the amount of source material for $CaCO_3$ is the same, regardless of annual rainfall.*

ACCUMULATION STAGES

Accumulation of secondary calcium carbonate follows a more or less fixed course of events, which can be recognised morphologically in four stages:

1. Accumulation of calcite as small, thread-like accumulations in the smaller biopores. In soil profiles, such thread-like accumulations are described as 'pseudo-mycelium' (Fig. 9.3). At the same time, small calcite crystals and thin coatings may be formed in voids. In coarse deposits, calcite accumulates at the bottom of individual pebbles.
2. Continued accumulation causes the formation of hard or soft calcite nodules. These may be arranged in vertical macropores or dispersed through the matrix, depending on the geometry of the pore system.
3. Upon further accumulation, nodules coalesce, and eventually calcite banks (the so-called 'petrocalcic horizon') are formed (Fig. 9.4; Plate Q, p.214).
4. When hardened calcite accumulations impede vertical water flow, a massive, laminar calcite accumulation may be formed on top of the nodular zone.

This description holds in the absence of groundwater at shallow depth. In arid soils with shallow groundwater, laminar calcite accumulations are also formed when calcite is precipitated by evaporation from groundwater, which is usually saturated with $CaCO_3$.

Question 9.4. *In coarse deposits, calcite accumulations start at the bottom of pebbles (Plate R, p.244). Why?*

The four stages may be found superimposed on each other in profiles with well-developed calcite accumulations. Such a model profile is given in Figure 9.5.

Question 9.5. *Figure 9.5 shows that calcite accumulation stage I, II, and III are found in opposite order above and below the massive laminar zone). Explain the formation of a vertical sequence (I-II-III; III-II-I) for a soil that developed in a sediment that was moderately calcareous throughout.*

In thin section, calcite accumulations exhibit a large variety of forms. Some of these can form in various environments, but some are very specific for one specific precipitation environment. The latter can be used to reconstruct soil genesis (Freytet & Plaziat, 1982; PiPujol & Buurman, 1997).

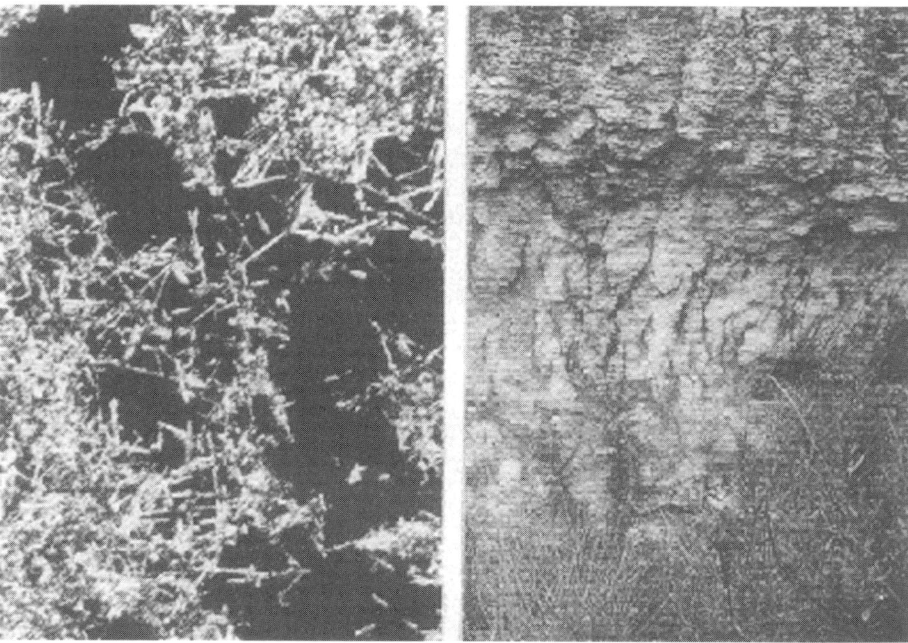

Figure 9.3. Pseudomycelium in a soil from Barbados. Width of photo is 0.5mm. From Esteban and Klappa, 1983.

Figure 9.4. Petrocalcic horizon. Upper part with cemented nodules, lower part with vertically arranged nodules. Height: 2.5 m. From Esteban and Klappa, 1983.

TRAVERTINE

Flowing groundwater may transport dissolved calcium carbonate over large distances. Precipitation may occur where the flowing water comes into contact with the atmosphere or with aquatic plants. Contact with the lower CO_2 concentrations of the atmosphere leads to a decrease of dissolved CO_2. This leads to precipitation of calcium carbonate. Plants growing in water cause an even stronger removal of CO_2. Precipitation in an aquatic vegetation leads to typically porous calcite precipitates with an abundance of plant remains and fresh-water molluscs. Such a calcite accumulation is called *travertine*. If degassing through contact with the atmosphere causes precipitation, layered and non-porous precipitates are formed, e.g. in post-volcanic calcite terraces.

Question 9.6. *Explain why calcite precipitates upon CO_2 degassing of $Ca(HCO_3)_2$-rich groundwater? Why would aquatic plants further stimulate calcite precipitation?*

GYPSUM ACCUMULATION

Gypsum ($CaSO_4.2H_2O$) is a common constituent of ancient marine evaporites. Such deposits are common from the Permian, Triassic, and Miocene. In recent sediments, gypsum is formed as a weathering product of pyrite in calcareous sediments (See Chapter 7), but this rarely leads to significant accumulations. Significant gypsum accumulations in soils are usually due to evaporation from groundwater and to redistribution in gypsiferous sediments. Gypsum is more soluble than calcium carbonates (2.6 $g.L^{-1}$ in pure water) and is therefore more mobile.

At temperatures above 40°C, gypsum dehydrates to bassanite ($CaSO_4.½H_2O$), which, because of its rapid rehydration upon addition of water, is used in commercial 'gypsum' plaster. Dehydration to bassanite is rapid, but because soils rarely heat up above 40°C, bassanite is not a common constituent of soils. Further dehydration of bassanite to anhydrite ($CaSO_4$) requires higher temperatures and hardly occurs in soils but is common in older metamorphic sediments. Because of its higher solubility, gypsum may be moved over a larger distance with percolating or capillary water than calcium carbonate.

Question 9.7. *What are relative vertical positions of accumulation horizons of gypsum and calcite in a) well-drained soils, and b) soils with evaporation from ground water?*

Gypsum accumulations have stages that are similar to those of calcite. Massive, laminar deposits are due to evaporation from ground water. Gypsum is usually found in the form of lenticular crystals, at high concentrations as fibrous aggregates or aggregates of lenticular crystals. In arid soils, gypsum is always found in combination with calcite, and frequently with soluble salts. Purely gypsiferous soils are scarce and are only found on old evaporitic sediments.

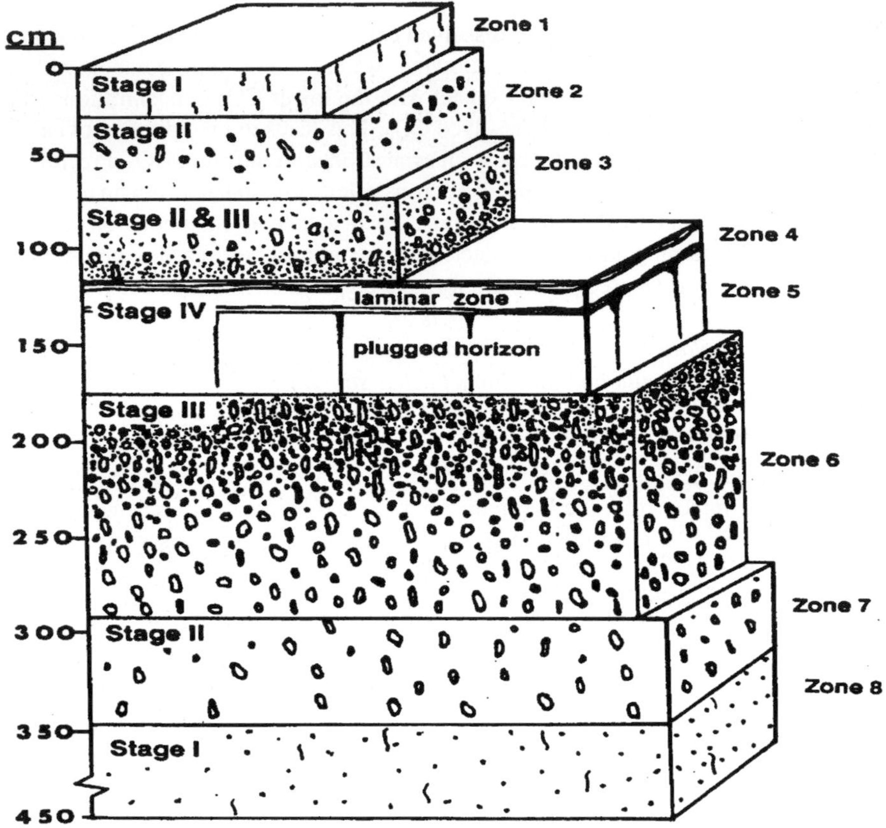

Figure 9.5. Stages of calcite accumulation in a soil from New Mexico, USA. From Monger et al., 1991.

Because of its high solubility and Ca content, gypsum strongly stabilises soil structure (See Chapter 2; Figure 8.1). For this reason, pure gypsum is sometimes applied for structure improvement on soils with low organic matter contents (e.g. Oxisols, Ultisols), and on sodic soils (see 9.3).

DIAGNOSTIC HORIZONS

If the accumulation of secondary calcium carbonate in the soil is sufficiently strong, it is recognised in soil classification as a *calcic* or *petrocalcic* horizon. The petrocalcic horizon is a calcic horizon that is cemented. Such horizons are also called *calcrete* or *caliche*.

In case of gypsum accumulation, the diagnostic horizons are *gypsic* and *petrogypsic*. The term *gypsite* is sometimes used for such accumulations, but should be avoided because of possible confusion with gibbsite ($Al(OH_3)$).

9.3. Soluble salts

Salts more soluble than gypsum are called *soluble salts.* These are normally combinations of the cations Ca^{2+}, Mg^{2+}, Na^+, K^+, and H^+, with the anions CO_3^{2-}, SO_4^{2-}, Cl^-, and NO_3^- and water. Different amounts of these ions lead to a large number of possible combinations. The most common of these are listed in the inset on the following page. The differences in solubility are illustrated in Table 9.2.

Salinization, i.e. accumulation of soluble salts in soils, occurs if evaporation/ evapotranspiration exceeds or equals effective rainfall (the rainfall that penetrates into the soil). Sources of soluble salts are local rocks and sediments, inundation by seawater, irrigation water, atmospheric deposition of so-called *'cyclic salts'* (salts that become airborne through evaporation of seawater spray), and dust deposition. Obviously, salinization by dust or atmospheric deposition is slow, while inundation by sea and evaporation from groundwater and irrigation water may cause rapid salinization. Salinization is a common process in arid and semi-arid areas, but may also occur locally in more humid climates.

Question 9.8. *Which of the sources of soluble salts do you expect to be most important in causing salinization in relatively humid climates?*

Different salt minerals dominate in subhumid and arid areas. The most soluble salts, such as magnesium and calcium chlorides, are only found under the most arid conditions; magnesium and sodium sulphates and sodium chloride are common in arid and semi-arid environments. With increasing humidity, the most soluble salts disappear and only gypsum and finally only calcium carbonate is found. Precipitation from a salt solution that is concentrated by evaporation proceeds in the opposite direction, with least soluble salts precipitating first. For a system with only simple salts, the precipitation sequence can be predicted from the relative solubilities of these salts.

Salt minerals in arid soils

Carbonates	*Formula*	*Abundance*
calcite	$CaCO_3$	*
aragonite	$CaCO_3$	
nesquehonite	$MgCO_3.3H_2O$	
dolomite	$CaMg(CO_3)_2$	
thermonatrite	$Na_2CO_3.H_2O$	**
nahcolite	$NaHCO_3$	
trona	$NaHCO_3.Na_2CO_3.2H_2O$	***
pirssonite	$Na_2Ca(CO_3)_2.2H_2O$	
gaylussite	$Na_2CO_3.CaCO_3.5H_2O$	*
Sulphates		
gypsum	$CaSO_4.2H_2O$	**
arcanite	K_2SO_4	
thenardite	Na_2SO_4	**
starkeyite	$MgSO_4.4H_2O$	
hexahydrite	$MgSO_4.6H_2O$	*
epsomite	$MgSO_4.7H_2O$	
mirabilite	$MgSO_4.10H_2O$	
aphtithalite	$K_3Na(SO_4)_2$	*
bloedite	$Na_2Mg(SO_4)_2.4H_2O$	**
konyaite	$Na_2Mg(SO_4)_2.5H_2O$	
schoenite	$K_2Mg(SO_4)_2.6H_2O$	
glauberite	$Na_2Ca(SO_4)_2$	*
eugsterite	$Na_4Ca(SO_4)_3.2H_2O$	*
loeweite	$Na_{12}Mg_7(SO_4)_{13}.15H_2O$	
Halides		
halite	$NaCl$	***
sylvite	KCl	
Nitrates		
niter	KNO_3	
sodaniter	$NaNO_3$	
Mixed anions		
burkeite	$Na_6(CO_3)(SO_4)_2$	*
darapskite	$Na_3(NO_3)(SO_4).H_2O$	
kainite	$KMgClSO_4.3H_2O$	
northupite	$Na_6Mg_2Cl_2(CO_3)_4$	

Relative abundance on the soil surface in 85 locations in Kenya and Turkey: *** more than 30 times; ** 15-30 times; * 5-15 times. Data from Vergouwen, 1981.

Table 9.2. Solubilities of some common simple salts ($mol.L^{-1}$). Simplified after Vergouwen, 1981.

Chemical formula	*Mineral name*	*solubility at 10°C*	*solubility at 30°C*
$CaCl_2.6H_2O$ (10°C) or $4H_2O$ (30°C)		5.9	9.2
NaCl	halite	6.1	6.3
$MgCl_2.6H_2O$		5.6	5.9
KCl	sylvite	4.2	5.0
$MgSO_4.7H_2O$	epsomite	2.4	3.3
$Na_2CO_3.10H_2O$		1.2	4.1
$Na_2SO_4.10H_2O$	mirabilite	0.66	3.0
$NaHCO_3$	nahcolite	0.96	1.3
K_2SO_4	arcanite	0.52	0.74
$CaSO_4.2H_2O$	gypsum	11.10^{-3}	11.10^{-3}
$CaCO_3$	calcite		14.10^{-5}

Assume water that contains only Ca^{2+} and SO_4^{2-} in (necessarily) equal concentrations. If such a water evaporates, concentrations will increase until the saturation of gypsum is reached ($K_{so} = 10^{-4.6}$). Further evaporation will cause precipitation of gypsum, but the concentration in solution remains constant. This sequence of events is illustrated by pathway 1 in Figure 9.6.

If the solution contains a second cation, e.g. Na^+, in addition to calcium, the concentration of sulphate will be larger than that of calcium, and upon evaporation, the concentrations will change according to pathway 2. Upon reaching the saturation concentration of gypsum, the constancy of the ionic activity product $(Ca^{2+}) * (SO_4^{2-}) = 10^{-4.6}$ must be maintained. The concentration of sulphate, which is in excess, continues to increase, while that of calcium decreases. Further evaporation would precipitate more gypsum until the equilibrium concentration of the more soluble Na_2SO_4 is reached, and sodium sulphate would precipitate. Alternatively, a solution containing Ca^{2+}, SO_4^{2-} and Cl^- would first produce gypsum and later precipitate $CaCl_2$ from a highly concentrated solution that is practically devoid of sulphate. Figure 9.7 shows simplified precipitation pathways in a solution with unequal amounts of Ca^{2+}, Na^+, CO_3^{2-}, SO_4^{2-}, and Cl^-.

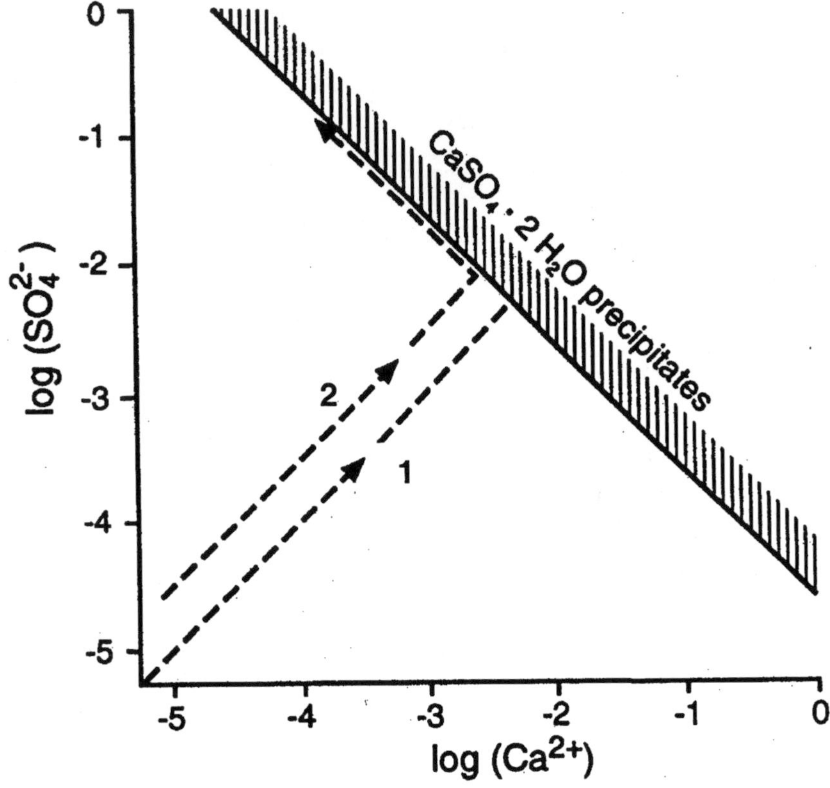

Figure 9.6. Changes in concentrations of Ca^{2+} and SO_4^{2-} upon evaporation. 1: (Ca) = (SO_4); 2: (Ca) < (SO_4).

Question 9.9. *Sketch in Figure 9.6 the composition pathway of a solution containing only* Ca^{2+}*,* SO_4^{2-}*, and* Cl^-*. Indicate the approximate location in the diagram where* $CaCl_2$ *would precipitate (consult Table 9.2).*

Question 9.10. *Why would sulphate have virtually disappeared when* $CaCl_2$ *starts to precipitate (consult Table 9.2 and Appendix 2).*

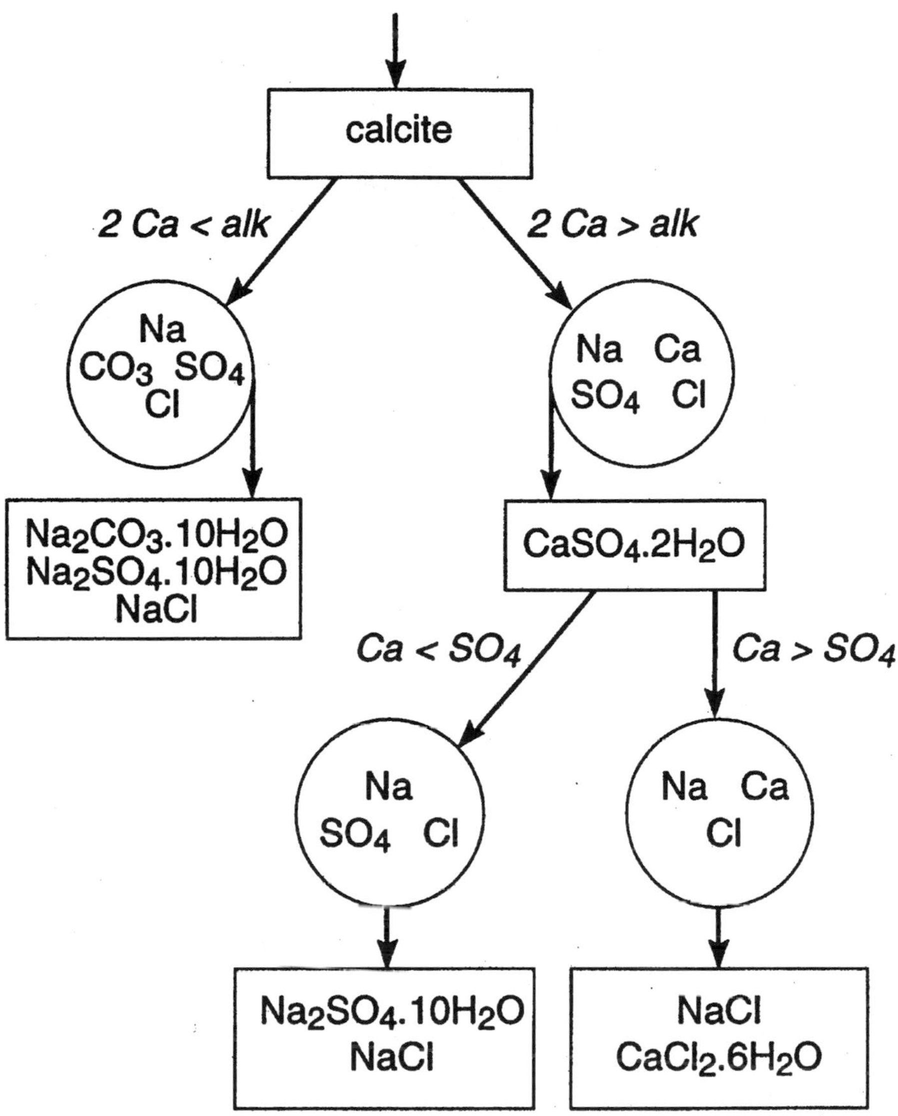

Figure 9.7. Different precipitation pathways during evaporation of a solution containing Ca^{2+}, Na^{+}, CO_3^{2}, SO_4^{2-}, and Cl^{-}.
'Alk' stands for alkalinity, defined as (HCO_3^-) + 2(CO_3^{2-}) + (OH^-) - (H^+). At pH < 9.5, 'alk' is very close to the concentration of HCO_3^- . From Vergouwen, 1981.

The precipitation model of Figure 9.7 does not account for the presence of complex salts. The list of salts clearly illustrates that many salts have a complex composition, and that precipitation pathways are complex. Moreover, which salts are precipitated depends not only on the composition of the water, but also on temperature at the soil surface (see Table 9.2).
The location of salt accumulation is governed by penetration depth of rainfall and, in case of precipitation from ground water, by capillary rise. In arid areas with shallow ground water, salt crusts can form at the soil surface.

Question 9.11. *a) Where do you expect soluble salts relative to the calcic horizon in Figure 9.1; b) how would the relative positions of calcic and saline horizons be in an arid soil with shallow ground water?*

SODICATION
Accumulation of sodium salts that leads to an increasing ratio of sodium to divalent cations in solution and at the adsorption complex is called *sodication* and soils with more than 15% of the exchange complex occupied by Na^+ are called *sodic* soils. Sodication may be accompanied by a strong rise in pH. In older literature, sodication and sodic soils were called 'alkalinization' and 'alkaline soils', suggesting high pH values. This is misleading, because sodication is not necessarily accompanied by high pH. Sodication is of practical importance because both high Na levels and high pH are harmful to crops. In addition, a high percentage of exchangeable Na may lead to easier peptisation of clay. This causes a decrease in structure stability, leading to physical soil degradation.

Question 9.12. *What would be the effect of high or low salinity level on peptisation of Na-saturated clay?*

Sodic soils can be formed by two different processes: accumulation of sodium sulphates or chlorides, and concentration of dilute solutions containing HCO_3^- in excess of $Ca^{2+} + Mg^{2+}$.

1. Accumulation of sodium salts
Supply by ground water of dissolved sodium chloride or sodium sulphate to a low-Na soil causes a direct increase of exchangeable Na^+ because of increased Na^+ in solution and exchange of other cations on the adsorption complex for Na^+. Sodic soils formed in this way usually have a pH around 8.5.
This type of sodication sometimes leads to the development of pH values larger than 8.5 in playas (ephemeral salt lakes) where Na_2SO_4 accumulates. In the presence of organic matter and under seasonal waterlogging, sulphate is reduced. This results in the formation of Na_2CO_3. The Na_2CO_3 persist, and thus leads to permanent alkalinization

if the reduced sulphur is not retained in the soil (e.g. as FeS), but leaves the system as gaseous H_2S (Van Breemen, 1987).

Question 9.13. *Why would the Na_2CO_3 not persist in seasonally flooded soils, unless reduced sulphur disappears?*

2. Concentration of dilute solutions with sodium carbonates

By evaporation of solutions with a calcium concentration that is lower than the HCO_3^- concentration, practically all Ca^{2+} is precipitated as calcite. The solution that remains is dominated by HCO_3^-, normally with Na^+ as the counter-ion. Upon further evaporation, concentrations of dissolved Na^+ and bicarbonate increase (left branch of Fig. 9.7). In equilibrium with calcite, the concentration of Ca^{2+} is depressed by the increase of dissolved HCO_3^-. Upon further evaporation, lowering of dissolved Ca^{2+}causes further displacement of exchangeable Ca by Na (forced exchange), while calcite precipitates:

$$Ca^{2+}_{(ex)} + 2\,Na^+_{(aq)} + 2HCO_{3\ aq)}^- \rightarrow CaCO_{3(s)} + CO_2 + H_2O + 2Na^+_{(ex)} \qquad (9.1)$$

Because the solubility of sodium carbonates is much higher than that of calcite, the concentration of CO_3^{2-} may increase strongly, leading to high pH values, according to:

$$CO_3^{2-} + H_2O \rightarrow HCO_3^- + OH^- \qquad (9\text{-}2)$$

In this system, pH values above 10 are possible.

SOLONCHAK, SOLONETZ, SOLOD

As long as the salt concentration is high, clay remains flocculated, even at high contents of exchangeable sodium (Figure 8.1). As a result, soil structural stability is high. Such soils with an accumulation of soluble salts, but without excess of bicarbonate over calcium + magnesium, are called *Solonchak*. This includes soils that are recently reclaimed from marine deposits.

When the soil solution is diluted by rainwater, however, sodium-saturated clay will be dispersed easily. This may lead to clay eluviation and the formation of a textural B-horizon with a high Na saturation, which is typical for *Solonetz* soils. Such B-horizons are characterised by conspicuous, prismatic elements with rounded tops (columnar structure). As stated in Chapter 8, also organic matter is mobile at high Na-saturation, and it moves with the clay. Eventually, most exchangeable sodium will be leached from the surface soil under the influence of dissolved CO_2:

$$Na^+_{(ex)} + H_2O + CO_2 \rightarrow H^+_{(ex)} + Na^+ + HCO_3^- \qquad (9\text{-}3)$$

This results in an acidic topsoil relatively low in clay, underlain by a Btn horizon. This is characteristic for so-called degraded Solonetz soils, known as *Solods*.

Dense, non-sodic soils are found in the marine clay area of Friesland (northern Netherlands). This so-called 'knip' -clay was probably deposited in a saline environment and have become strongly compacted as a result of desalinisation and the resulting destabilisation of soil structure. Even though these soils have lost their high levels of exchangeable sodium through slow replacement by Ca and Mg, they are generally still very compact and have a low hydraulic conductivity. Their structure has improved where better drainage has increased earthworm activity.

Where saline marine soils originally contained appreciable amounts of calcite and some pyrite, empoldering resulted in oxidation of pyrite (see 7.4). The sulphuric acid produced by this oxidation dissolves calcite, thus providing a good supply of Ca^{2+} in the soil solution. Such soils never became sodic. In highly acid soils (pH<4.5; large amounts of pyrite and insufficient calcite for buffering), an appreciable part of the exchange complex is saturated by Al, which also causes stable structures. Therefore, the conventional limit of 15% exchangeable sodium for sodic soils, has no practical significance in acidic coastal soils. Many saline and sodic soils owe their formation to human action. Millions of hectares of potentially fertile land in arid and semi-arid areas were once made productive by irrigation, but eventually turned into salt deserts. Irrigation, even if the water used is low in salts, will inevitably lead to salinization and sodication, unless excess water (over evaporative demand) is supplied and removed by drainage.

Question 9.14. *In Figure 8.1, the lower right field (high pH, low exchangeable* Na^+*) is marked as 'non-existent. Explain this.*

SALT AND CONDUCTIVITY

Accumulation of soluble salts in the soil leads to an increase in electrical conductivity and osmotic pressure of the soil solution (Figures 9.8 and 9.9). An increase in osmotic pressure decreases the availability of water for plants. Because electrical conductivity is easy to measure in the field, this parameter is commonly used to classify saline soils. The most sensitive plants are already affected when the electrical conductivity exceeds 2 $dS.m^{-1}$ (2 $mmho.cm^{-1}$).

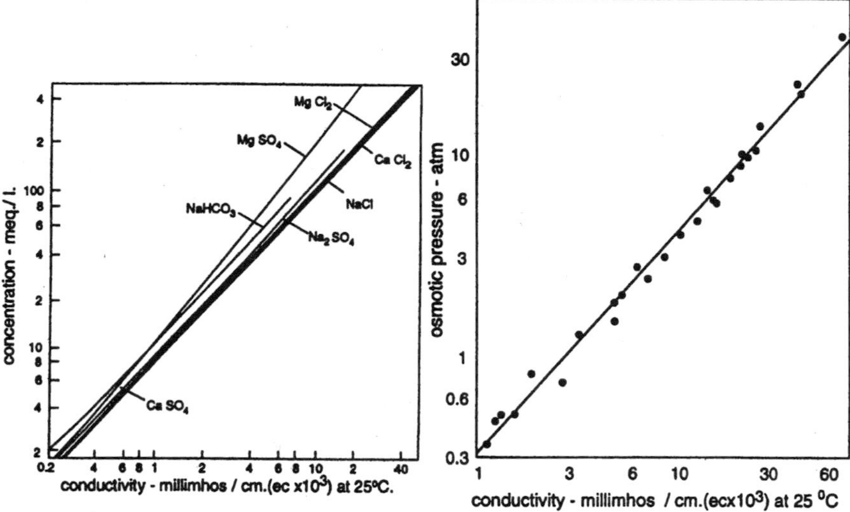

Figure 9.8. Relation between electrical conductivity and salt concentration in solutions of simple salts. From USSLS, 1954.

Figure 9.9. Relation between electrical conductivity and osmotic pressure. From USSLS, 1954.

DIAGNOSTIC HORIZONS

The following diagnostic horizons are related to salinization and sodication:

Natric horizon: a clay illuviation horizon with >15% exchangeable Na and columnar structure.

Salic horizon: a horizon of accumulation of salts which are more soluble than gypsum, with an electrical conductivity (1:1 extract) of >30 $dS.m^{-1}$

9.4. Clay minerals specific to arid soils

Ground- and pore waters of soils in arid regions frequently have high concentrations of Mg^{2+} and H_4SiO_4. This may lead to formation of specific clay minerals. Depending on the activity of aluminium, either Mg-smectite, palygorskite, or sepiolite may form. Palygorskite and sepiolite resemble 2:1 sheet silicates, but differ from these in that the tetrahedral layer reverses after a specific number of atoms (Figure 9.10). Dissolved silica and Mg^{2+} are commonly set free by weathering in wetter areas upstream, and are provided by ground- and surface water draining from such areas into arid basins. Landscape relationships show that smectites form closest to weathering zones, while sepiolites form farther away at higher solute concentrations and lower aluminium availability. It is unclear what exactly determines the formation of either Mg-smectite or palygorskite. In fossil environments, they alternate at a small spatial and time scale. Palygorskite is a common constituent of (petro)calcic horizons and duripans (see Chapter 14), and Mg-smectite appears to be more common in lacustrine environments. Sepiolite and palygorskite are important as molecular sieves for industrial use, because of the fixed dimension of the channels in the crystallographic structure.

Question 9.15. *Indicate the tetrahedral and octahedral layers in Figure 9.10. Expand the figure by completing/adding octahedra and tetrahedra.*

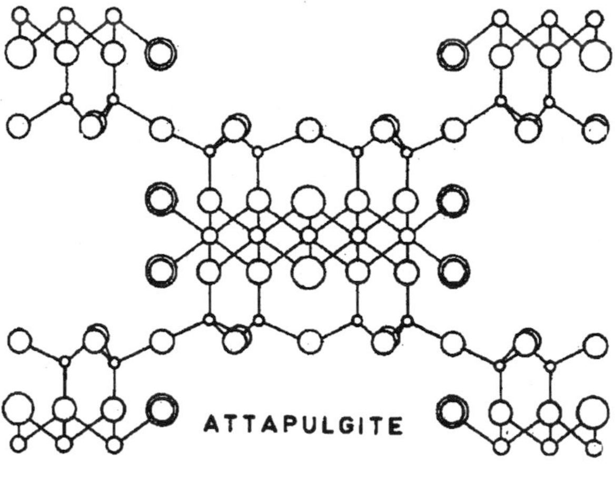

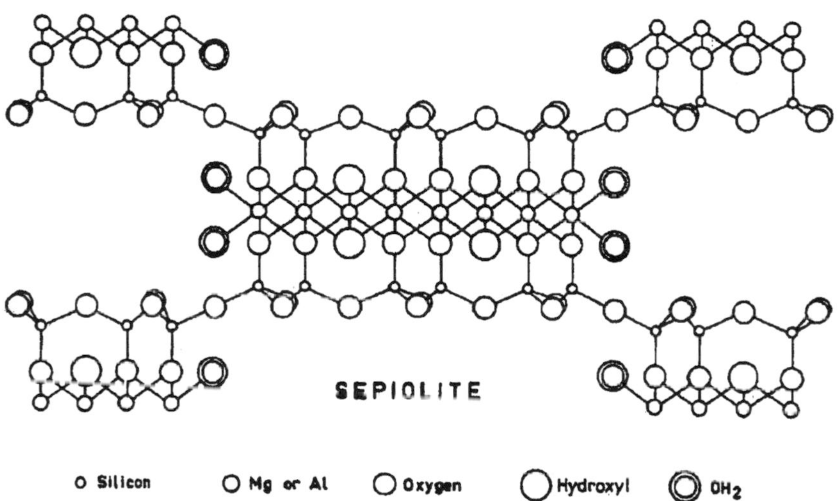

Figure 9.10. Structure of palygorskite (attapulgite) and sepiolite.
From Millot, 1970. Reprinted with permission of Springer Verlag GmbH & Co., Heidelberg.

9.5. Problems

Problem 9.1

Calculate the amount of $CaCO_3$ (mass fraction, %) accumulated over 100 years in a 50 cm thick soil horizon of bulk density 1.4 kg dm^{-3}, by evapo-transpiration (5 mm day^{-1}, throughout the year) from shallow groundwater containing 100 mg of $CaCO_3$ per litre.

Problem 9.2.

What sequence of salt precipitation would you expect upon evaporation of each of the following waters? Evaporation of which water could cause high-pH sodic soils?

water nr.	equivalent concentrations in mol(+/-)/m^3				
	Na^+	Ca^{2+}	HCO_3^-	Cl^-	SO_4^{2-}
1	5	1	3	1	2
2	2	4	3	1	2
3	3	3	2	0	4

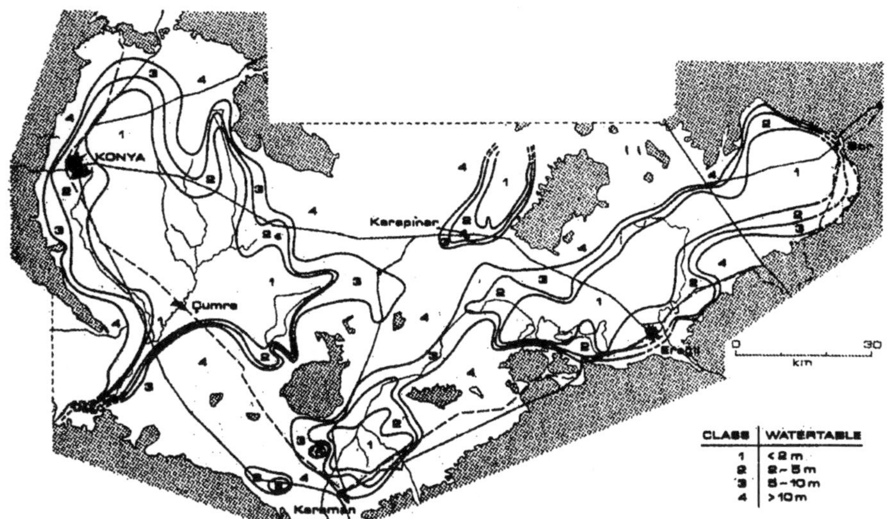

Figure 9A. Depth of water table in May in the Great Konya Basin, Turkey. From Driessen, 1970.

Problem 9.3

The Konya basin in Turkey is a lacustrine plain in an area with less than 200 mm precipitation per year. Figures 9A (previous page) and 9 B (below) show water table depths in the wettest part of the year and salt contents (expressed in terms of the electrical conductivity of the saturation extract in mmho.cm^{-1}; class 1 indicates non-saline land, class 5 is extremely saline) in the driest season. What can you conclude about the relationships between hydrology and salinity as indicated by the two maps?

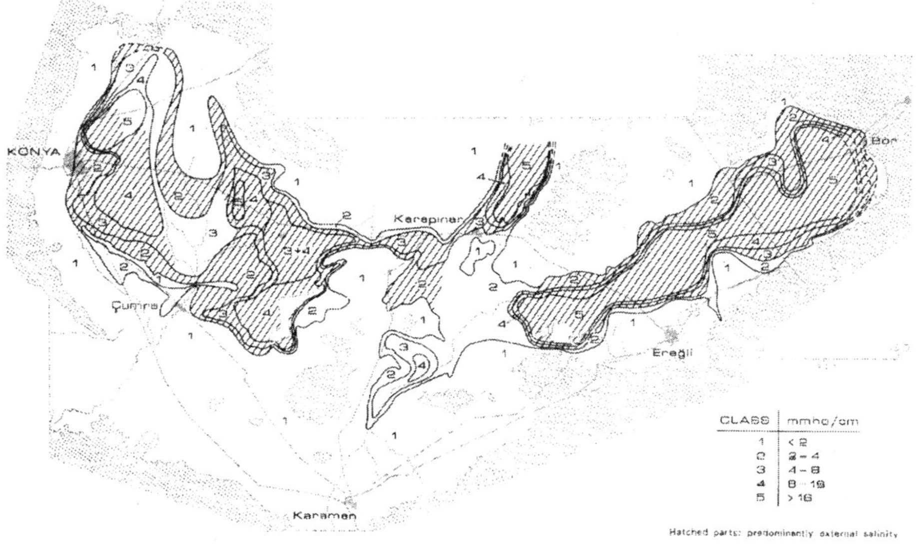

Figure 9B. Distribution of electrical conductivity in soils of the Great Konya Basin, Turkey. From Driessen, 1970.

Problem 9.4

Figure 9C shows salt profiles (from analyses of the aqueous extracts of water-saturated soil samples from each 10cm layer of soil) in three soils from a toposequence (Figure 9D) in the Konya area. Site 6 is well drained, sites 7 and 8 have shallow groundwater. The bar diagrams below the salt profile diagrams of sites 7 and 8 refer to the composition of that groundwater. Note the different concentration scales.

a. Concentrations of cations and anions have been plotted cumulatively right and left of the O-line. Why are the plots symmetrical?
b. Discuss the data in terms of the aspects of salinization treated in this chapter. Discuss which anions are most mobile, CO_3^{2-} and HCO_3^-, or SO_4^{2-} and Cl^-?

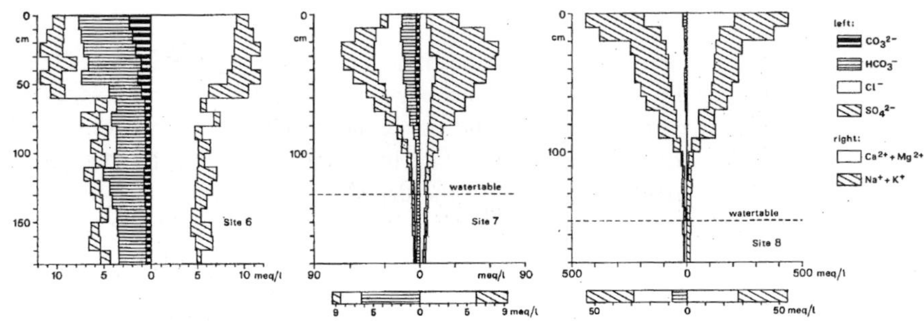

Figure 9C. Salinity profiles in three soils from Konya. From Driessen, 1970.

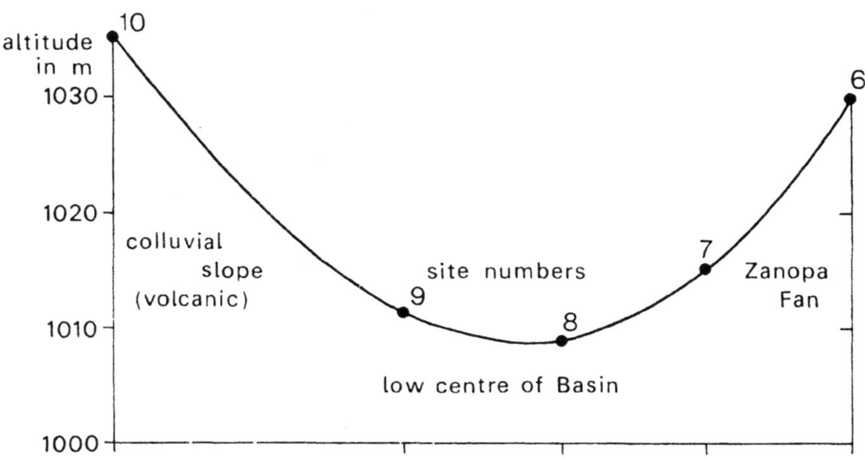

Figure 9D. Topographic relation between the profiles of Figure 9C. From Driessen, 1970.

Problem 9.5

Lake Chad is one of the world's largest closed evaporative basins. Fig. 9E shows the changes in concentration of the major solutes in the water of Lake Chad polders during evaporation. The vertical axis shows the log(molar concentration), the horizontal concentration the log of the concentration factor. Secondary minerals observed include Mg-smectite, $CaCO_3$, and amorphous silica.

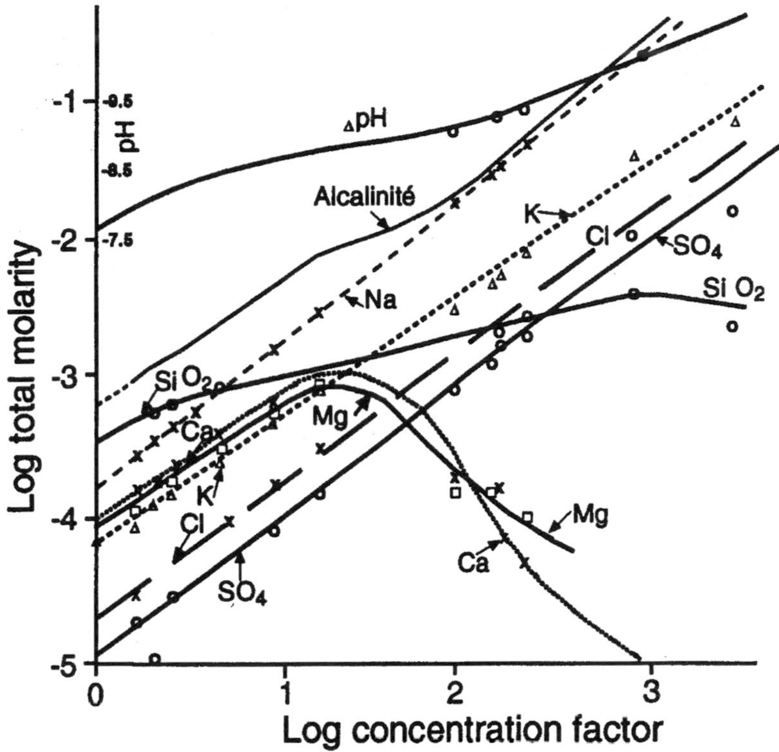

Figure 9E. Change of concentration of solutes upon evaporation of water from a Lake Chad polder. Drawn lines connect measurements. From Al-Droubi, 1976.

a. Explain the changes in alkalinity, and in the concentrations of Ca, Mg and SiO_2 in the course of concentration of the water. At which pH do Mg-smectite and calcium carbonate precipitate?
b. Which of the two sodication processes described on pages 226-227 would take place in a soil in an arid area, irrigated with Chari river water?

Problem 9.6
Write the reaction equation for the alkalinization of a soil by bacterial reduction of sulphate from Na_2SO_4 to H_2S.

Problem 9.7

Figure 9F illustrates how certain soil properties of a sodic soil changed after planting two species of trees in 1979. The original silty surface soils (0-15 cm) had a pH of 10.5, contained 0.7 % $CaCO_3$, and had an exchangeable sodium percentage of 96. Between 1982 and 1984, average litter production (in kg/ha.yr) was 1.1 x 10^3 for Eucalyptus and 3.7 x 10^3 for Acacia. a) Give a possible explanation for the changes in soil properties. b) Are these data solid proof for an ameliorative effect of these trees on this sodic soil?

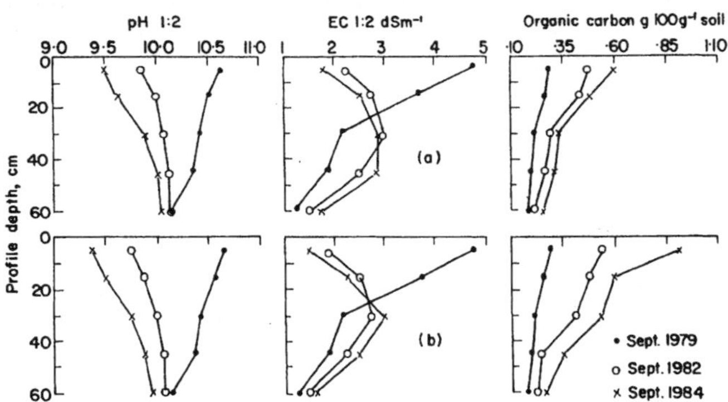

Figure 9F. Change with time of pH, soluble salts, and organic C in soils planted with (a) Eucalyptus, and (b) Acacia. From Gill and Abrol, 1986.

9.6. Answers

Question 9.1

$CaAl_2Si_2O_8 + 3H_2O + 2CO_2 \rightarrow Ca^{2+} + 2HCO_3^- + Al_2Si_2O_5(OH)_4$

Upon evaporation, both the Ca^{2+} and the HCO_3^- concentration increase, and calcite is formed by: $Ca^{2+} + 2HCO_3^- \rightarrow CaCO_3 + CO_2 + H_2O$

Question 9.2

The equilibrium between $CaCO_3$, HCO_3^- and CO_2 is described in Chapter 3. The dependence on pH is clear from the reaction $CaCO_3 + 2H^+ \rightarrow Ca^{2+} + H_2O + CO_2$.
The precipitation of $CaSO_4$ is described by: $Ca^{2+} + SO_4^{2-} + 2\ H_2O \rightarrow CaSO_4.2H_2O$
This reaction is virtually independent of pH, because sulphuric acid is completely dissociated at normal soil pH values (>2).

Question 9.3

The frequency of rainfall events that penetrate deeper than the carbonate accumulation horizon probably increases with increasing annual rainfall. At each of such an event, a small amount of calcite is removed from the soil profile.

Question 9.4

In a coarse material, water flows around, and accumulates at the bottom of pebbles, if a pebble-bottom does not touch other soil particles. Whenever the water dries before it drips, solutes will precipitate on the bottom of pebbles. The first precipitate to form if normally calcite.

Question 9.5

Leaching causes redistribution in the upper horizon, but there is not enough calcite to form concretions (or time has been too short). The gradual removal from the topsoil causes a gradient in calcite content from the top to the horizon of maximum accumulation. In the subsoil, penetration to great depth occurs only occasionally, and therefore, the intensity of calcite accumulation decreases below the zone of maximum accumulation (which is the zone of mean rainfall penetration depth).

Question 9.6

Contact with the atmosphere and especially photosynthesis by growing plants remove CO_2 from ground water. Removal of CO_2 causes precipitation of calcite (see reaction equation 2 in Question 9.1).

Question 9.7

In well-drained soils, gypsum can be transported further down the profile because its saturation is reached at a later stage in the evaporation of the percolating water than that of calcite. This implies that it can be removed from the soil in a drier climate than calcium carbonate (In Figure 9.1, the gypsum accumulation horizon would lie below the lime horizon, and would disappear earlier).

In soils with groundwater, capillary rise would transport gypsum further upward than calcite and the gypsum accumulation would be found above the calcite accumulation.

Question 9.8

In relatively humid climates, inundation by seawater is the most likely cause of salinization. Fossil salt deposits may also play a role.

Question 9.9

If $Ca^{2+} > SO_4^{2-}$, the first part of the reaction path is parallel to, but to the right of, the reaction path for $Ca^{2+} < SO_4^{2-}$. When gypsum starts to precipitate, SO_4^{2-} becomes exhausted, while the Ca^{2+} concentration increases, so the path follows the gypsum line towards the lower right hand corner. The approximate location of the Ca^{2+} concentration when $CaCl_2$ starts to precipitate (at 30°C) can be derived from the solubility of calcium chloride. At a solubility of 9 $mol.L^{-1}$ of $CaCl_2$, the Ca^{2+} concentration is 9 $mol.L^{-1}$, so $\log(Ca^{2+})$ is close to 1.

Question 9.10
At a Ca^{2+} concentration of 9 $mol.L^{-1}$, the SO_4^{2-} concentration can be derived from the solubility product of gypsum: $(Ca^{2+}) * (SO_4^{2-}) = 3.4*10^{-5}$ (see Appendix 2). If $(Ca^{2+}) = 9$, (SO_4^{2-}) is about $7*10^{-6}$ $mol.L^{-1}$ (activity corrections neglected).

Question 9.11
Accumulation of soluble salts would occur below the calcite (and below the gypsum) accumulation of Figure 9.1. It would disappear from the solum at lower precipitation than calcite. In case of evaporation from groundwater, accumulation of soluble salts is above the calcite accumulation.

Question 9.12
Clay would easily disperse at low salinity levels (expanded double layer) and be flocculated at high salinity levels (compressed double layer).

Question 9.13
If reduced sulphur stays in the soil, it will oxidise to H_2SO_4 in the next dry season, transforming Na_2CO_3 again to Na_2SO_4: $Na_2CO_3 + H_2SO_4 \rightarrow Na_2SO_4 + H_2O + CO_2$.

Question 9.14
The situation in the lower right-hand corner of Figure 8.1 is unlikely, because high pH values are usually caused by the presence of $NaHCO_3$, which would automatically result in a high percentage of exchangeable Na^+.

Question 9.15
The tetrahedral sheet comprises the silica atoms *and* the surrounding oxygens. The octahedral sheet consists of the magnesium atoms *and* the surrounding oxygens and hydroxyls. In expanded drawings, the whole tetrahedral sheet should flip over every 4 (palygorskite) or 6 (sepiolite) Si atoms, so that a network of tubes is formed.

Problem 9.1
Evaporation of $5mm.day^{-1}$ on a surface of 1 dm^3, during 100 years, gives a total amount of evaporation of: $0.05 * 365 * 100\ L = 1825\ L$
This contains $1825 * 100\ mg\ CaCO_3 = 182.5\ g\ CaCO_3$.
The mass of a column of 50 cm depth and 1 dm^2 surface equals $5*1.4\ kg = 7.0\ kg$.
So the total addition of $CaCO_3$ during 100 years is $0.1825/7.0 * 100\% = 2.6\ \%$

Problem 9.2
To solve the problem, we can make use of a decision tree as in Fig 9.7. The least soluble salt crystallises first. Note that concentrations are in charge equivalents, so $CaCO_3$ is

formed with 1 Ca^{2+} and 1 HCO_3^-. After precipitation of calcite (1 mol in 1, 3 mol in 2, 2 mol in 3), the water compositions are:

water nr.	equivalent concentrations in mol_c / m^3				
	Na^+	Ca^{2+}	HCO_3^-	Cl^-	SO_4^{2-}
1	5	0	2	1	2
2	2	1	0	1	2
3	3	1	0	0	4

After precipitation of remaining Ca^{2+} as gypsum (0, 1, and 1 mol, respectively), the solutions look as follows:

water nr.	equivalent concentrations in mol_c / m^3				
	Na^+	Ca^{2+}	HCO_3^-	Cl^-	SO_4^{2-}
1	5	0	2	1	2
2	2	0	0	1	1
3	3	0	0	0	3

The next precipitate is $NaHCO_3$ (2,0,0 mol), after which the solutions are:

water nr.	equivalent concentrations in mol_c / m^3				
	Na^+	Ca^{2+}	HCO_3^-	Cl^-	SO_4^{2-}
1	3	0	0	1	2
2	2	0	0	1	1
3	3	0	0	0	3

The next component is $Na_2SO_4.10H_20$ (2,1,3 mol), and the remaining solutions are:

water nr.	equivalent concentrations in mol_c / m^3				
	Na^+	Ca^{2+}	HCO_3^-	Cl^-	SO_4^{2-}
1	1	0	0	1	0
2	1	0	0	1	0
3	0	0	0	0	0

And finally, NaCl crystallises from solutions 1 and 2. Only water No. 1 can lead to high concentrations of dissolved $NaHCO_3$, and therefore to high pH.

Problem 9.3

Salinity is closely related to high ground water levels.

Problem 9.4

a) The plots are symmetrical, because cationic and anionic charges must balance, and concentrations of OH^- and H^+ are negligible.

b) At site 6, which is beyond reach of the water table, there is some downward transport of soluble salt, while calcite is predominantly concentrated at the surface. Chloride accumulation occurs between 50 and 60 cm depth, gypsum slightly higher (30-40 cm). At site 7, there is a combination of evaporation from ground water and slight downward transport with precipitation. Chlorides are accumulated below 10 cm. The soil is strongly dominated by NaCl and Na_2SO_4, and there appears to be some excess alkalinity.

At site 8, evaporation from ground water dominates, and accumulation is strongest at the soil surface. This soil does not have excess alkalinity.

In general, salt accumulation increases towards the depression.

Problem 9.5

a) The increase in alkalinity temporarily slackens at concentration factors between 10 and 100, which means that a compound containing HCO_3- is precipitated. Because both Ca^{2+} and Mg^{2+} concentrations start to decline from this point, precipitation of Mg-and Ca-carbonate is likely. After precipitation of these carbonates, the alkalinity continues to increase because the remaining bicarbonates are much more soluble.

Silica concentration first increases with the concentration factor, but the increase is less steep above Log[concentration factor] = 0.5. Because none of the other components is affected at this point, precipitation of silica (opal) is possible (Al-Droubi suggests the presence of a biological buffering mechanism). Concentration of silica increases further, 1) because opal precipitation is a relatively slow process, which involves strongly hydrated gel-like phases and 2) because of the dissociation of silicic acid. Decrease in the Si concentration above concentration factor 3 marks the beginning of Mg-smectite precipitation. At this moment, the pH is strongly alkaline.

b) Irrigation with Chari River water causes sodication through the presence of Na^+ plus HCO_3^- in excess over $Ca^{2+} + Mg^{2+}$, leading to high-pH sodic soils.

Problem 9.6

$Na_2SO_4 + 2CH_2O \rightarrow H_2S + 2Na^+ + 2HCO_3^-$

Problem 9.7

a) In both cases, we observe an increase in organic carbon throughout the profile. The EC of the topsoil has decreased, but there is an increase in the subsoil. This points to downward transport, but not removal, of soluble salts. The drop in pH may be due to increased P_{CO2}, caused by root respiration and decomposition of plant litter. This will increase the solubility of $CaCO_3$, (which is very low at high pH) and decrease the exchangeable Na percentage. Sodium saturation may have increased in the horizon of

highest conductivity.
b) Because there is no untreated control plot (showing profiles of pH, EC, and organic C in 1979, 1982, and 1984 in the absence of trees), there is no way to prove that the soil changes were induced by the trees.

9.7. References

Al-Droubi, A., 1976. *Géochimie des sels et des solutions concentrés par évaporation. Modèle thermodynamique de simulation. Applications aux sols salés du Tchad.* Mém. No. 46, Univ. Louis Pasteur de Strasbourg, Inst. de Géologie.

Bocquier, G., 1973. *Genèse et évolution de deux toposéquences de sols tropicaux du Chad.* Mém. ORSTOM. No. 62, 350 pp. ORSTOM, Paris.

Driessen, P.M., 1970. *Soil salinity and alkalinity in the Great Konya Basin, Turkey.* PhD Thesis, Wageningen, 99 pp.

Freytet, P., and J.C. Plaziat, 1982. *Continental carbonate sedimentation and pedogenesis.* Contr. Sediment., 12. Schweitzerbart'sche Verlag, Stuttgart, 213. pp.

Garrels, R.M., and C.L. Christ, 1965. *Solutions, minerals and equilibria.* Harper and Row, New York, 450 pp.

Gill, H.S., and I.P. Abrol, 1986. Salt affected soils and their amelioration through afforestation. In: R.T. Prinsley and M.J. Swift (eds): *Amelioration of soil by trees, a review of current concepts and practices.* Commonwealth Scientific Council, London, pp 43-53.

Jenny, H., 1941. *Factors of soil formation.* McGraw-Hill, New York, 281 pp.

Millot, G., 1970. *Geology of clays - weathering, sedimentology, geochemistry.* Springer, New York, 429 pp.

Monger, H.C., L.A. Daugherty, and L.H. Gile, 1991. A microscopic examination of pedogenic calcite in an Aridisol of southern New Mexico. In: W.D. Nettleton (ed.): *Occurrence, characteristics, and genesis of carbonate, gypsum, and silica accumulations in soils.* SSSA Special Publication No. 26, pp. 37-60. Soil Science Society of America, Madison.

Pipujol, M.D., and P. Buurman, 1997. Dynamics of iron and calcium carbonate redistribution and water regime in Middle Eocene alluvial paleosols of the SE Ebro Basin margin (Catalonia, NE Spain). Palaeogeography, Palaeoclimatology, Palaeoecology, 134:87-107.

USSLS (United States Salinity Laboratory Staff), 1954. *Diagnosis and improvement of saline and alkali soils.* USDA Agriculture Handbook No., 60, 160 pp.

Van Breemen, N., 1987. Effects of redox processes on soil acidity. Netherlands Journal of Agricultural Science, 35:275-279.Vergouwen, L., 1981. *Salt minerals and water from soils in Konya and Kenya.* PhD. Thesis, Wageningen, 140 pp.

Vergouwen, L., 1981. *Salt minerals and water from soils in Konya and Kenya.* PhD Thesis, Wageningen, 140pp.

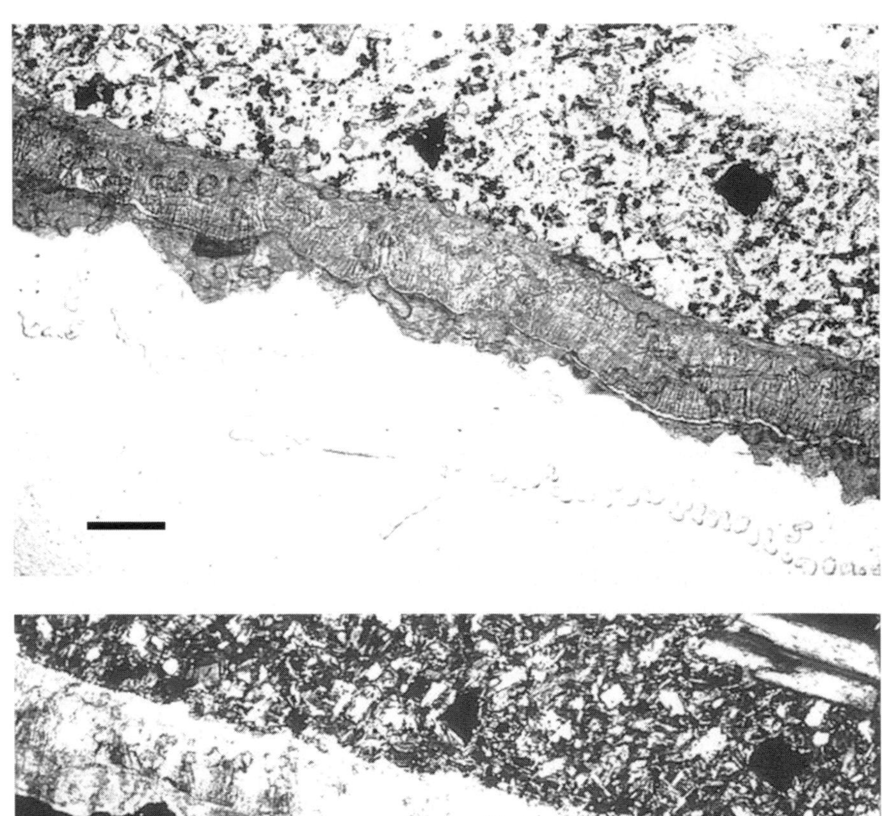

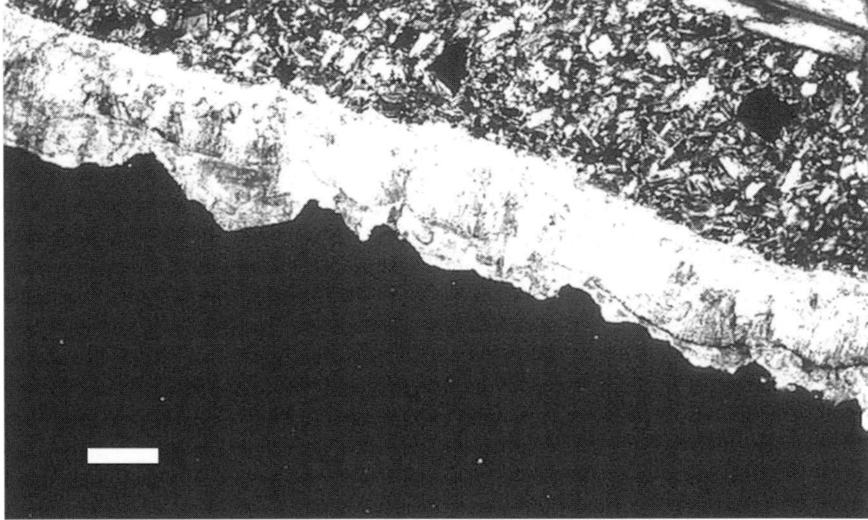

Plate R. Calcite pendant at the bottom of a basaltic pebble. Top: plain light; bottom: crossed polarisers. Note strong birefringence and perpendicular growth of the calcite crystals at the pebble surface. Origin:France. Scale bar is 215 μm. Photographs A.G. Jongmans

CHAPTER 10

FORMATION OF VERTISOLS

10.1 Introduction

Soils dominated by smectitic clay minerals that are found in a strongly seasonal climate usually exhibit 'vertic properties'. Vertic properties are a combination of (1) deep cracks when dry (Fig. 10.1); (2) intersecting slickensides in the subsoil (polished and grooved shiny surfaces, produced by one mass of soil sliding past another); (3) wedge-shaped structural aggregates in the subsurface soil (25 to 100 cm deep); and (4) a strong, nutty structure at the soil surface. Often, these soils show a so-called *gilgai* microrelief: rounded mounds and depressions or series of ridges and inter-ridge depressions, with distances between highs and lows of 2 to 8 m and vertical differences of 15 to 50 cm. The combination of these properties characterises Vertisols. Two sets of processes related to the genesis of Vertisols will be discussed: the formation of smectitic clays, and the development of the typical physical characteristics.

10.2. Formation of smectite clays

Smectites are 2:1 clay minerals with relatively little isomorphous substitution, resulting in a low layer charge (0.2 to 0.6 per unit cell; see Chapter 3.2.). Because the clay plates are only loosely bound together, water can enter the interlayer space, and cause an increase in volume of more than 100% compared to the dry clay (see Chapter 2).

Question 10.1. *How does the composition of the exchange complex influence the swelling capacity? (See Chapter 2).*

Smectite clays are formed at relatively high concentrations of constituent cations (Mg in case of montmorillonite) and silica in the soil solution (Figure 3.1). Conditions favourable for the formation of smectites occur when base-rich aluminium silicate minerals weather under moderate leaching. Mafic and intermediary rocks in semi-arid and mediterranean climates usually produce smectites upon weathering. Smectites also form where groundwater containing Mg^{2+} and H_4SiO_4 drains from higher areas and collects in lower areas where it becomes more concentrated by evapo-transpiration. This is common in tropical monsoon regions, where strong weathering in higher parts of a catena is accompanied by smectite formation in the lower parts. Many alluvial deposits of tropical lowlands are rich in smectite as well, because rivers erode upstream areas with smectite formation.

Figure 10. Cracks in a Vertisol from Trinidad. From Ahmad, 1983; reproduced with permission of Elsevier Science, Amsterdam.

Question 10.2. *What factors could explain why Vertisols are more common in seasonal tropical than in seasonal temperate climates?*

Question 10.3. *Explain the common presence of calcium carbonate and relatively high levels of exchangeable Na in most Vertisols (consider factors that Vertisols and soils with secondary calcite have in common)?*

10.3. Physical deformation

Alternation of dry and wet seasons and corresponding desiccation and rewetting of the soil, causes strong changes in volume of smectite clay with its strong potential to swell and shrink (see Chapter 2). Such changes cause strong pressures in the soil. The release of pressure leads to the characteristic structures of Vertisols. The development of these structures as envisaged by Wilding and Tessier (1988) is illustrated in Figure 10.2.

STRUCTURE FORMATION

When a wet, smectitic soil dries out, the clay minerals lose interlayer water, which results in strong shrinkage. The shrinkage is translated into a subsidence of the soil

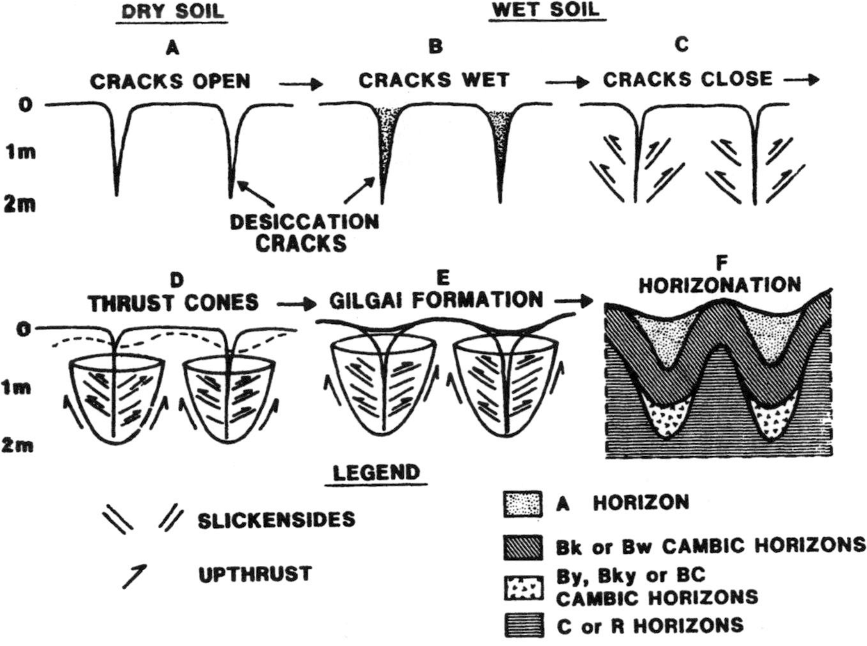

Figure 10.2. Stages in the formation of Vertisol structure. From Wilding and Tessier, 1988.

surface (which is not readily observed) and the formation of cracks (Figure 10.2,A). Depending on the depth of the solum, the content of smectite clays and the severity of the dry season, cracks may be between 0.5 and 2 m deep. Strong fragmentation caused by cracking in the top layer leads to the development of fine, nutty structural elements. This layer may be up to 30 cm thick. Part of the fine fragments falls into the cracks, partially filling the latter (Figure 10.2,B).

Upon rewetting, the soil swells again. This results in both vertical and lateral pressures that counter the former subsidence and cracking. Because the cracks have been partially filled with topsoil material, there is now an excess soil volume in the subsoil. The combination of vertical and lateral pressure causes a failure plane at about 45°. The soil material slides along these failure planes, causing grooved surfaces: slickensides (Figure 10.2,C; Plate S, p.256).

Repeated rearrangement of soil particles along a failure plane, during each stage of contraction and expansion, makes such a plane more or less permanent. The zone close to the vertical cracks becomes a topographic low, while the inter-crack area becomes a high: formation of *gilgai* (Figure 10.2,DE). In the long run, strongly different soil profiles develop in highs and lows, as is shown in Figures 10.3 and 10.4. Figure 10.5 gives an aerial view of the gilgai relief.

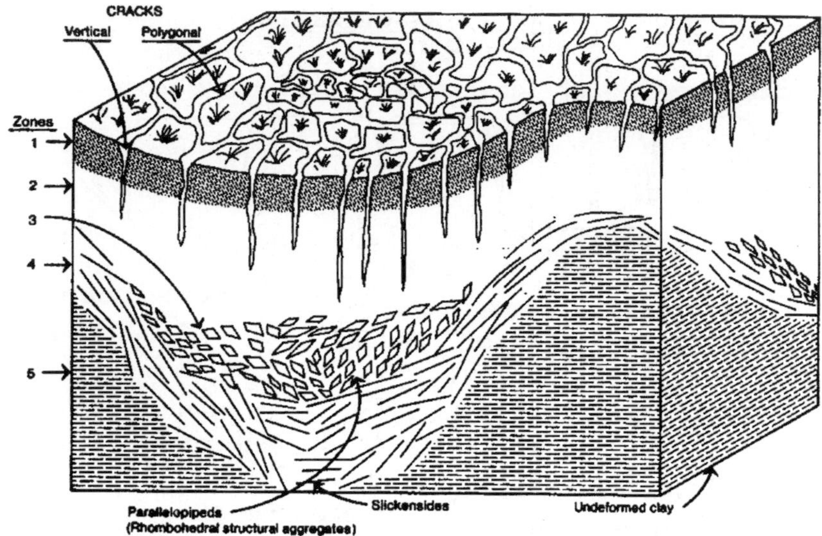

Figure10.3. Block diagram of a gilgai microrelief. 1. nutty structure; 2. vertical cracks and prisms; 3. rhombohedral elements; 4. slickensides; 5. stable subsoil.
From Dudal and Eswaran, 1988.

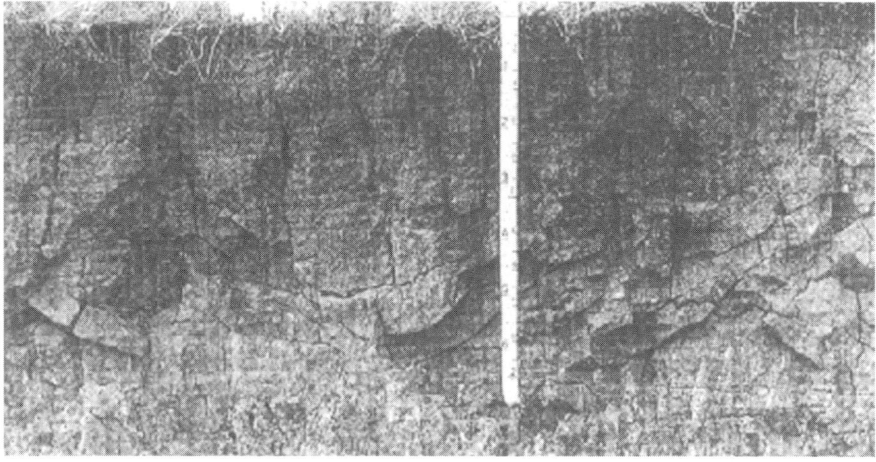

Figure 10.4. Vertisol profile (Typic Pelludert) from Texas, USA. Tape measure is 2.1m long. Topsoil is black, subsoil reddish brown. From Wilding and Puentes, 1988.

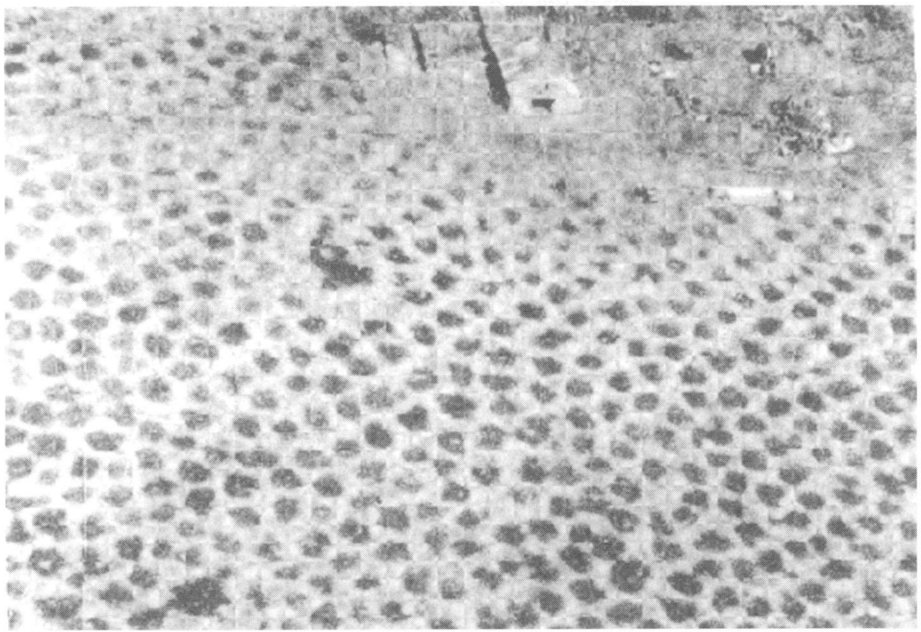

Figure 10.5. Aerial photograph of gilgai microtopography. Dark patches are heavily vegetated lows. From Wilding and Puentes, 1988.

The intensity of swelling of soils is expressed in the Coefficient of Linear Extensibility, or COLE (see Glossary). The COLE of a soil sample depends on the clay content, the clay mineral, and the saturating cations. Of the smectites, montmorillonite has the strongest swelling capacity. Most of the swelling and shrinkage occurs at relatively high matric potentials (between 33 and 2000 kPa; see Figure 2H), so directly upon moistening of a dry soil.

Question 10.4. *Explain why Vertisols do not form in perhumid tropical or humid temperate climates. Why is surface mulch (nutty elements) less pronounced in a udic than in an ustic climate?*

Question 10.5. *Sketch in Figure 10.4: a. The land surface, emphasising the location of highs and lows; b. the approximate upper limit of the stable subsoil, and c. The position and direction of slickensides.*

The interaction between vertical and lateral forces causes an optimum depth for the formation of slickensides (Figure 10.6). Below this optimum depth, seasonal moisture differences are too small to cause sufficient pressure, while above the optimum depth the porosity is higher, resulting in a higher elasticity. In the zone above maximum slickenside development, oriented rhombohedral elements are formed by cracking

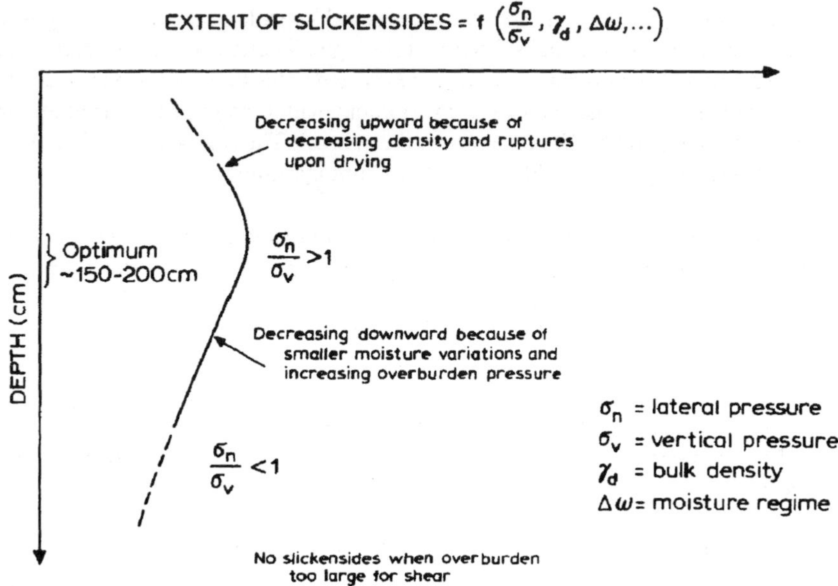

Figure 10.6. Depth of maximum slickenside development as a function of lateral and vertical pressure From Ahmad, 1983; reproduced with permission of Elsevier Science, Amsterdam.

across large structural elements (Figure 10.3,3). Normal gilgai with round depressions, as depicted in Figure 10.5, is typical of flat, homogeneous terrain. In gently sloping terrain, gilgai usually consists of ridges and inter-ridge depressions, with a length axis perpendicular to the slope. Irregularities in the subsoil, e.g. due to hard bedrock at variable depth, cause irregular gilgai patterns.

PROFILE DYNAMICS

Many Vertisols have deep, dark topsoils. This dark colour reflects the presence of organic carbon, but does not indicate high levels of organic matter. In fact, the organic C levels are generally rather low, in the order of 1 % (mass fraction). The dark colour is traditionally explained by a highly stable association between smectite and part of the organic carbon, but the presence of black soot (char) particles from regular burning of the vegetation is probably also a factor..

Before the monograph by Wilding and Puentes (1988), the deep dark coloured A horizons were seen as signs of a strong homogenisation by shrinkage and swelling, causing a mass movement towards gilgai mounds, followed by erosion into cracks in the depressions. Vertisols (from Latin *vertere* = to turn) were in fact named after the presumed strong pedoturbation. If this process of pedoturbation were rapid, with a turnover time of, e.g., centuries, one would expect Vertisols to be very homogeneous with depth between lows and highs, both morphologically and chemically. Radiometric age determination of the organic matter, however, indicates that the turning over of the soil is relatively slow. Measured ages of the organic matter may increase from less than 1000 years in the topsoil to 10 000 years in the subsoil. The formation of slickensides and gilgai is probably more rapid than vertical homogenisation.

Question 10.6. *How would addition of a small amount of young organic matter into the subsoil by pedoturbation affect the ^{14}C age of old organic matter in the subsoil (see Chapter 6)? What does this imply about the rate of vertical mixing in a Vertisol that has subsoil organic matter with a MRT of 10.000 years?*

As the climate becomes moister, contents of calcium carbonate and exchangeable Na decrease, while organic matter contents increase. In a given climate, Vertisols of flat topographies are relatively grey, while Vertisols of undulating terrain have browner and redder colours.

Question 10.7. *How can you explain these differences in colour?*

10.4. Problems

Problem 10.1

Figure10A shows depth functions for organic C, $CaCO_3$, exchangeable Na and conductivity of Vertisols (Udic Pellustert, Texas, USA) at high and low spots in the gilgai relief. Without considering possible causes of the differences between high and low spots, what do these differences imply about the hypothesis that Vertisols are strongly homogenised over an appreciable depth?

Problem 10.2

Figure 10B presents soil water contents with depth at various times in gilgaied Vertisols in a mound position and in an adjacent depression, as illustrated in the block diagram of Figure 10.3. The first curve (08/09) is just after heavy rain following a dry season. The next three curves illustrate increasing desiccation, while the last three are from the next rainy period and show rewetting. The curve for 27/10 occurs in both the right hand and left hand diagrams.

a) Which set of curves (ab or cd) belongs to the mound position, and which to the depression?
b) What are the main differences in wetting pattern between mound and depression? Refer to Figure 10.3 in your answer. Consider the speed and depth of penetration of the water.
c) Why would cracking be stronger in the depression, and how would the differences in hydrology contribute to the formation of slickensides and the gilgai relief?

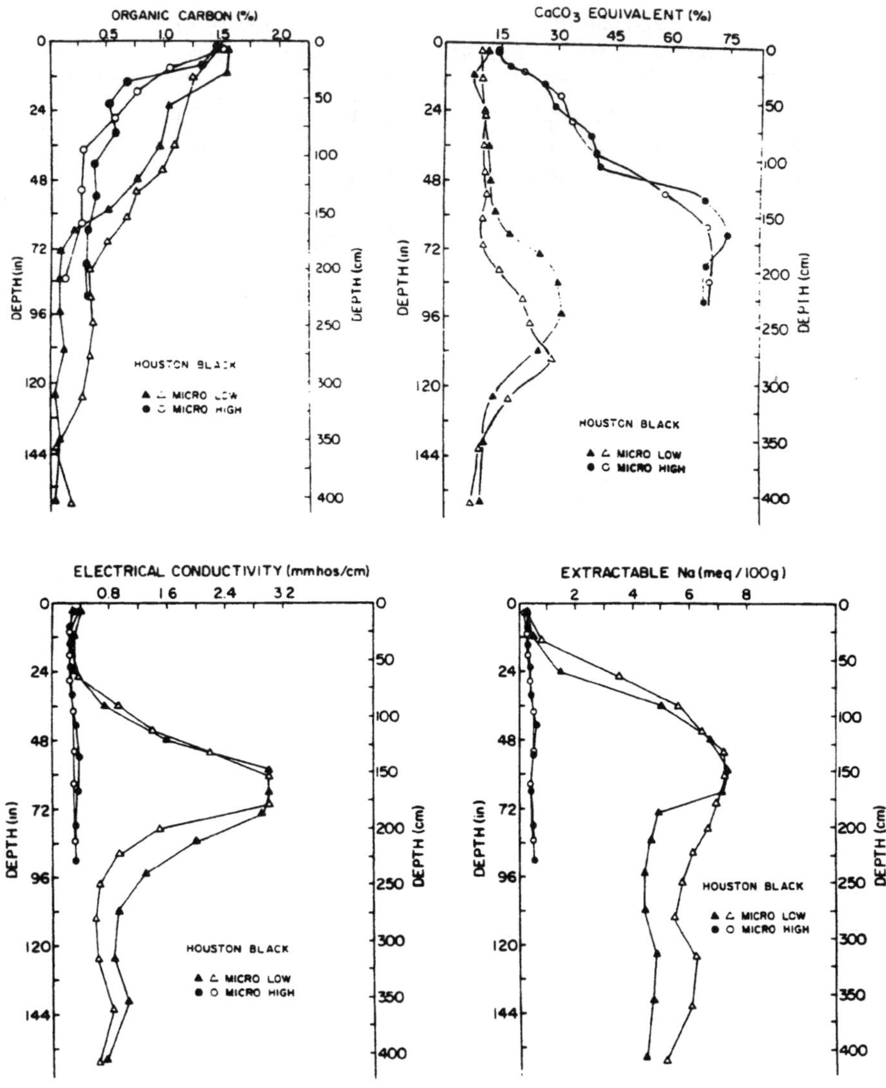

Figure 10A. Chemical characteristics in low and high spots of a Vertisol from Texas. From Wilding and Tessier, 1988.

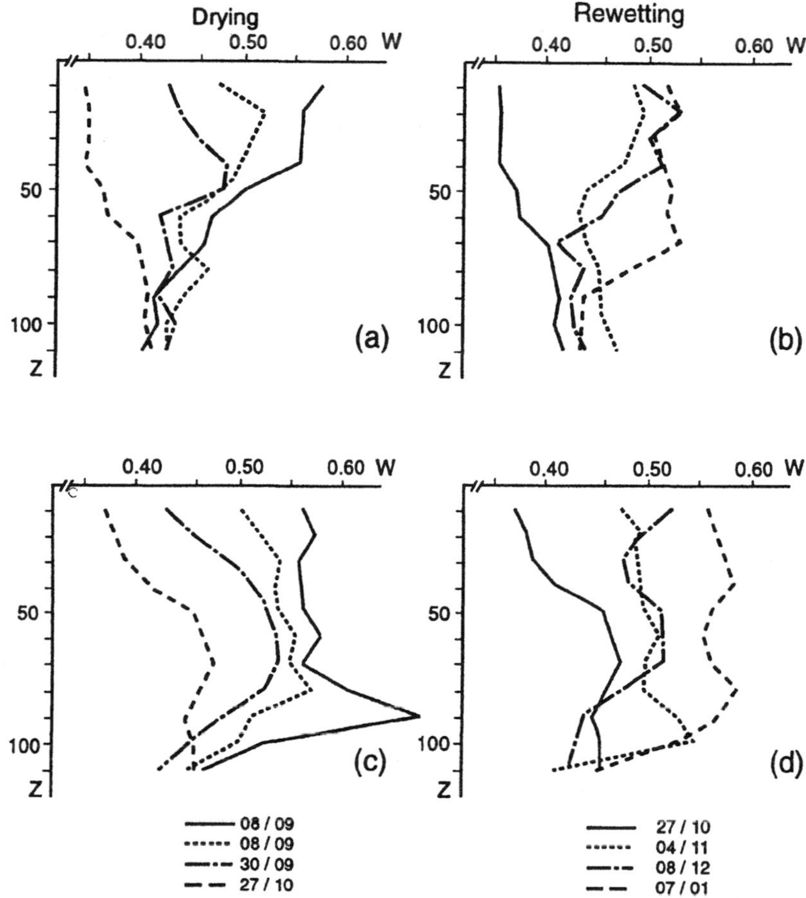

Figure 10B. Soil water content in a mound and in a depression position in a gilgaied landscape. Explanations see text. From Jaillard and Cabidoche, 1984.

10.5. Answers

Question 10.1

Sodium clays have a larger amplitude in swelling and shrinking than Ca-clays; especially the rehydration is more reversible (Figure 2J).

Question 10.2

1) The formation of smectite clays requires considerable weathering (either in the soil profile itself, or in an adjacent part of the same landscape), and/or time. Liberation of

cations and silica is slower in temperate than in tropical climates. 2) Young soils abound in many temperate areas (glaciations!). 3) Concentration of solutes by evaporation is required to cause precipitation of smectite from the soil solution. Alternating wet and dry seasons are a characteristic of the monsoonal tropics.

Question 10.3

Both Vertisols and calcite accumulation are common in seasonally dry climates. Evaporation of (pore) water leads to a strong increase in Na concentration, while Ca concentration is kept low by precipitation of $CaCO_3$. As a result, Na saturation of the adsorption complex increases.

Question 10.4.

Vertisols do not form in perhumid climates because the seasonal desiccation that stimulates formation of smectite clay, and that is needed for swelling and shrinking, does not take place. Surface mulch consists of small dense aggregates. These are formed through fragmentation of larger aggregates. In a udic moisture regime, organic matter contents are higher, and therefore aggregates become more porous. In addition, desiccation, and therefore fragmentation, is weaker.

Question 10.5

See indications in figure below.

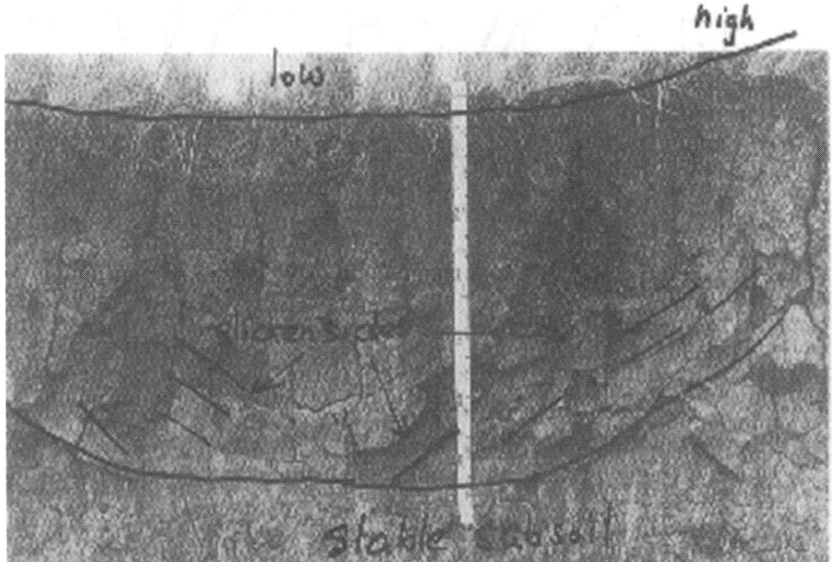

Question 10.6.

Admixtures of small amounts of young carbon with high ^{14}C activity would considerably decrease the mean age ^{14}C age of subsoil organic matter. A high ^{14}C age of organic matter in the subsoil implies therefore that there is little vertical mixing.

Question 10.7

Redder or browner colours are caused by iron oxides (goethite and hematite), which form more readily in the better drained soils of undulating landscapes than in the poorly drained soils of flat lands or depressions, where part of the iron released during weathering of Fe-silicates is removed as soluble Fe^{2+}.

Problem 10.1

The rate of physical homogenisation by vertical mixing due to shrink and swell processes is clearly slower than the vertical differentiation of the properties shown in Figure 10A. Because the variations in organic C with depth as well as the steep profiles of $CaCO_3$ and differences in $CaCO_3$ contents between micro highs and micro lows must develop slowly (centuries), vertical mixing cannot be a very fast process.

Problem 10.2

a. After a rainstorm on dry soil, we expect a stronger wetting of the deeper soil in the depressions, than in the mounds. The curve of 08/09 shows a very strong peak at 90 cm in graph (c). This suggests that graphs (c) and (d) belong to the depression.

b. Desiccation (a and c) starts in the topsoil. In the depression, water below 50 cm depth is preserved longer than in the mound, and there appears to be some upward movement of water. On 27/10, water contents in the depression are still higher than in the mound. This may be due to a higher clay content and higher initial water content in the depression. Upon rewetting, water contents in the subsoil of the depression increase more rapidly than in the mound and also final water contents are higher in the depression.

c. In the subsoil of the depression, the differences in water content between dry and wet season are larger than in the mound. Consequently, shrinking and swelling is stronger in the depression. The fact that depressions gather more water increases the difference in moisture between mound and depression.

10.6. References

Ahmad, N., 1983. Vertisols. In: L.P. Wilding, N.E. Smeck, and G.F. Hall (eds). *Pedogenesis and Soil Taxonomy. II. The Soil Orders*. pp. 91-123. Developments in Soil Science 11B. Elsevier, Amsterdam.

Bocquier, G., 1973. *Genèse et évolution de deux toposéquences de sols tropicaux du Chad.* Mém. ORSTOM no 62, 350 p., ORSTOM, Paris.

Dudal, R. and H. Eswaran, 1988 Distribution, properties and classification of vertisols. In: L.P.Wilding and R.Puentes (eds). Vertisols: their distribution, properties, classification and management. Pp.1-22. Technical Monograph no 18, Texas A&M University Printing Center, College Station TX, USA

Jaillard, B., and Y.-M. Cabidoche, 1984. *Etude de la dynamique de l'eau dans un sol argileux gonflant: dynamique hydrique*. Science du Sol, 1984: 239-251.

Wilding, L.P., and R. Puentes (eds), 1988. *Vertisols: their distribution, properties, classification and management*. Technical Monograph no 18, Texas A&M University Printing Center, College Station TX, USA.

Wilding, L.P. and D. Tessier, 1988. Genesis of vertisols: shrink-swell phenomena. In: L.P. Wilding and R. Puentes (eds): *Vertisols: their distribution, properties, classification and management*. Pp.55-81. Technical Monograph no 18, Texas A&M University Printing Center, College Station TX, USA.

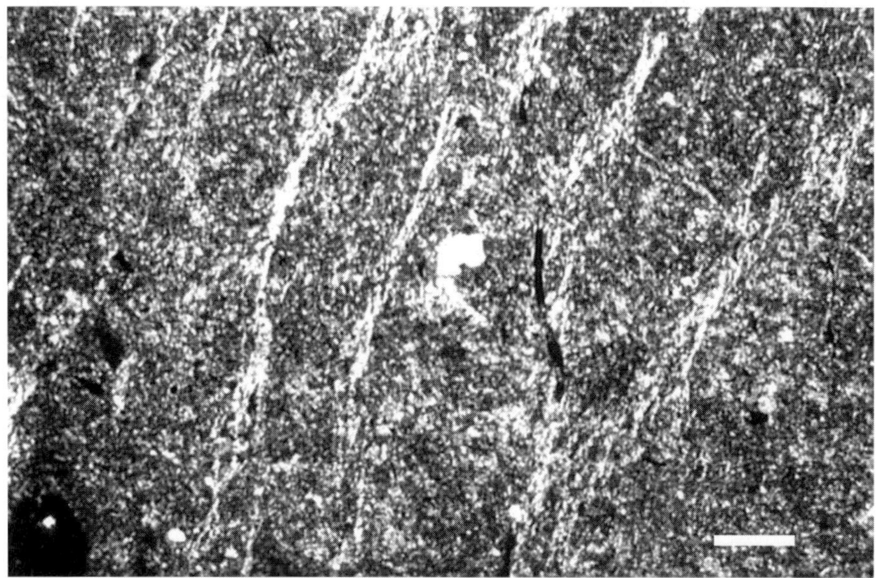

Plate S. Pressure faces (striasepic fabric) in a Vertisol. Crossed polarisers. Pressure faces are visible because clay has roughly parallel orientation. Note the difference with clay illuviation cutans (Figures 8.2-8.4). Scale bar is 345μm. Photograph A.G. Jongmans.

CHAPTER 11

PODZOLISATION

11.1. Introduction

The striking and dramatic soil profiles of podzols have intrigued generations of soil scientists. This is reflected by a voluminous literature on the subject.
From top to bottom, a podzol profile consists of up to five major horizons, two of which may be absent (in italics):

- a litter layer (O);
- *a humose mineral topsoil (Ah), caused by biological mixing;*
- a bleached, eluvial layer (E);
- an accumulation layer of organic matter in combination with iron and aluminium (Bh, Bhs, Bs);
- *thin bands with organic matter accumulation in the subsoil.*

In thin section, podzol O- and A-horizons show the normal transformation of plant litter into humic substances (see Chapter 4). In addition, organic remnants develop diffuse boundaries upon decomposition, which indicates the transformation of solid to soluble organic substances.
E horizons usually have thin and scattered remnants of coating-like material. Organic matter in B-horizons is present either in polymorphic or in monomorphic form. Polymorphic organic matter (Figures 6.2; 11.8) consists of mesofauna excrements in various sizes, and is frequently linked to decomposing root remnants. Monomorphic organic matter (Figure 11.7) consists of amorphous coatings around sand grains and in pores. Polymorphic organic matter may grade into monomorphic, but not vice-versa. Organic matter in the thin bands below the B-horizon usually consists of monomorphic organic matter. Polymorphic organic matter in B-horizons is associated with de decomposition of root material.

Podzols occur as zonal soils associated with a boreal climate, mainly on glacial, mixed parent materials, including those with relatively high amounts of weatherable minerals. Outside the boreal zone, podzols occur on poor, siliceous parent materials, both in temperate zones (e.g. on periglacial sand deposits) and in the tropics (e.g. on uplifted coastal sands).

THREE THEORIES
Podzols form through transport in solution, from the surface to deeper horizons, of

organic matter, iron and aluminium. The process consists of a phase of mobilisation and one of immobilisation of these compounds. There are various theories concerning the mechanism of podzolisation. The main three are:
a) *The fulvate theory*
b) *The allophane theory*
c) *The low molecular weight acids theory*
The three theories ignore the influence of root-derived organic matter accumulation in the B-horizon. In the following, we will integrate this aspect and show that it is dominant in some podzols.

The *fulvate theory* (e.g., Petersen, 1976) postulates that unsaturated fulvic acids in the topsoil dissolve Fe and Al from primary and secondary minerals, and that the dissolved metal-organic matter complexes precipitate upon saturation (charge compensation) of the organic ligand. The precipitate has a specific C/metal ratio (Mokma and Buurman, 1982; Buurman, 1985,1987).

The *allophane theory* (Anderson et al., 1982; Farmer et al., 1980) postulates that Fe and Al are transported to the B horizon as positively charged silicate sols, where they precipitate as amorphous allophane and imogolite (see Chapter 12) through an increase in pH. After this, organic matter would precipitate on the allophane and thus cause a (secondary) enrichment in the B-horizon.

The *low molecular weight (LMW) organic acid theory* postulates that LMW acids are responsible for the transport of iron and aluminium to the subsoil, and that precipitation of Fe and Al is caused by microbial breakdown of the carrier (Lundström et al., 1995).

Buurman and Van Reeuwijk (1984) have made clear that amorphous Si-Fe-Al sols cannot be stable in an environment with complexing organic acids. Nowadays, podzolisation is mostly regarded as the result of a combination of process a) and c), whereas the formation of allophane in process b) may precede or accompany podzolisation but is not essential to it. This view will be detailed in the following.

11.2. Conditions favouring podzolisation

Conditional to the formation of podzols is the production of soluble organic acids, in the absence of sufficient neutralising divalent cations, such as Ca^{2+} and Mg^{2+}. Conditions in which production of soluble organic acids is high are usually related to impeded decomposition of plant litter and increased exudation of organic acids by plants and fungi, resulting from one or more of the following factors:

- Low nutrient status of the soil and unpalatable litter.
  Sandy parent materials with low amounts of weatherable minerals have insufficient buffer capacity to neutralise organic acids that are produced during litter decomposition. In addition, low availability of nutrients such as N and P (1) strongly

decreases litter decomposition because nutrient-stressed plants produce poorly decomposable substances such as tannins and other phenolic products, and because decomposers (microbes) are nutrient-limited too; (2) favours conifers and ericaceous plants with litter high in tannins and phenolic substances, and (3)causes plants to allocate more carbon to mycorrhizal fungi, which exude organic acids.

- Low temperatures and high rainfall during the growing season.
  The period of litter decay is too short and biological activity is too low for efficient mixing of organic and mineral matter. High rainfall favours weathering, loss of nutrients, and transport of Al, Fe, and organic solutes.

- Poor drainage.
  Wet conditions retard mineralization of plant litter to CO_2 and so favour the production of water-soluble, low-molecular-weight organic substances.

- Impeded biological mixing of organic and mineral material.
  Both low nutrient status and defective drainage decrease biological activity in the soil because the animals need both food and oxygen to function. Nutrient-poor soils and soils of cold climates tend to lack burrowing soil fauna. Low biological activity impedes vertical mixing of the soil and favours vertical differentiation. Introduction of earthworms to podzols lacking burrowing animals, and increase in nutrient status have in some cases led to the disappearance of podzol morphology by vertical homogenisation (Alban and Berry, 1994; Barrett and Schaetzl, 1998).

Question 11.1. *In which of the ecosystems of Table 4.1 do you expect conditions favourable for podzolisation? What other information, which is not given in Table 4.1, do you need to better predict the occurrence of podzolisation?*

11.3. Transport and precipitation of organic and mineral compounds

MOBILIZATION

Breakdown of litter and exudation by plants and fungi under the conditions sketched above, results in a large variety of organic compounds. Part of the decomposition products is soluble - or suspendible - in water and can be transported with percolating rainwater to deeper parts of the soil. The remaining organic matter at the soil surface forms the humified part of the litter layer (H layer). It can be mixed with the mineral soil to form an Ah horizon, and is further broken down by microflora, also supplying soluble organic substances.

'Soluble' organics include high-molecular weight (HMW) substances with various carboxylic and phenolic groups, low-molecular weight (LMW) organic acids, such as simple phenolic and aliphatic acids, and simple sugars. Formulas of some acids are given in Chapter 4. 'Fulvic acids' and LMW organic acids such as citric and oxalic acid are particularly effective in complexing trivalent ions, and therefore in extracting such ions from crystal lattices, e.g. from clay minerals, pyroxenes, amphiboles, and feldspars.

This means that percolating organic acids cause strong weathering. Low-molecular-weight organic acid - metal complexes remain soluble, while larger organic acids, e.g. of the 'fulvic acid' fraction, may become insoluble when their charge is neutralized by metal ions.
The discovery that weatherable minerals in podzol E horizons are perforated by numerous micropores presumably from ectomycorrhyzal fungi, via exudation of LMW acids at their hyphal tops (Jongmans et al., 1997), emphasises the importance of LMW acids in podzolisation. It indicates that ectomycorrhyzal plants, which are typical of most boreal podzols (in Europe: *Pinus silvestris* and *Picea abies)* have a direct impact on weathering and podzolisation (Van Breemen et al., 2000). The more nutrient-poor the soils, the more carbon the trees will provide to the fungal symbiont, hence the greater the production of LMW acids.

Question 11.2. *What would be the causes of differences between the Podzol and the Cambisol in terms of amounts of organic acids presented in Figure 11.1? Consider effects of litter layer, nutrient status, pH, temperature.*

IMMOBILIZATION OF ORGANIC MATTER AND SESQUIOXIDES
Several processes, a combination of which will play a role in all soils, may halt the movement of dissolved organic matter and any aluminium and iron associated with it.

- The LMW organic carrier is decomposed by micro-organisms.
  LMW acids belong to the compounds that are readily decomposed. As a result, any complexed metal ions are released. The released metals can precipitate as (hydr)oxides or, together with silicic acid, as amorphous silicates (Lundström et al., 1995). Metal ions released from LMW organic complexes may also remain in solution if they are complexed by larger organic molecules.

- The metal-saturated HMW organic complex may precipitate.
  Precipitation of metal-saturated complexes of larger organic molecules has been demonstrated in the laboratory (Petersen, 1976) and saturation is probably a major cause of immobilisation of organic matter and attached metals. Dissolved HMW organic acids may become saturated through dissolution of primary or secondary minerals during transport, especially when they hit a layer with relatively large amounts of (oxy)hydrates or amorphous silicates of iron or aluminium. Such layers may have formed prior to podzolisation and include iron-stained B-horizons of brown soils and allophanic horizons (for allophane, see Chapter 12), etc. Part of the organic matter does not form complexes and may percolate further through the soil. Desaturation of complexes by addition of unsaturated organic compounds from overlying horizons may cause redissolution of metals. C/metal atomic ratios at which complexes appear to precipitate are around 10-12, depending on pH (Mokma and Buurman, 1982; Buurman, 1985, 1986).

- The water transport may stop.
  Transport of water through the soil depends on amount and intensity of rainfall and

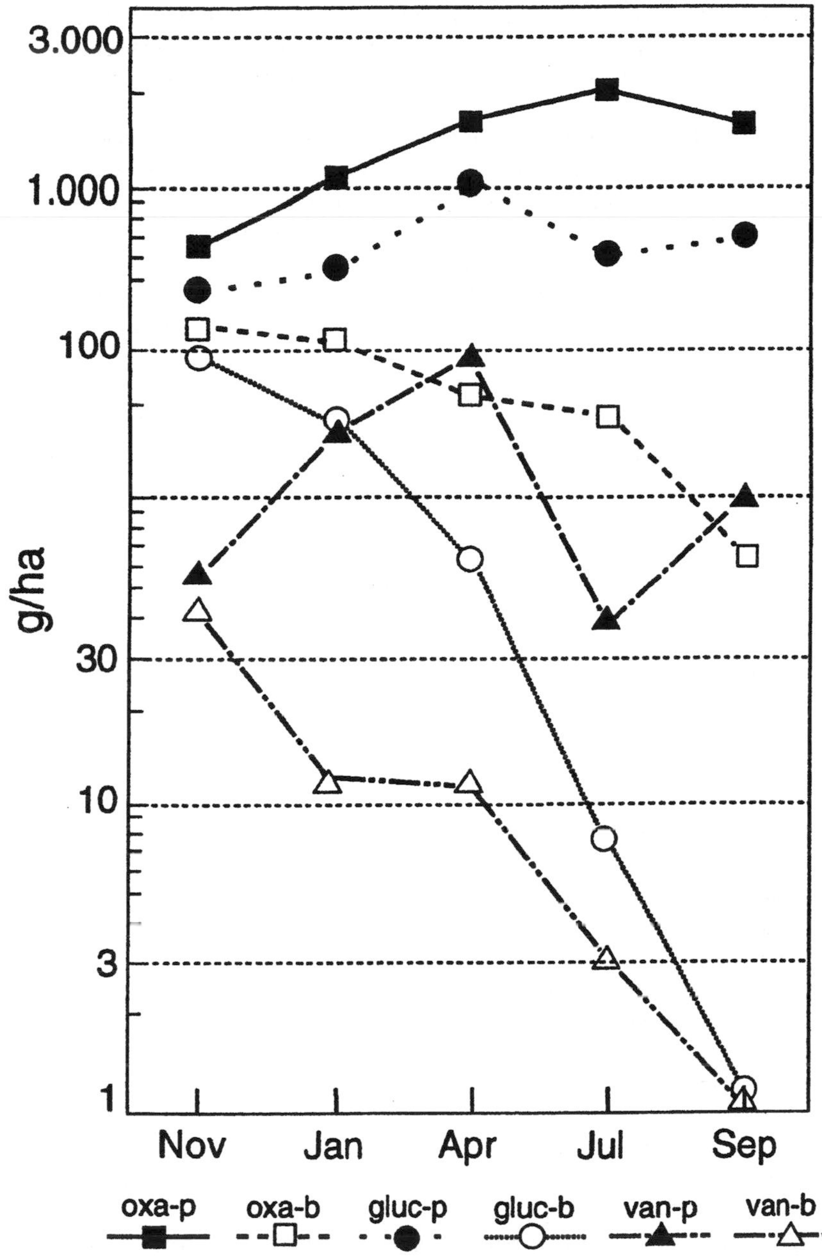

Figure 11.1. Seasonal variation in amount of three low-molecular-weight organic acids in litter of a Podzol (closed symbols) and a Dystric Cambisol (open symbols). Oxa = oxalic acid; gluc = glucuronic acid; van = vanillic acid. Data from Bruckert, 1970.

on evapo-transpiration. A precipitation event on relatively dry soil has a certain penetration depth, beyond which dissolved and suspended material cannot be transported. If the soil dries out again from above, dissolved and suspended material accumulate at the moisture front. Especially non-complexing substances, substances that are not precipitated by complexed metals and suspended organics are immobilised this way. The suspended organics are not easily remobilised because they are predominantly hydrophobic and will be difficult to rewet once precipitated on a grain surface or on a pore wall.

Question 11.3. *Show, from the stoichiometry of the reaction, how consumption of Al-oxalate by aerobic microbes can lead to precipitation of $Al(OH)_3$. Use the 2:1 complex $Al(C_2O_4)_2^-$.*

INFLUENCE OF pH

Complexation of Fe and Al is pH-dependent for two reasons: 1) these metals form various complexes with OH^-, depending on pH (e.g., dominant aluminium species shift from Al^{3+} to $Al(OH)^{2+}$, $Al(OH)_2^+$, $Al(OH)_3^o$, and finally $Al(OH)_4^-$, as pH increases from <3 to >8, and 2) soil organic matter has acidic groups with a range of pK values so that the number of dissociated groups increases with increasing pH. Therefore, complexation of metals by, e.g., fulvic acids, is dependent on pH and preferential binding may change with pH (Schnitzer & Skinner, 1964-1967).

At pH 4.7: $Fe^{3+}>Al^{3+}>Cu^{2+}>Ni^{2+}>Co^{2+}>Pb^{2+}=Ca^{2+}>Mn^{2+}>Mg^{2+}$

At pH 3.0: $Cu^{2+}>Al^{3+}>Fe^{3+}$

The two sequences indicate that at pH 3, aluminium is more strongly bound than iron, while at pH 4.7 the reverse is true. Elements other than Al and Fe play a negligible role in complexation in podzols.

Question 11.4. *If organic matter precipitates upon saturation with metal ions, this means that at a fixed pH, the precipitate will have one typical C/metal ratio (e.g. 10, as stated above). What is the effect of a decrease in pH on the C/metal ratio at which the complex precipitates? Consider the effect of pH on dissociation of organic acids and on hydroxylation of metal ions.*

Immobilisation of soluble organic compounds in a B-horizon leads to monomorphic organic matter. Actually, polymorphic organic matter dominates the B-horizons of most zonal podzols and of many well-drained intrazonal ones. This suggests that decay of roots is important in the accumulation of organic matter in many podzol-B horizons. The following paragraphs describes this effect as part of the podzolisation process, following Buurman and Jongmans (2002).

11.4. The podzolisation process in stages

Podzolisation involves simultaneous production, transport, precipitation, redissolution, and breakdown of substances in different parts of the soil profile. The process itself changes with time, because pH of the soil tends to decrease with continuing depletion of weatherable minerals (acid neutralising capacity, see Chapter 7). This causes a change in microflora and microbial activity, and in the dissociation of organic acids. The changes with time and the resulting profile morphology are best understood when we describe the podzolisation process in various stages. It should be clear, however, that processes acting in early stages may continue throughout the development.

a. Decomposition of plant litter and roots, and exudation by roots and fungi, results in humic substances and LMW organic acids. Both are effective in weathering of primary minerals, and percolating water transports part of the organic substances. Weathering results in liberation of metal ions and silica. Most of the dissolved silica (H_4SiO_4) is removed from the profile by percolating water, together with monovalent and divalent cations.

b. During downward transport, the LMW acids are largely decomposed. Trivalent metal ions bound by these acids can be precipitated as hydroxides or silicates, or transferred to organic larger molecules. This is probably the source of allophane or imogolite in the B-horizon of podzols on rich parent materials in northern Scandinavia (Gustafsson et al., 1995). Allophane, however, is only found in podzols formed in relatively rich parent materials. The metal - HMW organic matter complexes migrate downward in the profile, taking up more metals on the way. The complexes precipitate upon 'saturation', forming an illuvial horizon that is enriched in organic matter and metals. Organic matter precipitates as monomorphic coatings. When topsoils still contain mobilisable Fe and Al to neutralise and precipitate the organic matter, migration of HMW acids is absent or shallow. Biological mixing also counteracts downward migration of other dissolved and suspended material. As a result, the initial surface horizon is enriched in organic matter plus associated Fe and Al (Ah-horizon). As the process continues, however, larger amounts of organic matter are produced and transported, the soil acidifies and biological homogenisation decreases both in intensity and in depth. Ah horizon formation is decreased while transport of soluble material reaches deeper into the soil. This results in the formation of an eluvial (E) horizon, just below a progressively thinner zone of biological homogenisation, and of an illuvial layer immediately below the E-horizon (see also Figure 11.2). Formation of an E-horizon at the site of a former B-horizon implies that the organic matter of the B-horizon must disappear (see c). Fe(III)-oxides usually coat the dominant soil minerals (quartz, feldspars), and cause most well-aerated soils to have yellowish-brown colours. The removal of these oxide coatings by complexing organic acids is responsible for the ash grey colour of the eluvial horizon of podzols.

c. As the production of transportable organic matter in the litter layer continues, the progressively weathered topsoil further loses its capacity to provide metal ions to neutralise the organic acids. Organic acids that reach the top of the illuvial (B) horizon are increasingly unsaturated. These acids redissolve part of the complexed metals from the top of the B-horizon. The removal of metals increases the degradability of the accumulated organic matter and results in microbial degradation of the organic matter in this part of the B-horizon: the organic matter is not redissolved, but removed by respiration. Reprecipitation of the fresh metal organic complexes occurs deeper in the B-horizon. The combination of processes results in a downward shift with time of the illuvial horizon and an increase in thickness of the eluvial horizon. Strong weathering in the E horizon eventually results in quartz-dominated, white sand.

Eluviation of the O, Ah, and E horizons removes nutrients and decreases water storage in these horizons, making their environment less favourable for roots. Roots therefore concentrate in the B-horizon. This implies that, with time, the zone of maximum rooting shifts downward in the profile. Dead roots in A and E horizons are decomposed, leaving hardly any trace. In the zone of maximum root activity, and root litter input, decay of root remnants by mesofauna causes accumulation of polymorphic organic matter.

d. In sandy soils with appreciable amounts of both Al and Fe, the illuvial horizon tends to separate with time into an upper (Bh) and lower (Bhs) horizon. The Bhs horizon is characterised by visible accumulation of iron hydroxides (yellowish brown colours), while the Bh horizon is black from a predominance of organic matter. The Bh-horizon is relatively rich in organically bound aluminium, because iron is preferentially removed from this horizon and transported further downward. This is probably due to a pH-dependent change in preferent complexation of Fe and Al.

   Profiles on fine materials, in which podzolisation is not the first profile development, may initially have an iron maximum above the aluminium maximum.

   Hydromorphic podzols, which have lost all iron (see 11.5), may have B-horizons that are coloured brown by organic matter, which may be confused with iron-rich Bs horizons.

   In the richer zonal podzols that also have higher pH throughout the profile, accumulation of organic matter in the B-horizon is much less expressed, and separation of Al and Fe in the B-horizon has not been documented.

e. Especially in sandy podzols, part of the dissolved organic matter that percolates through the soil is not precipitated by metal ions and may move through the illuvial horizon. Such organic matter precipitates where the water front stops moving or when it reaches the groundwater. In well-drained podzols, this organic matter is responsible for the formation of thin bands (fibres) below the B-horizon. Because the penetration of water in the soil is very irregular, such bands also have an irregular morphology, which is influenced by discontinuities in the pore system of the soil: root channels, sedimentary stratification, burrows, etc. Dissolved organic matter that is not precipitated by metals appears to be more abundant in hydromorphic podzols than in well-drained ones (Figure 11.4). If transported to open water by lateral groundwater flow, such organic matter is a cause of 'black-water' rivers.

Question 11.5. *Explain how heating a soil sample from a brown B-horizon to 900°C will distinguish iron-rich and iron-poor/organic-rich podzol-B horizons.*

ILLUVIAL ORGANIC MATTER AND ROOT-DERIVED ORGANIC MATTER
Podzol B-horizons may contain both polymorphic (root-derived) and monomorphic (illuviated) organic matter. The two forms are illustrated in Figures 117, 11.8, and Plate T (page 284). Which kind of organic matter dominates depends on the dynamics of the two forms.
In soils on relatively rich parent materials, under boreal climates, considerable amounts of low-molecular weight acids and 'fulvic acids' are produced in the litter layer These compounds are broken down relatively rapidly. Also root-derived organic matter is mineralised rapidly. Because the relative contribution of roots to the organic matter in the B-horizon is larger than that of illuvial organic matter, polymorphic (root-derived) organic matter dominates in the B-horizon. Rapid dynamics causes little overall humus accumulation (see Chapter 6), but the accumulation of sesquioxides is cumulative over time. Because parent materials of zonal podzols largely consist of glacial tills, iron contents in the parent material are relatively high. In such situations the accumulation of iron (and aluminium) in the B-horizon far exceeds that of organic matter. In extreme cases, iron podzols (Ferrods) are formed.

Intrazonal podzols, by contrast, are always formed on very poor parent materials. In spite of higher soil temperatures than in the boreal zone, organic matter turnover is strongly slower because of lower pH, high exchangeable aluminium, and low nutrient status. The organic acids produced by the litter are respired at greater depth, and at a slower rate, than in zonal podzols. Because the illuvial organic matter decays slowly, it contributes considerably to organic matter accumulation in the B-horizon, mainly in the form of monomorphic organic coatings. As in zonal podzols, root-derived polymorphic organic matter is present, but in lower concentrations than illuvial organic matter. The soils usually have low contents of weatherable minerals and therefore absolute accumulation of sesquioxides is low; (upper) illuvial horizons are dominated by organic matter.

In hydromorphic podzols, illuvial organic matter is even more dominant. Rooting in water-saturated soil is usually restricted, while the decay of organic acids is further slowed down. Soluble organic components are more mobile, because the soil does not dry out, and because iron is usually removed from the soil by reduction and lateral transport, which reduces chemical precipitation. Organic matter and aluminium dominate B-horizons, and large part of the organic matter remains mobile and is eventually removed to surface water. B-horizons of hydromorphic podzols have organic matter compositions that resemble Dissolved Organic Carbon (DOC). DOC contains appreciable amounts of polysaccharides, which cannot be precipitated through complexation of metals. It is likely, therefore, that a considerable part of the organic matter in B-horizons of hydromorphic podzols is precipitated physically. The strong relation of humus precipitation zones with sedimentary stratification also points in this direction (Figure 11.4).

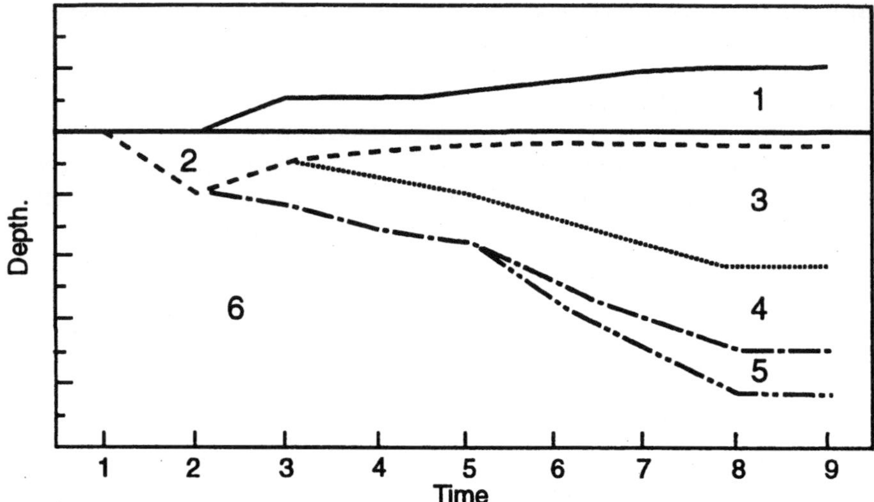

Figure 11.2. Changes with time in the thickness (depth) of six different horizons (labelled 1-6) in a podzol. Arbitrary scales.

Question 11.6. *Figure 11.2 depicts the horizon development with time of a podzol in a porous medium with good drainage. Give horizon designations (consult Appendix 1) for the horizons 1-6.*

Question 11.7. *How can podzolisation occur even when the A- and E-horizons still contain appreciable amounts of weatherable minerals? Consider the kinetic aspects of different processes involved in podzolisation.*

11.5. Influence of parent material, nutrients, and hydrology on the podzol profile.

Figure 11.2 suggests that the gradual deepening of the profile with time stops after a certain period, after which the profile morphology does not change further. The final thickness of the E horizon and the expression of the B horizon depend on many factors, such as composition of the parent material, nutrient status of the subsoil (decomposition of organic matter) and composition of the organic fraction itself. It is possible that the maximum depth at which dissolved organic matter is decomposed would determine profile equilibrium with respect to depth of the E horizon. The maximum content of organic matter is determined by its Mean Residence Time (see Chapter 6).

Question 11.8. *Explain why equilibrium may be reached under the circumstances outlined above.*

Parent material influences podzolisation through its content of weatherable minerals, Fe and Al oxide contents, and nutrient content. Content and kind of weatherable minerals determine the neutralising capacity of the soil for organic acids, and thereby its buffer capacity against podzolisation.

Question 11.9. *How would the development of a podzol profile be influenced by a) large amounts of easily weatherable Al-silicate minerals; b) large amounts of slowly weatherable Al-silicate minerals; and c) low amounts of weatherable minerals.*

As discussed above, availability of nutrients strongly influences organic matter dynamics. Higher levels of nutrients cause faster decay. Faster decay leads to a lower Mean Residence Time of the organic fraction. A lower MRT implies less accumulation of organic matter. Mean residence times of organic matter in B-horizons of zonal podzols tend to be much lower than that in intrazonal podzols. Mean residence times of B-horizon humus vary from a minimum of 400-500 years in some Scandinavian (zonal) podzols through 2000-2800 years in cemented B-horizons of (intrazonal) podzols in SW France, up to 40.000 years for some tropical podzols (Schwartz, 1988). In the Netherlands, MRT's measured so far range between 2000 and 3000 years.

Question 11.10. *Parent materials of zonal podzols are often relatively rich in weatherable minerals and lithogenic nutrients. Give the (geological) reason for this phenomenon.*

Question 11.11. *Two podzols have equal net production of organic acids in the O/Ah horizon, but organic matter in soil (1) has a lower MRT than that of soil (2). a) What do you expect about the relative accumulation of organic matter and sesquioxides in the B-horizons of profiles 1 and 2? b) Does the MRT influence the <u>total</u> amount of sesquioxides that is accumulated in the B-horizon?*

In a given climate, the speed of podzol formation strongly depends on parent material. On poor parent materials it is a fairly rapid process. In Dutch inland sand dunes, which do not contain calcium carbonate and are very low in weatherable silicates, clear podzol morphology is already present after a hundred years of vegetation development.
Periodic high groundwater in porous soils favours podzolisation through three main factors: inhibited decomposition, inhibited mixing, and reduction/removal of iron. Extreme wetness may fully inhibit podzolisation if downward water movement is absent.

In coarse-textured podzols, hydromorphism (reduction of iron) combined with lateral groundwater flow usually leads to complete depletion of iron. Dissolved Fe^{2+} is hardly bound by organic matter and is easily removed by flowing groundwater. In such podzols, aluminium is the dominant complexed cation. A hydrosequence of sandy podzols is shown in Figure 11.3.

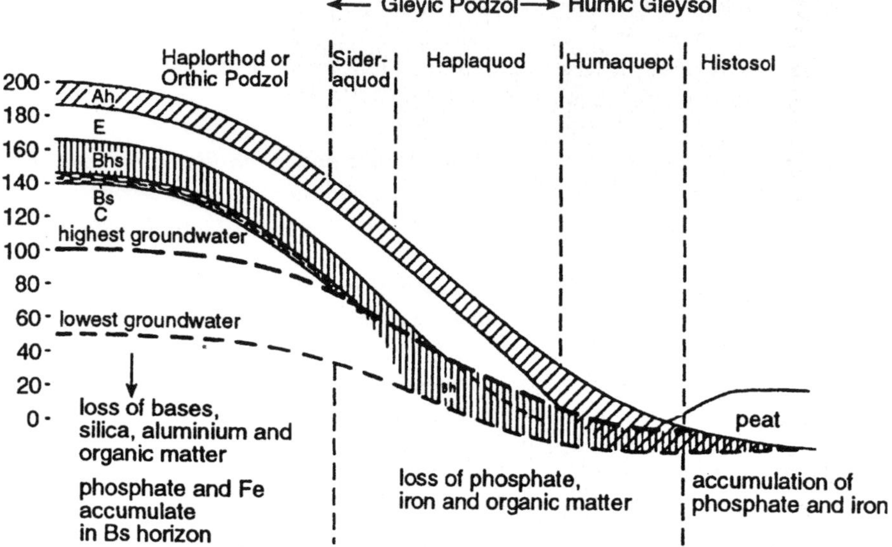

Figure 11.3. A podzol hydrosequence in cover sand of the Netherlands. Explanation see text. From Buurman, 1984.

At the left-hand side, a well-drained podzol exhibits Ah, E, Bhs and Bs horizons. Where the highest groundwater level touches the Bs-horizon, a concretionary Bs(g) horizon is formed. Further downward, the Ah horizon becomes peatier and the Bh merges with the Ah, while the E-horizon disappears. Iron and iron-bound phosphate, that are solubilised by reduction, are leached from the soil and transported to depressions where they may accumulate depending on aeration (iron hydroxide at anoxic/oxic interfaces, blue vivianite, $Fe_3(PO_4)_2$, or siderite, $FeCO_3$, in anoxic zones). In poorly drained podzols, lateral transport of organic matter may cause accumulation of dense humus pans. If this organic matter reaches open water, it causes 'black water rivers'. Part of the laterally transported organic matter accumulates as bands in the subsoil, where it accentuates lithological discontinuities (Figure 11.4).

In fine-textured parent materials, such as glacial tills, stagnant groundwater causes reduction and redistribution of iron, but seldom leads to iron depletion. In loamy soils, there is a whole range of transitions between podzols and gley soils in which iron concretions can be found either in the topsoil, the albic horizon, the spodic horizon, or below the spodic horizon, depending on groundwater fluctuations and parent material characteristics.

Figure 11.4. Horizontal accumulation bands in the subsoil of a hydromorphic podzol. Wavy bands in E are dissolved Bh of former well-drained podzol. Photo P.Buurman.

Question 11.12. *Is podzolisation in iron-depleted soils faster or slower than in their well-drained equivalents? Why?*

TROPICAL PODZOLS

In tropical lowlands, podzols are usually found on uplifted sandy, quartzic, coastal sediments. Such podzols are usually hydromorphic. Because of old age, coastal uplift, high rainfall and long MRT's of soil organic matter, many tropical podzols are very deep. When the soil is so strongly leached that no aluminium remains to precipitate organic matter and when percolation is deep enough to remove all organic matter to the ground water, the B-horizon eventually disappears and only an E horizon remains (giant podzols). The full breakdown of the illuvial layer appears to be strongest in locations with strong lateral water movement. Figures 11.5 and 11.6 depict such an environment. The tonguing transition from the E to the Bh horizon is due to dissolution of the Bh.

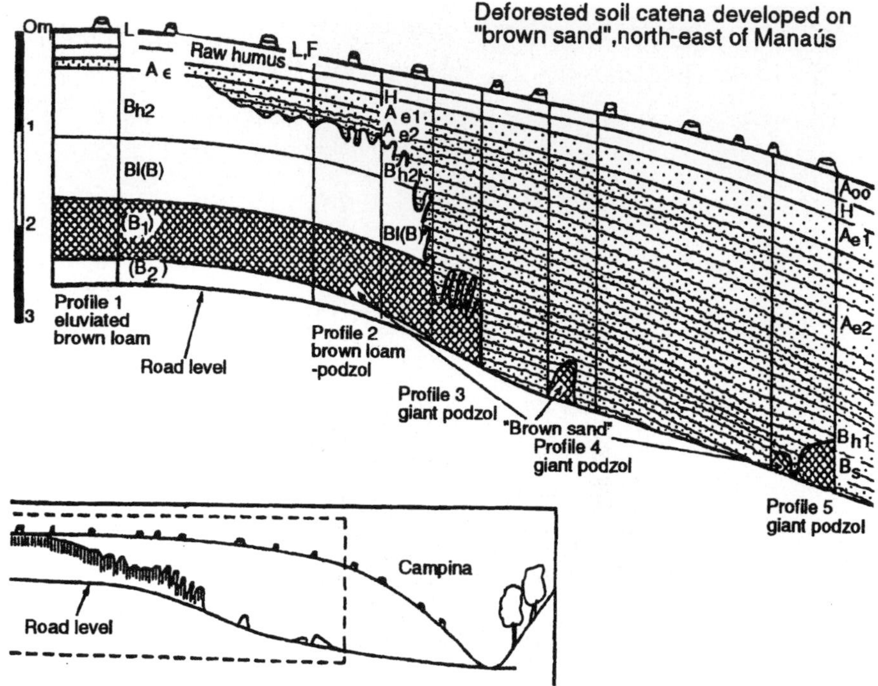

Figure 11.5. The formation of giant podzol in Brazil by dissolution of the B-horizon of a shallower podzol ('eluviated brown loam'). Ae stand for the E-horizon.
From Klinge, 1965; reproduced with permission of Blackwell Science Ltd., Oxford.

11.6. Mineral transformation in podzols

Podzolisation profoundly affects both primary and secondary minerals. The most resistant minerals (e.g. quartz (SiO_2), zircon ($ZrSiO_4$), rutile and anatase (TiO_2), ilmenite ($FeTiO_3$), and tourmaline) are residually accumulated in eluvial horizons (A and E), while minerals that are more susceptible to chemical attack are dissolved and disappear.

Physical and chemical weathering in eluvial horizons (e.g. mycorrhizae tunnelling into feldspars, see Chapter 4) results in more fine sand and silt than in the parent material. Illuvial horizons may show a higher content of silt and clay. The clay fraction may consist of allophane and imogolite, which was accumulated from overlying horizons previous to or during podzolisation, or of phyllosilicates that were formed by weathering *in situ*.
Strong weathering in the A- and E-horizons cause formation of beidellite (a smectite) at the expense of chlorites, micas and vermiculites (see Chapter 3.2).

Figure 11.6. Dissolution features in the top of a podzol from the coastal lowlands of eastern Malaysia. The E-horizon is about 70 cm thick. Photo P. Buurman.

Question 11.13. *Which processes cause the transformation from mica to beidellite (see Chapter 3.2)?*

The fraction of expanding 2:1 minerals in the A- and E-horizons tends to increase with time, temperature, drainage, and acidity. In very strongly weathered profiles, however, smectite may be removed completely by weathering.

Clay weathering is stimulated by low concentrations of free, uncomplexed Al^{3+}. In the eluvial horizon, complexation and removal by leaching depress Al^{3+} concentrations, and weathering rates are high. In the B-horizon, most of the aluminium occurs in the form of organic complexes, but microbial degradation increases the concentration of free Al^{3+}. If the pH in B-horizons is sufficiently high, aluminium may polymerise and form Al-hydroxy interlayers in vermiculites and smectites (formation of 'soil chlorites').

In tropical and subtropical podzols, where weathering is very intense, kaolinite may dominate the illuvial horizon. It is not clear whether this kaolinite was formed during podzolisation.

11.7. Recognising podzolisation

DIAGNOSTIC HORIZONS

Because of the intense colour contrast between Ah, E, and B-horizons, podzols are easily recognised in the field. If an eluvial horizon is absent, as in incipient podzols and in in transitions to Andisols and some Cambisols, micromorphologal and chemical criteria are used to identify podzolisation.

The *spodic* and *albic* horizons are used to classify Podzols (Spodosols). The presence of a *spodic* horizon reflects the dominance of the podzolisation process over other soil forming processes, such as weathering, ferralitization, and clay illuviation. The spodic horizon is a B-horizon that shows a certain accumulation of humus, iron and aluminium (sesquioxides). Al and Fe arrive in podzol B-horizons as organic complexes. Upon decay of the organic carrier, the metal is set free and forms an amorphous component, which may crystallise with time. Therefore, we can expect organically bound, and amorphous and crystalline sesquioxides of Fe and Al as well as amorphous Al-silicates in podzol B-horizons.
Chemical criteria (see below) for this diagnostic horizon are minimal because of the difficulty of defining appropriate boundaries between Andisols and Spodosols. The present definition (Soil Survey Staff, 1994; Deckers et al., 1998) assures that almost any recognisable podzol B-horizon is a spodic horizon.
The World Reference Base and SSS have the following criteria in common:

Colour:	B-horizons can be either black or reddish brown
Organic matter:	more than 0.6% C accumulation
pH-water:	≤ 5.9
Sesquioxides:	$Al_{ox} + 1/2Fe_{ox} \geq 0.5$
Optical density oxalate extract:	≥ 0.25
Thickness:	≥ 2.5 cm

In addition, SSS has the following criteria:

Cementation:	A horizon cemented by organic matter and Al, without Fe.
Coatings:	10% or more cracked coatings on sand grains.

Question 11.14. *What aspect of the podzolisation process is reflected by each of these criteria?*

The albic horizon in podzols, which is defined as a bleached horizon with a colour determined by uncoated mineral grains (without iron coatings), is virtually equivalent to the E-horizon.

Question 11.15. *E horizons often contain small remnants of organic matter resembling amorphous coatings typical of B-horizons. Why?*

MICROMORPHOLOGY

Weatherable minerals in E-horizons of podzols under coniferous forest are criss-crossed by tunnels with a diameter of about 5 μm, formed by (presumably ectomycorrhyzal) fungi. Tunnels are rare or absent in young podzols and tunnel frequency regularly increased with time over 8000 years of soil formation (Hoffland et al., 2002). Tunnels appear to be rare or absent in related soils (e.g. Dystric Cambisols) that lack and E-horizon.

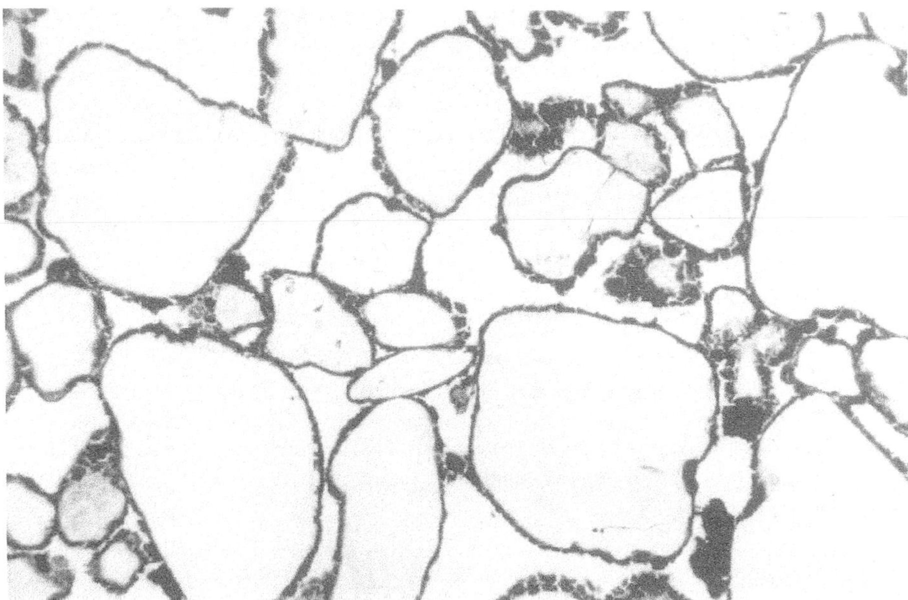

Figure 11.7. Amorphous humus coatings (cutans) in the B-horizon of a Humod, the Netherlands. Height of picture 1 mm. Photo A.G. Jongmans.

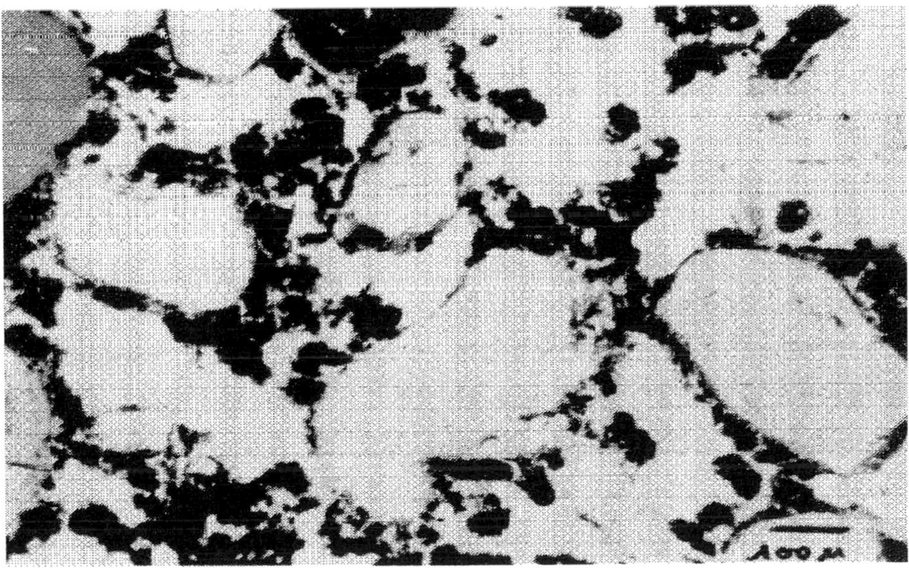

Figure 11.8. Polymorphic organic matter from a Humod on a river dune, Blitterswijk, the Netherlands. Height of picture 1 mm. Photo A.G. Jongmans.

Thin sections of podzol B-horizons show two dominant morphologies of organic matter: 1) amorphous coatings around mineral grains or in pores (monomorphic organic matter, Figure 11.7), and 2) pellets of organic matter in pores (polymorphic organic matter, Figure 11.8. The coatings are precipitated and dried gels, dominant in very poor and hydromorphic horizons and in bands below the B-horizons of well-drained podzols, while the pellets, which are mainly insect droppings, predominate in well-drained and richer soils. Both forms qualify for the spodic horizon.

Question 11.16. *Which kind or organic matter has a longer MRT, coatings of pellets?*

11.8. Problems

Problem 11.1

Table 11A gives data from a podzol profile from Nyänget, in northern Sweden. The soil is developed on glacial till rich in weatherable minerals. In the B-horizons, imogolite-like materials are encountered.

Table *11.A*. Analyses of soil and soil solution of the Nyänget profile, Sweden. From Karltun et al., 2000. DOC = dissolved organic C.

Hor.	C %	N %	pH $CaCl_2$	Fed	Feo	Fep	Ald	Alo	Alp	Sio
				------	------	------	mmol/kg	------	------	------
O	51	1	3.2							
E1	1.42	0	3.0	4	8	1			5	
E2	0.57	0	3.3	5	9	2	5	17	5	3
B1	2.32	0	4.4	307	252	2	10	17	8	2
B2	1.30	0	5.0	222	166	32	306	530	135	145
B3	0.79	0	5.1	127	98	5	206	478	54	163
B4	0.42	0	5.0	68	61	2	120	327	37	128
B5	0.34			36	36	1	70	235	25	91
C14	0.27			48	21		25	44		17

Soil solution June 1996; from Riise et al., 2000.

Hor.	DOC	Si	Al	Fe	DOC	Si	Al
	mmol/L	µmol/L			% in HMW component		
O1	50	85	40	6	60	5	50
O2	32	115	48	9	60	10	55
E1	8	270	56	8	50	8	65
E2	4	325	70	13	55	8	65
B1	3	265	32	11	35	5	60
B2	1	220	1	1	30	2	10
B3	0.5	170	-	-	35	1	-

a. Draw a diagram that depicts the changes of carbon content, Fe_d, Fe_p, Al_o, and Al_p with depth. Where are the maxima of each of these components? What part of free Fe, resp. amorphous Al, is bound by organic matter? What do the ratios Fe_p/Fe_d and Al_p/Al_o suggest about the MRT of the organic matter (compare question 11.10)?
b. Test the three different theories of podzolisation (11.1), using the data on soil solution chemistry and soil chemical data. Assume that Si_o and (Al_o - Al_p) represent allophane with 14% Si by mass, and that the solute concentrations represent a weighted annual mean, and that percolation rates through the bottom of the O, E, B1, B2, and B3 horizons are 200, 150, 120, 100, and 90 mm/year, respectively.

Problem 11.2

Figures 11B and 11C give organic matter fractionation and ^{14}C ages of a podzol hydrosequence in Les Landes du Médoc in SW. France. The profiles are developed in sandy coastal deposits. New horizon codes are: A1 = Ah; A2 = E; B21h = Bh1; B22h = Bh2.

The soils in the hydrosequence are, from left to right:

- An Aquod (Gleyic Podzol)) with loose B21h and cemented B22h and B3 horizons,
- An Aquod with cemented B2h horizon,
- An Aquod with a loose B2h horizon,
- An Aquent (Gleyic Arenosol)

The lowest ground water level is just below the B22h horizon; at high ground water, the soil at the right hand side is just flooded.

Assume that 'fulvic acids' and 'humic acids' represent LMW and HMW components, respectively, and that 'humin' in these profiles is mostly undecomposed plant litter. Ignore the 'acetyl bromide fraction'.

a. Explain the lateral changes in profile development. Consider vertical and lateral water movement.
b. What form of organic matter do you expect in the various B-horizons?
c. Study the trends in humus fractions (for the meaning of the fractions, see Chapter 1.5). Explain differences between A and B-horizons.

A ground water sample from below the B22h horizon was analysed for organic matter, aluminium and iron. The find out whether the metals were transported as organic matter complexes, the water was filtered over two different gels (Figure 11D). Small molecules move faster through the gel than large molecules, and therefore, the first eluent (away from the Y-axis) carries predominantly small molecules, while the larger molecules follow later (peaks close to the Y-axis). Each type of gel has a so-called 'cut-off value', which indicates which size of molecules can pass. Molecules larger than the cut-off value are retained in the gel. For gel G10, the cut-off value is 700 Daltons; molecules of weights between 100 and 5000 D pass through G25. The vertical axis in Figure 11D gives the amount of a component in the eluent; the horizontal axis is the eluted volume.

d. Discuss the composition of dissolved organic matter (DOC, DOM). Explain the different behaviour of Fe and Al on one hand and Ca on the other hand.

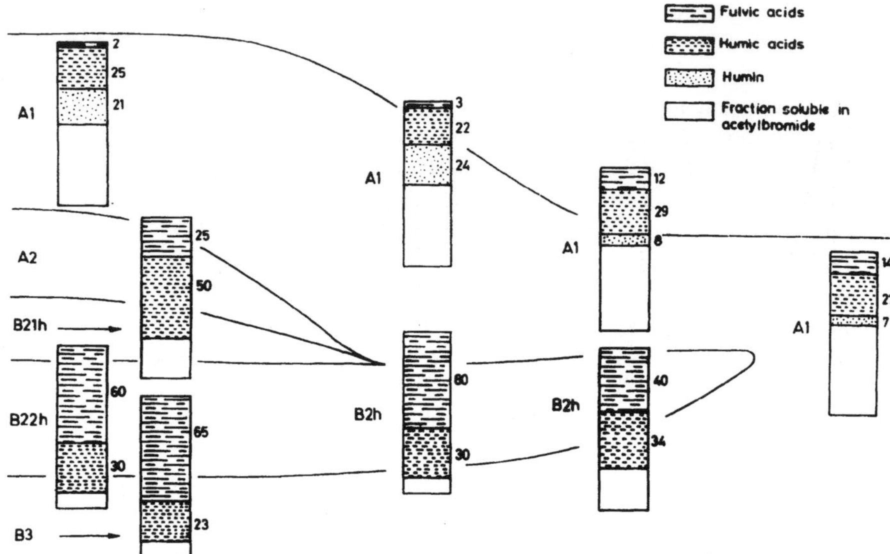

Figure 11B. Organic matter fractions in a podzol sequence of Les Landes (France). From Righi, 1977.

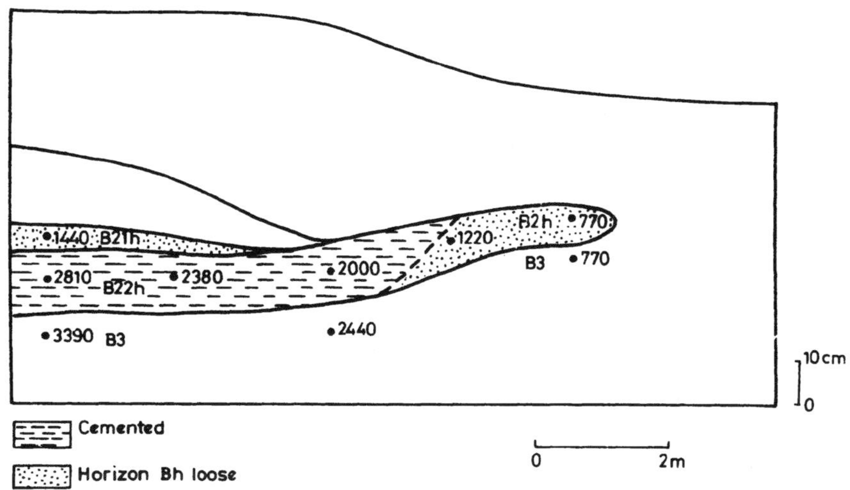

Figure 11C. ^{14}C ages of organic matter in the sequence of Figure 11B. Numbers indicate years BP. From DeConinck, 1980b.

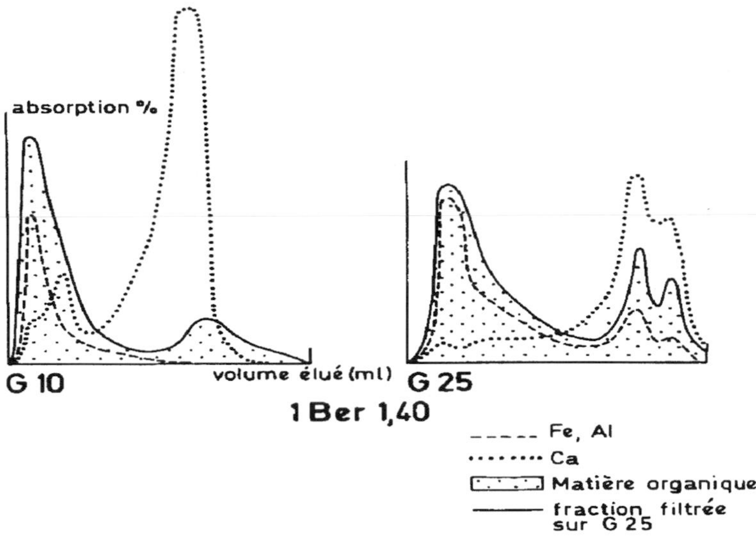

Figure 11D. Elution curves of groundwater components through gels Sephadex G10 and G25. From Righi, 1976.

11.9. Answers

Question 11.1

Because hampered decay of litter is one of the prerequisites of podzolisation, we should look for ecosystems with a litter layer (brushwood tundra, spruce forest, oak forest). For better prediction, data on percolation and parent material should be available.

Question 11.2

Because of a permanent litter layer, production in the podzol may continue throughout the year. In addition, decay is probably faster in the brown soil because pH and nutrient content are probably higher.

Question 11.3

$Al(C_2O_4)_2^-(aq) + O_2 + 2\ H_2O \rightarrow Al(OH)_3 + HCO_3^- + 3\ CO_2$

Question 11.4

'Upon saturation' means that all negative charge is compensated by metal ions. Because dissociation decreases at lower pH, lower amounts of metal can be bound. This causes an increasing C/metal ratio of the precipitate upon decreasing pH. On the other hand, hydroxylation of metal ions decreases with decreasing pH, so that fewer metal ions are necessary to compensate the negative charge of organic matter. Also this increases the C/metal ratio of the precipitating organic matter.

Question 11.5
Heating at 900°C causes oxidation of organic matter, and colours due to organic matter disappear. Iron hydroxides are converted to red hematite (Fe_2O_3).

Question 11.6
Horizon codes are: 1- O; 2- Ah; 3- E; 4- Bh; 5- B(h)s; 6- C.

Question 11.7
As long as removal of weathering products is faster than supply, the topsoil provides insufficient buffering against podzolisation. It is a matter of reaction kinetics of supply and removal rather than potential buffering capacity.

Question 11.8
Organic matter contents in the B horizon are - per definition - in equilibrium as soon as the MRT of the organic matter fraction is reached. As long as the topsoil still contains weatherable minerals, supply of Al and Fe to the B-horizon may continue. Therefore, further accumulation of Al and Fe only stops when the topsoil is depleted and when the top of the B-horizon does not move further downwards. If sesquioxides are transported as organic complexes, the depth to which these complexes can be transported determines the accumulation depth of the sesquioxides. Podzol dynamics predict that the depth necessary for the decay of the organic carrier will increase with increasing podzolisation and decreasing pH.

Question 11.9
Weatherable minerals provide Ca, Al and Fe, which can stabilise organic matter. Little buffering (stabilisation) results in fast podzolisation. Large amounts of slowly weathering primary minerals will, with time, cause a strong accumulation in the B horizon - but not necessarily to thick E horizons, while low amounts of such minerals hardly counteract podzolisation and may cause formation of very deep E horizons. Large amounts of easily weatherable minerals will counteract podzolisation.

Question 11.10
Zonal podzols are found in boreal climates. Parent materials in these areas are usually glacial deposits (a.o. moraines), which consist of finely ground fresh rock.

Question 11.11
a. The organic matter accumulation would be less in soil A, but the sesquioxide accumulation should be the same. Sequioxide accumulation is cumulative over time, while organic matter accumulation is determined by its mean residence time.
b. The MRT should not influence the total amount of accumulated sesquioxides, because the latter depends on the amount of weatherable minerals.

Question 11.12
Iron-depleted soils have less capacity to precipitate and hence retain soluble HMW organic acids. Podzolisation should therefore be faster in such soils.

Question 11.13

Mica-vermiculite changes are due to removal of interlayer cations. In the vermiculite to smectite transformation, the removal of interlayer cations is complete. The chlorite-vermiculite transformation is a removal of the second octahedral layer. The main agents in these transformations are organic acids, which remove metals from clay minerals by complexation.

Question 11.14

- The colour reflects dominant accumulation of either organic matter or sesquioxides. The first represents the chelate mechanism; the second the chelate/root mechanism.
- Organic matter reflects the role of DOC, either as carrier for sesquioxides, or as non-complexed component.
- pH reflects the zone in which buffering of organic acids is restricted. The pH is usually (much) lower than 5.9.
- The sesquioxide accumulation represents the effect over time of the accumulation (and decomposition) of organo-mineral complexes (compare Chapter 11). Accumulated Fe may be absent in hydromorphic sandy podzols.
- The optical density of the oxalate extract reflects the amount of organic matter in the solution. Because the oxalate extract has a pH of 3, the colour is mostly due to 'fulvic acids' (see Chapter 4).
- The thickness is used to define a minimum development.
- Cementation is a typical feature in podzols with strong cutan development (DOC+Al), such as hydromorphic sandy podzols.
- Cracked coatings also reflect the chelation/DOC/precipitation mechanism.

Question 11.15

The E horizon was first a B-horizon (gradual deepening of E horizon; decomposition of organic matter at the top of the B-horizon). Because coatings have a longer MRT, remnants of coatings are to be expected.

Question 11.16

Coatings are very compact, so that water and oxygen cannot easily penetrate (see Figure 11.7) and microbes can only access the outside. Therefore, organic matter in coatings has a longer MRT than that in pellets.

Problem 11.1

a) The diagram (not shown) indicates that maximum accumulation of Fe_d and Fe_o are in the B1, while the maxima of Al_o and Al_p are in the B2 horizon. The relative proportion of free Fe and Al bound by organic matter (Al_p/Al_o and Fe_p/Fe_d, respectively -see Appendix 3) is:

	B1	*B2*	*B3*	*B4*	*B5*	*C14*
Fe	0.01	0.14	0.04	0.03	0.03	-
Al	0.47	0.44	0.11	0.11	0.11	-

These are relatively small fractions, especially for Fe. This suggests that the organic carrier is broken down rapidly, so the MRT is probably low. Relatively small amounts of Al and especially Fe are (still) bound to organic matter.

b) The different theories predict different fates for dissolved substances and allophane formation:

i) The fulvate theory predicts that 1) HMW-Al and -DOC are produced below the O horizon, and 2) are transported downwards to be precipitated in the B horizon
ii) The allophane theory predicts that 1) HMW-Si is produced below the O horizon, and 2) are transported downwards to be precipitated in the B horizon as allophane or imogolite
iii) The LMW acids theory predicts that 1) LMW-Al and -DOC are produced below the O horizon, and 2) are transported downwards to be decomposed in the B horizon where Al precipitates, as amorphous product, e.g. allophane or imogolite
iv) The LMW acid/roots theory predicts that sesquioxides are transported by organic carriers and precipitate when the carriers are broken down.

The hypotheses can be tested by determining the fluxes of these different solutes as a function of depth. If soil solution concentrations would represent annual flux-weighted means, calculating them with the percolation rates gives solute fluxes in mmol/m^2.yr: (mmol/l) * mm/yr = mmol/m^2.yr.. The resulting solute fluxes at the bottom of a number of horizons are given below. The difference between total and HMW-DOC, -Si and -Al would represent LMW species:

horizon	water flux mm/yr	fluxes, mmol/m^2.yr						
		DOC tot.	DOC HMW	Si tot	Si HMW	Al tot	Al HMW	Fe tot
O2	200	6400	3840	23	2.3	9.6	5.3	1.8
E2	150	600	330	49	3.9	10.5	6.8	2.0
B1	120	360	126	32	1.6	3.8	2.3	1.3
B2	100	100	30	22	0.44	0.1	0.001	0.1
B3	90	45	16	15	0.15	0.009	0.001	0.001

The data are consistent with all three mechanisms: both HMW and LMW species disappear from solution as it moves from the E into the B-horizon. However, the contribution of HMW-Si (which may be inorganic colloidal Si) appears to be very small (3.9-0.15 = 3.75 mmol/m^2.yr of HMW Si are removed against 34 - 3.75 = 30.25 mmol/m^2.yr of LMW-Si). Therefore, ii) would be relatively unimportant if it exists at all. Possibly, HMW-Si is partly organically bound, so further work would be needed to test this. LMW and HMW forms of dissolved organic matter and of Al are about equally important, indicating that both i) and iii) apply. Note first, that the drop in DOC concentration is much larger between the O and E1, than between the E2 and B1 horizons, although accumulation of solid C is larger in the B1 than in the E1 horizon

C %). This indicates that most of the decrease in DOC with depth must be attributed to decomposition to CO_2, not to precipitation as solid organic material! This is also compatible with the LWM/root model.
Note, furthermore, that most of the soluble Al and Fe and about half of the soluble Si that percolate from the E into the B horizons are produced already in the organic O horizon, not in the E horizon as postulated by these two theories. A likely explanation is that mycorrhizal fungi bring Fe, Al and Si, mobilised by them in the E horizon (see Figure 4.12), towards mycorrhizal roots in the O horizon (where most of them reside). These substances (unwanted by plants, contrary to P, Ca, Mg and K that are taken up by these fungi too) are exuded or released, and next percolate back into the E horizon. Alternatively, Si may be produced by weathering of mineral components mixed with the litter in the O-horizon.
Theories ii) and iii) predict the formation of allophane or other forms of amorphous Al.

Allophane contents are calculated using a Si mass fraction of 14%. Therefore, Si_o concentrations have to be converted to weight percentages. The Al that can be attributed to allophane is Al_o-Al_p. Because both Al and Si are given in mmol.kg^{-1}, the molar ratio can be calculated directly: $(Al_o-Al_p)/Si_o$.
Thus, allophane contents and Al/Si ratios in allophane are:

	B1	*B2*	*B3*	*B4*	*B5*	*C14*
Allophane %	0	2.9	3.3	2.6	1.8	0.3
Al/Si ratio	4.5	2.7	2.6	2.3	2.3	2.6

Because of low contents, molar ratios for B1 are not reliable. Allophane may be formed by recombination of aluminium that is set free by decay of organic complexes with silica in solution. Note that the distribution of allophane with depth does not agree with that expected from the changes with depth of the Al and Si fluxes. These fluxes would result in higher allophane contents in the B1, and much lower contents in the B3 horizon. Probably, the fluxes based on solution concentrations in June 1996 are not representative for centuries of annual fluxes.

Problem 11.2
a) In the left part of the drawing, the podzol has an E horizon, which disappears towards the right. In addition, the B-horizon is higher in the profile at the right hand side. This suggests that there is more vertical percolation at the left side, and that restricted vertical water movement and increased influence of ground water fluctuations dominate at the right hand side. The B21h and B2h-horizons are loose, which means that they are aerated part of the year. However, it is illogical that the loose B2h disappears *under* the cemented B22h. This may be an error in the drawing.
b) The friable parts will have dominant polymorphic (excrement) organic matter, while the cemented parts have monomorphic (cutan) organic matter.
c) There is a strong shift in composition from the A to the B horizons. The most obvious differences are:
- Higher humin contents in the A;
- Much higher fulvic acid contents in the B;
- A smaller acetyl-bromide soluble fraction in the B.

Humin consists of strongly mineral-bound and of incompletely decomposed organic matter. In sandy soils, the latter is low. The difference in humin is therefore mainly due to incompletely decomposed plant material. The fraction of undecomposed plant material is, of course, mainly restricted to the topsoil and to the friable B-horizons, where roots decompose. Roots are probably much scarcer in the cemented horizons. Fulvic acids are water soluble, and are therefore easily transported to the B horizon, where they precipitate with metal ions (in this case Al, because it is a hydromorphic sequence).
d) Both elution curves indicate a dominance of high-MW organic molecules. In the G10 curve, all Fe and Al appears to be bound to the largest fraction, while in G25 it is bound both to the smallest and larger fractions (the smallest fractions on G25 are the largest on G10). The movement of Ca in G10 seems to be related to the low-MW organic fraction (adsorption), but it may also be independent of organic matter. The G25 curve, however, suggests that Ca is adsorbed to the low-MW organic fraction.

11.10. References

Alban, D.H., and E.C. Berry, 1994.Effects of earthworm invasions on morphology, carbon and nitrogen of a forest soil. Applied Soil Ecology, 1:243-249.

Anderson, H.A., M.L. Berrow, V.C. Farmer, A. Hepburn, J.D. Russell, and A.D. Walker, 1982. A reassessment of podzol formation processes. Journal of Soil Science, 33:125-136.

Bal, L., 1973. *Micromorphological analysis of soils*. PhD Thesis, University of Utrecht, 174 pp.

Barrett, L.R., and R.J. Schaetzl, 1998. Regressive pedogenesis following a century of deforestation: evidence for depodzolization. Soil Science, 163:482-497.

Bruckert, S., 1970. Influence des composés organiques solubles sur la pédogénèse en milieu acide. I. Etudes de terrain. Annales Agronomiques, 21:421-452.

Buurman, P. (ed), 1984. *Podzols*. Van Nostrand Reinhold Soil Science Series. 450pp. New York.

Buurman, P., 1985. Carbon/sesquioxide ratios in organic complexes and the transition albic-spodic horizon. Journal of Soil Science, 36:255-260.

Buurman, P., 1987. pH-dependent character of complexation in podzols. In: D. Righi & A. Chauvel: *Podzols et podzolisation*. Comptes Rendus de la Table Ronde Internationale: 181-186. Institut National de la Recherche Agronomique.

Buurman, P., and A.G. Jongmans (2002). Podzolisation - an additional paradigm. Proceedings 17th International Congress of Soil Science, Bangkok. In press.

Buurman, P., and L.P. van Reeuwijk, 1984. Allophane and the process of podzol formation - a critical note. Journal of Soil Science, 35:447-452.

De Coninck, F., 1980a. The physical properties of spodosols. In: B.K.G. Theng (ed): *Soils with variable charge*. 325-349. New Zealand Society of Soil Science, Lower Hutt.

De Coninck, F., 1980b. Major mechanisms in formation of spodic horizons. Geoderma, 24:101-128.

Deckers, J.A., F.O. Nachtergaele, and O. Spaargaren, 1998. World Reference Base for Soil Resources. Introduction. ISSS/ISRIC/FAO. Acco, Leuven/Amersfoort, 165 pp.

Farmer, V.C., J.D. Russell and M.L. Berrow, 1980. Imogolite and proto-imogolite allophane in spodic horizons: evidfence for a mobile aluminium silicate complex in podzol formation. Journal of Soil Science, 31:673-684.

Flach, K.W., C.S. Holzhey, F. de Coninck & R.J. Bartlett, 1980. Genesis and classification of andepts and spodosols. In: B.K.G. Theng (ed): *Soils with variable charge*. 411-426. New Zealand Society of Soil Science, Lower Hutt.

Guillet, B., 1987. L'age des podzols. In: D. Righi & A. Chauvel: *Podzols et podzolisation*. Comptes Rendus de la Table Ronde Internationale: 131-144. Institut National de la Recherche Agronomique.

Gustafsson, J.P., P. Battacharya, D.C. Bain, A.R. Fraser, and W.J. McHardy, 1995. Podzolisation mechanisms and the synthesis of imogolite in northern Scandinavia. Geoderma, 66: 167-184.

Hoffland, E., R. Giesler, A.G. Jongmans, and N. van Breemen, 2002. Increasing Feldspar Tunneling by Fungi across a North Sweden Podzol Chronosequence. Ecosystems 5:11-22 (2002)

Jongmans, A.G., N. Van Breemen, U. Lundström, P.A.W. van Hees, R.D. Finlay, M. Srinivasan, T. Unestam, R. Giesler, P.A. Melkerud, and M.Olsson, 1997. Rock-eating fungi. Nature, 389: 682-683.

Karltun, E., D. Bain, J.P. Gustafsson, H. Mannerkoski, T. Fraser and B. McHardy, 2000. Surface reactivity of poorly ordered minerals in podzol-B horizons. Geoderma, 94:265-288.

Klinge, H., 1965. Podzol soils in the Amazon Basin. Journal of Soil Science, 16:95-103.

Lundström, U.S., N. van Breemen and A.G. Jongmans, 1995. Evidence for microbial decomposition of organic acids during podzolization. European Journal of Soil Science, 46:489-496

Melkerud, P.A., D.C. Bain, A.G. Jongmans, and T. Tarvainen, 2000. Chemical, mineralogical and morphological characterization of three podzols developed on glacial deposits in Northern Europe. Geoderma, 94:125-148.

Mokma, D.L., and P. Buurman, 1982: *Podzols and podzolization in temperate regions*. ISM Monograph 1: 1-126. International Soil Museum, Wageningen.

Petersen, L., 1976. *Podzols and podzolization*. DSR Forlag, Copenhagen, 293 pp.

Righi, D., 1977. *Génèse et évolution des podzols et des sols hydormorphes des Landes du Médoc*. PhD Thesis, University of Poitiers.

Righi, D., and F. DeConinck, 1974. Micromorphological aspects of Humods and Haplaquods of the 'Landes du Médoc', France. In: G.K. Rutherford (ed.): *Soil Microscopy*. Limestone Press, Kingston, Ont. Canada, pp 567-588.

Righi, D., T. Dupuis, and B. Callame, 1976. Caractéristiques physico-chimiques et composition des eaux superficielles de la nappe phréatique des Landes du Médoc (France). Pédologie, 26:27-41.

Riise, G., P. Van Hees, U. Lundström and L.T. Strand, 2000. Mobility of different size fractions of organic carbon, Al, Fe, Mn, and Si in podzols. Geoderma, 94:237-247.

Robert, M., 1987. Rôle du facteur biochimique dans la podzolisation. Etudes expérimentales sur les mécanismes géochimiques et les évolutions minéralogiques. In: D. Righi & A. Chauvel: *Podzols et podzolisation*. Comptes Rendus de la Table Ronde Internationale: 207-223. Institut National de la Recherche Agronomique.

Ross, G.J., 1980. The mineralogy of spodosols. In: B.K.G. Theng (ed): *Soils with variable charge*. pp 127-146. New Zealand Society of Soil Science, Lower Hutt.

Schnitzer, M., and S.I.M. Skinner, 1964. Organo-metallic interactions in soils, II. Properties of iron and aluminium organic matter complexes, prepared in the laboratory and extracted from soils. Soil Science, 98:197-203.

Properties of iron and aluminium organic matter complexes, prepared in the laboratory and extracted from soils. Soil Science, 98:197-203.

Schnitzer, M., and S.I.M. Skinner, 1966. Organo-metallic interactions in soils, V. Stability constants for Cu^{2+}, Fe^{2+}, and Zn^{2+} fulvic acid complexes. Soil Science, 102:361-365.

Schnitzer, M., and S.I.M. Skinner, 1967. Organo-metallic interactions in soils, VII. Stability constants of Pb^{2+}, Ni^{2+}, Mn^{2+}, Co^{2+}, Ca^{2+}, and Mg^{2+} fulvic acid complexes. Soil Science, 103:247-252.

Schwartz, D., 1988. *Histoire d'un paysage: Le Lousséké - paléoenvironnements Quaternaires et podzolisation sur sables Batéké*. Etudes et Thèses, ORSTOM, Paris, 285pp.

SSS (Soil Survey Staff), 1994. Keys to Soil Taxonomy, 6th Edition. US Dept. of Agriculture, Soil Conservation Service, 306 pp.

Van Breemen, N., U.S.Lundström, and A.G. Jongmans, 2000. Do plants drive podzolization via rock-eating mycorrhyzal fungi? Geoderma, 94:163-171.

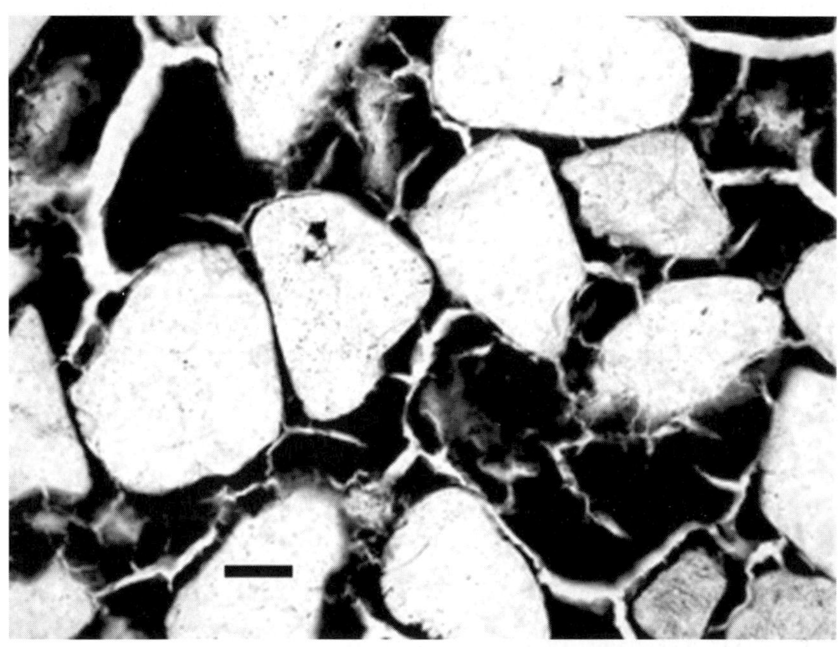

Plate T. Monomorphic organic matter filling the pores of a hydromorphic podzol B-horizon on marine sands, New Zealand. Scale bar is 100μm. Photograph A.G. Jongmans.

CHAPTER 12

FORMATION OF ANDISOLS

12.1. Introduction

DEFINITION
Andisols are soils that are dominated by amorphous (or short-range order) aluminium silicates and/or Al-organic matter complexes. They usually have an Ah - Bw - C horizon sequence. The Ah horizon is dark-coloured and normally very high in organic matter (often more than 10%) stabilised by aluminium. The B-horizon is usually dominated by amorphous aluminium silicates (allophane, imogolite). Andisols form mainly on volcanic ashes, but can also be found on other highly weatherable rock such as amfibolites, arkoses, etc. We will use the term 'andisol' loosely, to indicate soils formed by processes that are typical of the *Andisols* of the USDA soil classification and the *Andosols* of the FAO classification. Andisols have many peculiar properties, such as a high phosphate fixation, aluminium toxicity, irreversible drying, high water retention with low water availability, and high erosion resistance. They usually have sedimentary stratification, with the most unweathered materials occurring on top. Because the organic and non-organic secondary products in andisols are of overwhelming importance for these properties, most attention in this chapter is paid to the characteristics of these weathering products.

12.2. Weathering and mineralogy

Andisols form when weathering is rapid and when soluble weathering products are removed from the soil relatively rapidly. Rapid weathering requires easily weatherable materials. Of the common constituents in volcanic deposits, the following sequence of decreasing weatherability is found:

> **coloured glass - white glass - olivine - plagioclases - pyroxenes - amphiboles - ferromagnesian minerals**

Question 12.1. *K-feldspars and micas are not included in the weathering sequence above, because they are scarce in volcanic deposits. Where would these minerals fit in the weathering sequence? (Compare Chapter 3).*

Weathering of easily weatherable minerals and volcanic glass produces relatively high concentrations of Ca, Mg, Al, Fe and silica in solution. At such high solute concentrations, solutions are strongly super-saturated with respect to minerals of low solubility, and insufficient time is available for proper arrangement into crystalline structures. As a result, Al, Fe, and Si precipitate as amorphous components, such as allophane, imogolite, opal, and ferrihydrite. Part of the dissolved H_4SiO_4 and most of the dissolved basic cations (Ca, Mg, K, Na) are removed with the drainage water, together with HCO_3^-.

WEATHERING AGENTS

The presence of dissolved HCO_3^- shows that the weathering agent is CO_2, and that the pH is above 5. At such pH, silica is more mobile than aluminium and iron (compare the ferralitization process, Chapter 13). In these conditions, silica is selectively removed from the profile and the remaining amorphous silicates become more aluminous.
Most active volcanoes produce large amounts of sulphur and HCl gases. Such gases combine with water to form strong acids. The first stages of weathering of volcanic ashes near an active crater may therefore be influenced by these strong acids, causing a pH below 5. Weathering is rapid, and large amounts of Al, Fe, and Si are dissolved. Upon evaporation of the solution, the saturation value of opal is reached earlier than those of Fe and Al-hydroxides. This may give rise to the precipitation of opal as coatings on the soil surface and at shallow depth in the soil (Jongmans et al., 1996). Strong acidity counters the formation of allophane.

ALLOPHANE AND IMOGOLITE

Whether andisols will be dominated by amorphous silicates (allophane and imogolite) or by Al-organic matter complexes, depends on the supply of organic matter and on pH. At relatively low pH, the formation of Al-organic complexes inhibits the polymerisation of aluminium hydroxide, and thus the formation of hydrous Al-silicates (allophane structure features a polymerised $Al(OH)_3$ structure, see later). A rough subdivision in pH classes (pH measured in water or in 1M KCl) results in the following groups of andisols (from Nanzyo et al., 1993; order changed):

Group	pH_{H2O}	pH_{KCl}	Composition	Al_p/Al_o
1	4.8-5.3	3.8-4.4	non allophanic (humic)	high
2	5.0-5.7	4.3-5.0	allophanic, humus-rich	medium
3	5.2-6.0	5.0-5.6	allophanic, humus-poor	low

BIMODAL DISTRIBUTION

These three groups form a continuum, but because groups 1 and 3 tend to dominate, the population of organically bound Al relative to total amorphous Al (Al_p/Al_o) in Andisols has a bimodal distribution (Figure 12.1).
If the pH is sufficiently high to allow polymerisation of aluminium hydroxide, allophane and imogolite are formed. Allophane is an essentially amorphous substance with a

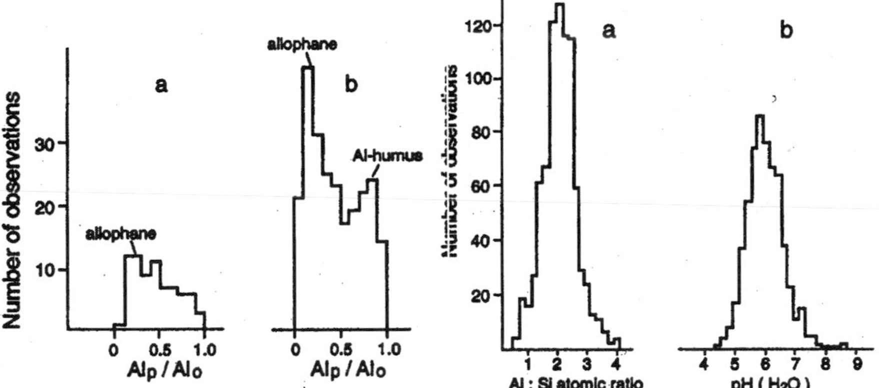

Figure 12.1. Histogram of the Al_p/Al_o ratio in a) New Zealand Andisols, and b) world Andisols. From Parfitt and Kimble, 1989.

Figure 12.2. Al/Si ratios and pH of a worldwide population of Andisols. From Parfitt and Kimble, 1989.

general formula of $Al_2O_3.SiO_2.nH_2O$. Oxygen and OH link aluminium and silica in a single layer (Figure 12.3). Most of the precipitated aluminium has sixfold co-ordination, as in the octahedral layer of phyllosilicates. The structure of Figure 12.3 is idealised; in fact it is neither as continuous, nor as neatly ordered. The Al-Si layer is curved and forms globules with outside diameters of 3.5 to 5 nm, and a wall thickness of 0.7-1 nm. The imperfections in the lattice appear as holes in the globules.

Question 12.2. *Pyrophosphate-extractable Al (Al_p) and oxalate-extractable Al (Al_o) and Si (Si_o) can be used to quantify amorphous phases typical of andosols. How will the Al_p/Al_o ratio shift from the Ah to the Bw horizon in an allophanic andisol?*

The Al/Si molar ratio for allophane and imogolite is usually around 2. Silica-rich allophanes (Al/Si = 1-1.5) contain polymerised silica units as in phyllosilicates, but part of the Al is in fourfold (tetrahedral) co-ordination. The structure of high-silica allophanes is still incompletely understood.

Imogolite has an idealised formula of $(OH)_3Al_2O_3SiOH$, or $Al_2O_3.SiO_2.2H_2O$. It has a tube-like morphology (Figure 12.4). Outer and inner diameter of the tubes are 2.0 and 1.0 nm, respectively. The tubes are up to several microns long. The tube form is caused by a particular arrangement of an $Al_2(OH)_6$ sheet and a hydrated $AlSiO_5$ sheet. The morphology of allophane and imogolite can be seen by transmission electron microscopy (Figure 12.5).

Question 12.3. *In Figure 12.4, the inner sheet is built up of silica atoms. If the idealised formula for imogolite and the structure of Figure 12.4 are both correct, what fraction of the Si atoms of the inner sheet must be replaced by Al?*

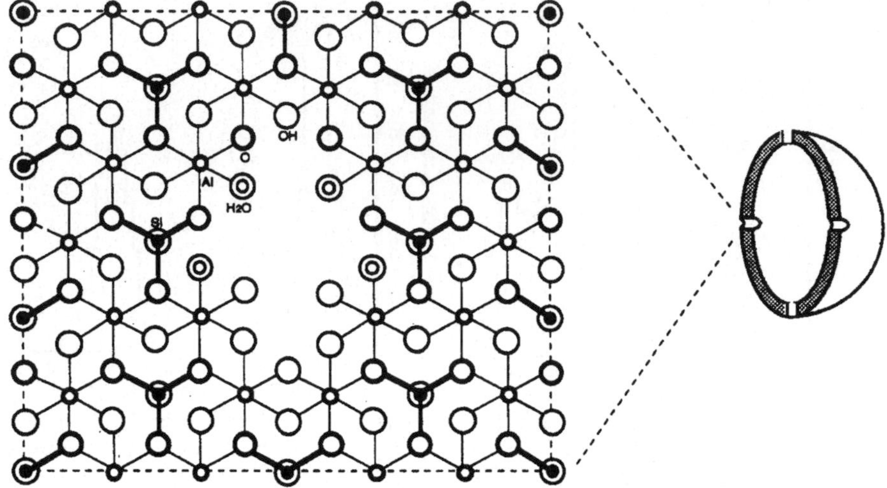

Figure 12.3. Schematic structure of allophane (left) and an allophane globule with micropores (right). From Dahlgren et al., 1993. Symbols are explained in the figure. Reproduced with permission of Elsevier Science, Amsterdam.

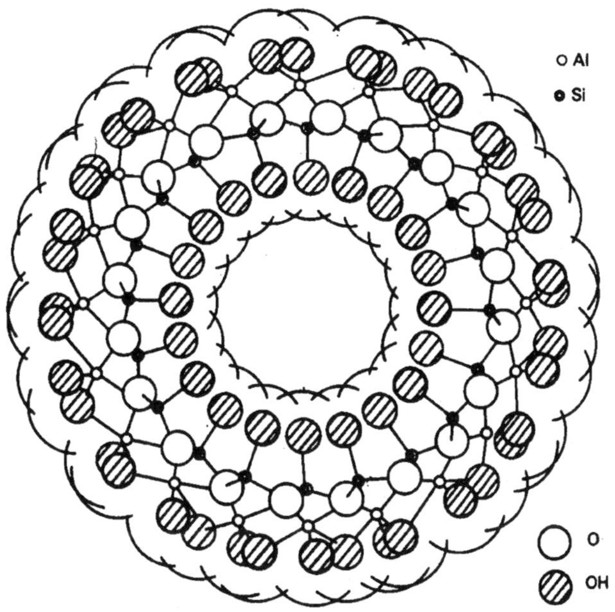

Figure 12.4. Cross-sectional structure of an imogolite tube. From Parfitt et al., 1980.

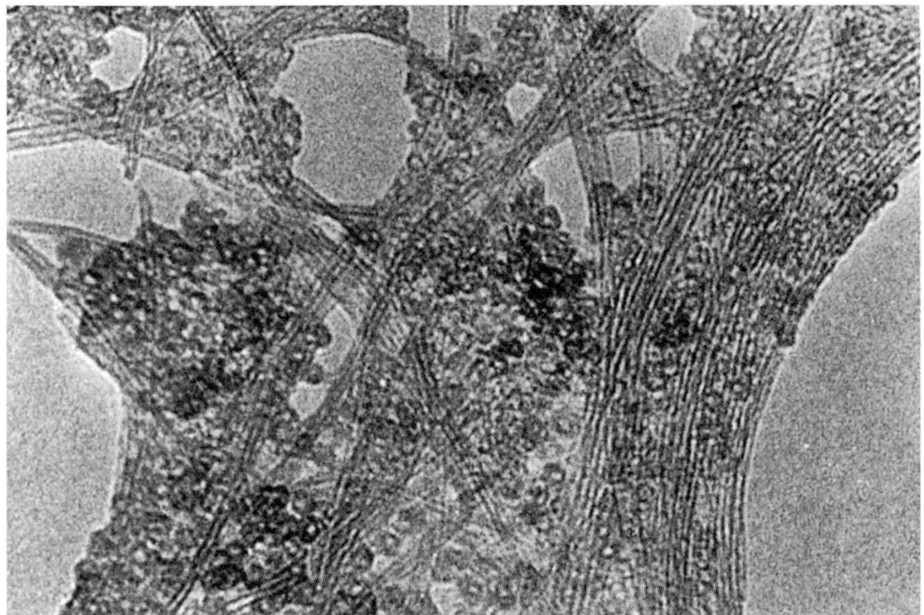

Figure 12.5. Transmission electron micrograph of allophane (globules) and imogolite (threads). Width of picture is approximately 0.2 μm. From Henmi and Wada, 1976.

Question 12.4. *Describe the differences between allophane or imogolite structure with those of phyllosilicates. Consider ordering (crystallinity), number of sheets, isomorphic substitution, surface composition*

Question 12.5. *Allophane is sometimes calculated from the amount of oxalate-extractable Si (Si_o) by assuming a 14% Si_o content. Calculate the water content of the allophane formula with an Al/Si ratio of 2. Express allophane as a sum of oxides and water, as above.*

Question 12.6. *What happens to silica in solution when allophane formation is inhibited by formation of Al-organic matter complexes?*

Allophane can be extracted by oxalic acid (see Appendix 3). The Al/Si ratio in the extract is taken to be the ratio in the amorphous silicates. Figure 12.2 gives the frequency distribution of Al/Si atomic ratios in the oxalate extract and of pH of a world-wide population of andosols.

OTHER SECONDARY MINERALS

Average compositions of a number of allophanic and non-allophanic andisols are given in Figure 12.6. The *crystalline clay minerals* in this figure are mostly halloysite.
Intermediary and mafic volcanic rocks contain iron-rich minerals. Some iron can

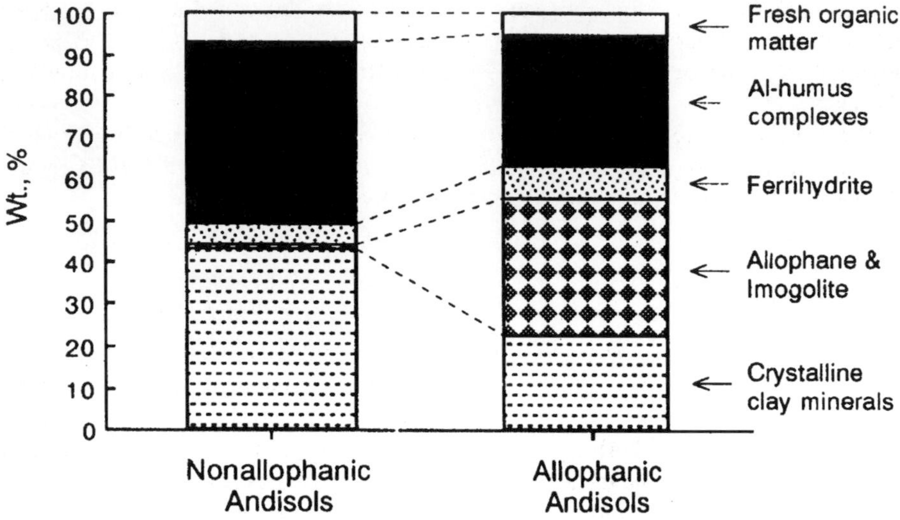

Figure 12.6. Average composition of the colloidal fraction of non-allophanic and allophanic andisols. From Nanzyo et al., 1993. Reproduced with permission of Elsevier Science, Amsterdam.

perhaps be accommodated in the allophane structure, but most of it forms a separate phase, the poorly crystalline *ferrihydrite*, with an approximate formula of $5Fe_2O_3.9H_2O$ (see also Chapter 3). Ferrihydrite is responsible for the yellowish-brown colour of many andisol B-horizons. Ferrihydrite content in soils is estimated by multiplying its Fe_o content by 1.7. Goethite and hematite are sometimes found in andisols, but their crystallinity is low. An amorphous iron-silicate phase (hisingerite), similar to allophane, has been proposed by various authors, but so far, little evidence has been found for its presence.

Gibbsite ($Al(OH)_3$) is a common mineral in strongly leached andisols. It forms by weathering of plagioclases, through desilication of allophane, or through precipitation from the soil solution in deeper parts of the profile.

Phyllosilicates are relatively scarce in andisols and these minerals may have various origins (see Dahlgren et al., 1993). Smectite clays may have been inherited from hydrothermally altered primary minerals such as pyroxene (Mizota, 1976; Jongmans et al., 1994a). Recrystallisation of allophane at low silica activity may lead to halloysite formation. In older andisols, halloysite may further transform to kaolinite.
Periodic desiccation of allophane (-coatings) may cause crystallisation to phyllosilicates (Buurman & Jongmans, 1987, Jongmans et al., 1994b). Finally, addition of *eolian dust* is a source of phyllosilicates in andisols that are found in the neighbourhood of sources

of dust, such as the andisols of Japan (close to the Gobi desert and the Chinese loess belt) and of the Canary Islands (close to the Sahara). In most andisols, the presence of phyllosilicates can be demonstrated by X-ray diffraction after complete removal of allophane.
In many andisols, stable primary minerals, such as cristobalite (SiO_2) and Fe and Fe-Ti minerals (magnetite, ilmenite) are relatively accumulated upon weathering. In dry climates, phytoliths of amorphous silica may be common in topsoils.

12.3. Organic matter

Dark Ah horizons, caused by high organic matter contents are a striking characteristic of many andisols. The high organic matter contents are probably caused by the large, positively charged surface area of allophane and by the stability of organic matter - allophane complexes and Al-organic matter complexes (both decay slowly and thus cause a build-up or organic matter content - see Chapter 6). Because andisols with high organic matter contents tend to have low allophane contents, organic ligands are supposed to inhibit allophane formation. However, if the ligands are saturated with Al, inhibition is probably zeroed. Little is known about the interaction between organic matter and allophane.

The darkest forms of these topsoils are called *melanic epipedon* (USDA). Such dark topsoils appear to be formed under grass vegetation that is frequently burned. The dark colour is usually ascribed to a specific group of humic acids, but the extremely high aromaticity of the organic matter in melanic horizons (50-70%; T. Yonebayashi, pers. com.) strongly suggests that it is due to charred grass, and not to a humic acid-allophane complex.
Organic matter binds both to phyllosilicate surfaces and to allophane. Phyllosilicates, which have a negative surface charge, bind mainly positively charged or non-polar organic groups (e.g. long-chain aliphatics). Allophane in the soil has a net positive charge and is therefore expected to have a preference for negatively charged (acid) organic groups, such as carboxyls. Molecular studies of allophane-bound organic matter are still very scarce.

12.4. Physico-Chemical properties

CHARGE
Andisols have a large component of fine material with a high surface area and surface charge, and large water binding capacity, due to the presence of Al-organic matter, allophane and imogolite, and ferrihydrite. All these components have a pH-dependent charge. Organic matter has a negative charge due to dissociation of acidic groups. This dissociation increases with increasing pH, and therefore, the CEC of the organic matter increases with pH. Amorphous silicates and ferrihydrite are amphoteric and have a pH value (the Point of Zero Net Charge, PZNC, see Chapter 13) above which they have a

negative charge, and below which the surface charge is positive so that they have an anion exchange capacity (AEC). Finally, phyllosilicates have a permanent negative charge. Most andisols have three or four of the above components, and show a distinct pH-dependent charge (Figure 12.7).

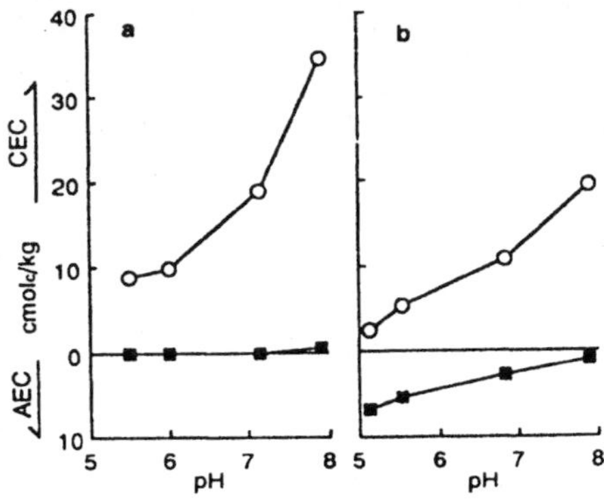

Figure 12.7. pH-charge relations for (a) a non-allophanic, and (b) an allophanic andisol. From Nanzyo et al., 1993. Reproduced with permission of Elsevier Science, Amsterdam.

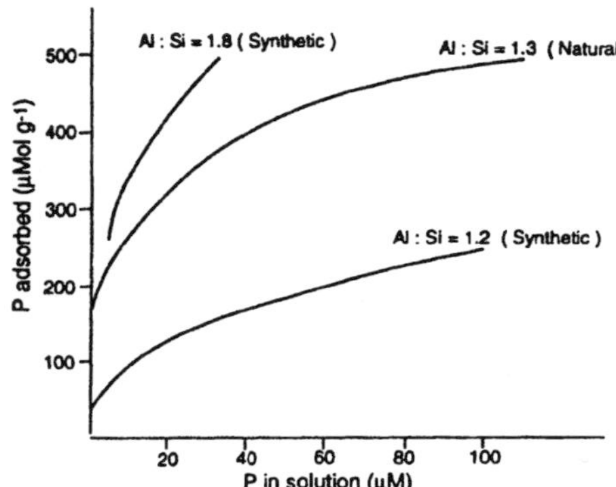

Figure 12.8. Phosphate adsorption in dependence of Al/Si ratio in synthetic allophanes. From Parfitt, 1990.

Question 12.7. *The pH-dependent charge of allophane or imogolite results from the chemical nature of their surface. Explain this, using Figures 12.3 and 12.4.*

Question 12.8. *Explain the difference in charge behaviour between the allophanic and non-allophanic andisols in Figure 12.7.*

Question 12.9. *Explain the effect of pH-dependent charge on the difference between pH_{water} and pH_{KCl} for the three andisol groups of paragraph 12.2.*

ANION ADSORPTION AND pH-NaF

The net positive charge of amorphous silicates at prevailing soil pH causes a strong adsorption of phosphate ions, a property that is used to identify andisols (phosphate retention, see Appendix 3). Phosphate adsorption capacity increases with the Al/Si ratio

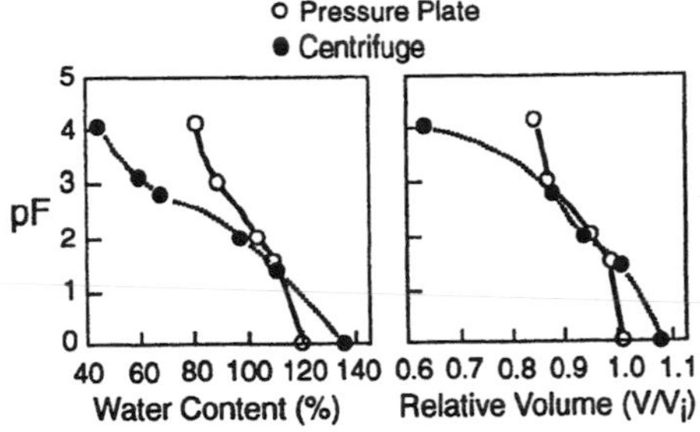

Figure 12.9. Water retention of a Japanese andisol measured by centrifuge and pressure plate. From Warkentin and Maeda, 1980.

of the allophane (Figure 12.8). Phosphate-binding groups are, Al-O-Si, R-COO-Al (organic ligands), and Al-OH groups.
High phosphate retention is due to (rapid) binding of $H_2PO_4^-$ to Al of the allophane structure. The amount of sorbed P varies between 50 and 200 μmoles per gram allophane, or 2-8 phosphate ions per allophane spherule.
Fluoride ions (F^-) form strong complexes with aluminium. Fluoride rapidly reacts with Al groups of the allophane surface, causing the release of OH^- groups, and consequently an increase in pH. Therefore, a strong rise in pH upon addition of a fluoride solution is sometimes used to signal the presence of allophane. If 1 gram of allophanic soil is dispersed in 50 ml 1M NaF, the pH usually rises to 10.5.

Question 12.10. *Explain why phosphate adsorption is stronger at higher Al/Si ratios in allophane (Figure 12.8).*

Because of the large surface area of allophane and the very finely porous nature of allophane aggregates, andisols can retain large amounts of water, even at high suctions. At pF = 4.2, water content in some allophanic soils may be as high as 200%. Water in allophanic soils is held in small voids rather than at mineral surfaces (in allophane-organic matter aggregates, we cannot properly speak of a particle 'surface').
This is illustrated by the difference in water content determined by pressure plate and by centrifuge method (Figure 12.9). Centrifugation destroys micropores, causing water retention at high suctions to decrease. Notwithstanding the high water binding capacity, plant-available water (held between pF 2-4.2) can be low in andisols: between 7 and 25 volume %.

Question 12.11. *Explain the relatively small amount of 'available water' in andisols in relation to the very high total water binding capacity.*

The fact that much water is held in small pores and that the pore geometry is easily disturbed by pressure explain the *thixotropy* of many allophanic soils. Thixotropic soil material appears moist or even dry when undisturbed, but when compressed, free water becomes visible and the soil becomes smeary (semi-fluid). This causes the specific slipperiness of andisol soil surfaces. Strongly weathered rock may also be thixotropic.

TEXTURE AND AGGREGATION

The high charge and colloidal nature of andisols makes dispersion difficult and the determination of particle-size distributions rather meaningless. This is illustrated by the effect of drying, which stabilises allophane aggregates, so that dried soil appears to have a coarser texture (Figure 12.10). Allophane, imogolite, ferrihydrite and organic matter have extremely small basic units, but they cluster strongly into aggregates of clay to silt-size (Buurman et al., 1997). The size of allophane aggregates appears to increase with allophane content (Figure 12.11), but the effect may also be due to a varying mixture of fine (ferrihydrite) and a coarser (allophane) aggregates.

Because the presence of allophane may inhibit dispersion, it is almost impossible to determine textures of both the allophanic and the non-allophanic mineral fraction. The non-allophanic fraction can be studied after removal of allophane through extraction. The resulting grain-size distribution gives insight in the size fractions of unweathered or partly weathered primary grains, and of phyllosilicate clays.

Question 12.12. *Explain the changes upon drying that are observed in Figure 12.10. What could be the practical relevance of this phenomenon?*

Question 12.13. *Ignoring the problem of grain-size determination, the calculated CEC of the 'clay' fraction is sometimes used to recognise allophanic soils. What is the effect of poor dispersion on this calculated CEC? (CEC_{clay}= CEC_{soil}*100/clay%).*

BULK DENSITY

The main components of andisols are porous, non-crystalline components with a high water retention capacity. As a result, andisols usually have a low bulk density, typically lower than 0.9 $g.cm^{-3}$, and sometimes even lower than <0.2 $g.cm^{-3}$. Because the soils may shrink strongly upon desiccation, the volume at pF2 (1 kPa) is taken as a reference. The combination of a large pore volume (high permeability) and a stable aggregation results in a low erodibility of Andisols on undisturbed slopes. Interruption of slopes may strongly increase mass wasting because of inherent sloping sedimentary contacts and lateral subsurface water transport.

12.5. Summary of properties and morphology

Summarising, andisols are formed during rapid weathering on rocks that have high amounts of weatherable materials. The very specific physico-chemical properties of the poorly crystalline secondary products (allophane and imogolite, ferrihydrite, organic complexes) dominate the soil properties.

Specific properties are: a high variable (pH-dependent) charge, strong adsorption of phosphates and reaction with fluoride, strong aggregation (low dispersion and clay contents), which increases upon drying, high water retention, high pore volume with non-rigid pores, and low bulk density.

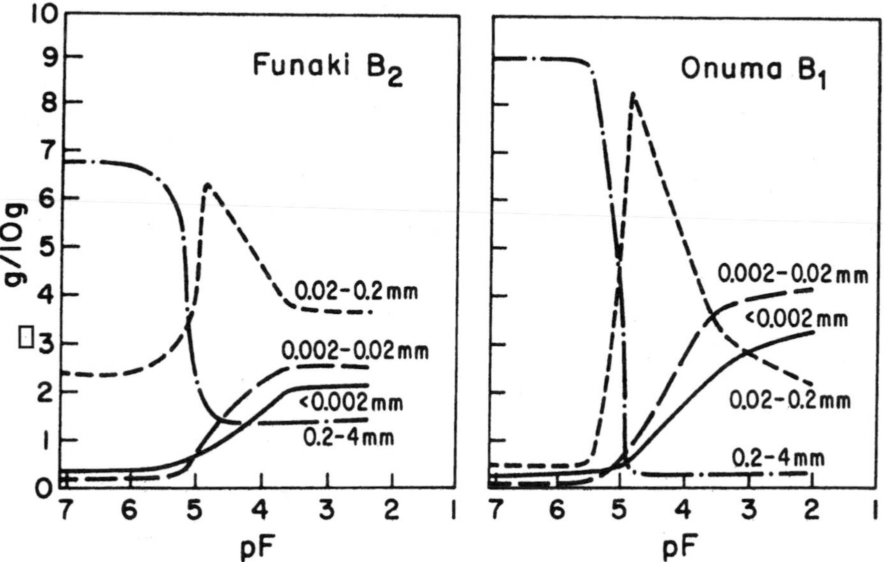

Figure 12.10. Measured particle size distributions of two allophanic soils after drying to different pF values. From Warkentin and Maeda, 1980.

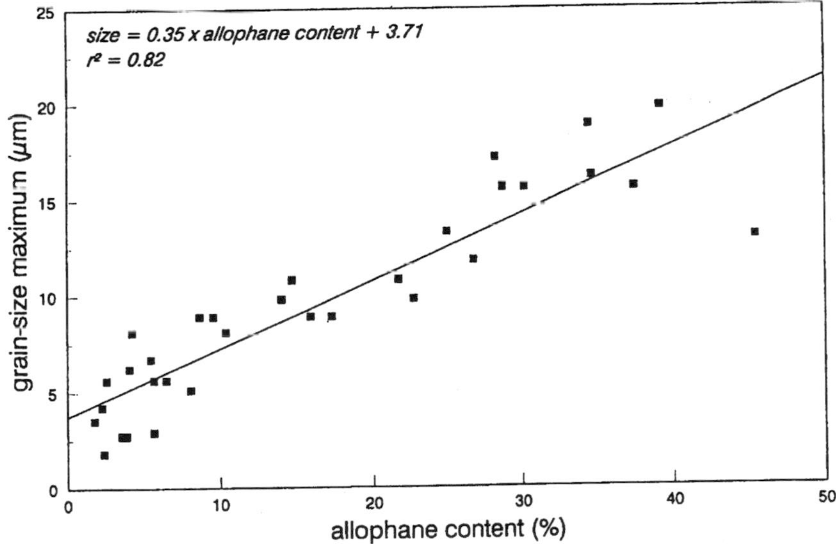

Figure 12.11. Relation between allophane content and aggregate sizes in Costa Rican andisols (field moist samples). From Buurman et al., 1997.

The strong interaction between organic matter and the abundant aluminium of andisols stabilises the organic matter so that organic matter contents are relatively high (up to 20%). Organic matter distribution with depth depends on the vegetation. Vegetations that bring much organic matter directly into the soil (root litter, grass vegetation) cause black, deep, high-organic matter A horizons, while regular burning may intensify the black colour and increase humus aromaticity. Stratification (deposition of fresh tephra during soil formation) causes irregular organic matter distribution with depth.

PROFILE MORPHOLOGY

Andisol profiles are built up of Ah, Bw and C-horizons (and transitions). Ah horizons increase in thickness and expression under grass and under humid climates; allophanic Bw horizons are also strongly expressed under humid conditions in well-developed soils.

Ah horizon development in andisols is related to climate and vegetation, and tends to be stronger in humid, cold than in warm and dry climates. Therefore, strongly humose Ah horizons are more common in Japan than in the tropics, while in the tropics humosity increases with altitude and rainfall.

Most andisols develop on volcanic deposits, especially on volcanic ejecta, which weather rapidly. Repeated eruptions cause burial of old soils and soil formation restarts on the new surface. Sequences of buried soils are therefore common. Sometimes dated eruptions allow estimates of the speed of soil formation.

Young andisols have little allophane. Allophane may become predominant after hundreds to thousands of years of weathering, and may in later stages convert to crystalline phyllosilicates, such as halloysite, kaolinite, and more rarely, smectites. Periodic desiccation favours crystallisation of allophane to phyllosilicates, which causes disappearance of the typical properties that accompany the presence of allophane. Andisols thus change into other soil types. Depending on climate, the next stages may be:

- Spodosols (Podzols), in cold and wet climates or on acid volcanics under strong leaching. Podzols may also form on rhyolitic tephras without prior andosol formation.
- Inceptisols and oxisols, in udic to ustic, warm climates.
- Vertisols, in accumulation positions in ustic, isohyperthermic climates.
- Mollisols, in ustic, isothermal climates

DIAGNOSTIC HORIZONS

For classification purposes, the general characteristics of andisols, such as high water retention, high contents of weatherable minerals and organic matter, low bulk density, high aluminium contents of aluminium, either as organic complexes or as amorphous silicates, have been combined into various diagnostic horizons. These reflect different stages of development (vitric horizon for young soils), the dominance of either aluminium organic complexes or amorphous aluminium silicates, and very dark topsoils.

Current criteria in the recognition of And*i*sols or And*o*sols (SSS, 1994; Deckers et al., 1999) are:

- an andic horizon
  In addition to a certain thickness and clay percentage, an *andic* horizon has a bulk density of < 0.9 kg.dm^{-3}, 2% or more oxalate-extractable (Al + 1/2 Fe); 70% phosphate retention, and less than 30% volcanic glass.
- a vitric horizon
  A *vitric* horizon is characterised by a dominance of volcanic glass over secondary weathering products.
- a fulvic horizon
  A *fulvic* horizon has, in addition to a specific thickness, a dark colour of the horizon, combined with a light colour of the oxalate extract, and at least 6% organic C.
- a melanic horizon
  A *melanic* horizon has a dark humus extract, and is otherwise similar to the fulvic horizon.
- A phosphate retention of more than 70 (85)%.
  The phosphate retention is described in Appendix 3.

Question 12.14. *Which aspects of soil genesis and properties of andisols are addressed by the various diagnostic horizons?*

12.6. Problems

Problem 12.1

The following table gives analyses of an andisol from Indonesia and one from Colombia (Mizota & Van Reeuwijk, 1989).

Profile CO 13, Colombia. Typic Hydrandept (Typic Hydrudand) or Humic Andosol

Depth cm	Hor.	Clay %	Org. %	pH H_2O	pH KCl	pH NaF	CEC	Bases	Al	Al_o %	Fe_o %	Si_o %	Al_p %	Fe_p %	water at pF 4.2 %	P ret. %
							cmol+ /kg									
0-12	Ah1	20	46.2	4.1	3.4	6.9	179	10	6.5	0.7	0.4	0.1	0.5	0.3	150	56
-30	Ah2	24	28.6	4.0	3.6	8.0	105	2	12.2	1.0	0.7	0.1	0.9	0.6	114	82
-51	AB	17	13.1	4.4	4.2	10.8	66	5	4.1	2.6	0.1	0.5	2.4	0.1	67	99
-78	Bh	32	15.9	4.4	4.2	10.9	122	0	4.5	7.1	0.5	1.6	2.7	0.4	106	99
-82	Bs	37	14.3	4.9	4.3	10.8	117	1	2.4	8.5	5.1	2.5	2.1	3.0	109	99
-98	BC	28	8.5	5.1	4.7	10.9	77	0	0.8	8.0	1.1	2.8	1.3	0.5	92	99
-120	C1	23	2.4	5.3	5.2	10.8	45	2	0.2	5.8	0.3	2.6	0.4	0.1	50	99
-150	C2	36	3.4	5.2	4.9	10.7	51	1	0.2	6.1	0.6	2.7	0.5	0.1	90	99

Profile INS 36, Indonesia, Typic Dystrandept (Hydric Hapludand) or Humic Andosol

Depth cm	Hor	Clay %	Org. C %	pH H_2O	pH KCl	pH NaF	CEC	Bases	Al	Al_o %	Fe_o %	Si_o %	Al_p %	Fe_p %	water 1500 kPa %	P ret. %
							cmol+ /kg									
0-30	Ah1	14	2.5	5.7	5.5	10.7	40	4	0	4.9	0.8	2.9	0.3	0.0	71	99
-44	C1	10	0.7	6.0	5.5	10.4	22	1	0	2.8	0.4	1.9	0.2	0.0		87
-100	B	22	1.9	5.6	5.4	10.5	55	2	0	5.5	1.0	3.6	0.3	0.0	91	99
-123	C2	10	0.4	5.6	5.2	10.2	26	1	0	3.5	0.4	2.5	0.1	0.0		99
-130	2A	28	2.4	5.6	5.2	10.3	64	5	0	6.2	1.9	3.9	0.8	0.3		99
-170	2B	34	1.6	5.4	5.1	10.2	89	5.	0	6.4	2.1	4.8	0.3	0.1	119	99

a. Check for each horizon whether it is allophane-dominated or Al-organic matter dominate
b. Allophane contents are calculated in two different ways. (1) By assuming 14% Si_o in alloph (Al/Si=2), and (2) by attributing all oxalate-extractable Al and Si (minus pyrophosph extractable Al) to allophane, and to change the Al/Si ratio of the allophane accordingly. T second procedures results in the following formula:

   allophane = $100*Si_o$ / $\{-5.1*(Al_o-Al_p)/Si_o)+23.4\}$

   Calculate allophane contents for all horizons in both ways. Put the results of both calculati in a graph with on the vertical axis allophane (2) and on the horizontal axis allophane (Discuss the difference between the results.
c. Calculate the Al/Si ratios for all allophanes of calculation (2). What is the reason for differences in Al/Si ratio in allophane (2) between the profiles and within one profile? Wh ratios are unlikely and what causes these deviations?
d. Suggest better horizon codes for both profiles, using all the chemical data and considering that both profiles may consist of two or more ash layers.

Problem 12.2

Contrary to other soils, physical properties of andisols do not change with the nature of the exchangeable cation. Explain this.

Problem 12.3

In many soils, allophane is present in coatings in pores and on peds. This allophane may later recrystallise to layer-lattice clays with strong parallel orientation which will show up in micromorphological studies of thin section as birefringence.
Such recrystallisation of allophane may cause confusion with a different soil forming process. Which process, and why?

Problem 12.4

Shaking 1 gram of allophanic soils with 50 ml 1M NaF increases the pH to 10.5.

a) Calculate the amount of OH^- replaced by F^- (in cmol/kg soil), assuming that the original pH of the NaF solution is 7.0.

b) What assumptions to you have to make with respect to 1) exchangeable acidity, 2) dissociation of the organic matter?
c) If the specific surface of allophane is 700 $m^2.g^{-1}$, what is the charge density ($cmol^+.m^{-2}$).

Problem 12.5
The following table gives the rate of release of silica and cations (10^{-9} $mol.kg^{-1}.s^{-1}$) of two fresh volcanic ash fractions in demineralised water and dilute HCl, and the pH of the extractant after 2 days (HCl) and 7 days (water). From Jongmans et al., 1996.

Fraction	Solution	pH	Si	Al	Ca	Mg	Na
<53 μm	10^{-3}M HCl	3.77	26.6	23.6	20.8	1.04	39.4
>500 μm	10^{-3}M HCl	3.48	17.6	18.5	12.2	0.64	35.9
<53 μm	water	5.81	2.35	0.03	2.05	0.02	8.27
>500 μm	water	5.88	1.12	0	1.67	0.05	5.15

a) Describe and explain the differences in pH and release elements in dependence of extractant and grain-size.
b) How many kilograms of Si would be released at pH 5.81 from 1 kilogram of fraction <53 μm ash in one year, provided that there is sufficient percolation and that the release would be constant in time?
c) And how much would be released in one year from the top 10 cm of one hectare of a silty soil (100% < 53 μm)? Assume a bulk density of the ash of 1.0 $kg.dm^{-3}$.

12.7. Answers

Question 12.1
K-feldspars and micas belong at the most resistant end of the weathering sequence.

Question 12.2
The Al_p/Al_o ratio is the ratio of organically bound Al to total amorphous Al, or the relative amount of Al that is in organic complexes. From the Ah to the Bw horizon in an allophanic soil, the ratio should decrease.

Question 12.3
The structure of Figure 12.3 has an Al/Si ratio of 1:1. To obtain a ratio of 2:1, 1/3 of the silica atoms should be replaced by Al.

Question 12.4
Allophane structure consists of one sheet, in which both Si and Al are placed. All the end atoms are OHs. The structure is idealised and not really repetitive. Apart from the

curvature, imogolite structure is somewhat similar to a 1:1 phyllosilicate structure, but the tetrahedra are reversed (top of tetrahedron points away from the octahdral sheet) and the top oxygen is an OH. The 'tetrahedral' sheet contains about 1/3 aluminium substitution.

Question 12.5

If the general formula is $Al_2O_3.SiO_2.nH_2O$, and the Si content is 14%, this means that the molecular weight must be 200 (Si=28, Al=27, H=1, O=16). The molar weight of the formula without water is 54+48+28+32=162, so the remaining 38 must be water. This equals 38/18=2.1 moles of water in the structural formula (=19 weight %).

Question 12.6

If silica cannot combine with Al to form allophane, it is either leached from the soil (good drainage) or precipitated as opal (impeded drainage).

Question 12.7

The surfaces of allophane and imogolite consist of OH-groups. These groups will associate or dissociate in dependence of pH.

Question 12.8

Allophanic andisols have both positive and negative charge. The positive charge is due to dissociation of OH-groups from the colloid surface, the variable negative charge is due to organic matter. The allophanic andisol has a PZNC of around pH 5.5.
The non-allophanic andisol does not have positive charge (no positively charged allophane or ferrihydrite surfaces) and only a pH-dependent negative charge due to organic matter. The soil non-allophanic soil does not have a PZNC.

Question 12.9

Non-allophanic andisols have both a low pH and a large difference between pH_{water} and pH_{KCl}. As seen in Figure 12.7, such soils have only negative charge, and the addition of KCl removes H^+ ions from the adsorption surface, but no OH^- ions.
In allophanic andosols, pH values are higher - but below the PZNC of allophane - and the difference between the two pH values is smaller. The system is less influenced by organic acids (higher pH), and addition of electrolyte removes both H^+ and OH^- ions from the adsorption surface - in this case more H^+.

Question 12.10

At low Al/Si ratios, part of the Al will be in tetrahedral co-ordination and cannot have a positive charge. At higher Al/Si ratios, the number of amphoteric groups increases, and consequently the amount of positive charge at soil pH.

Question 12.11

The high water contents are mainly due to abundance of very small pores. These pores do not yield their water between pF 2 and 4.2.

Question 12.12

In both soils, the fraction < 2μm disappears completely when the soil is dried to pF 5. At tensions higher than pF=3.5, small fractions aggregate to for a fraction of 0.02-0.2 mm. At tensions above pF=5, also this fraction disappears, and larger aggregates (0.2-4 mm) dominate.

NB. Air-dry samples - the normal preparation before analysis in the laboratory, have pF values in excess of 5. So, if one is interested in the natural state of the soil, andisol samples should not be dried before analysis.

Irreversible drying to coarser aggregates leads to a loss in water-binding capacity.

Question 12.13

Incomplete dispersion results in an underestimate of the clay fraction. Calculated CEC_{clay} will therefore be too high, which is frequently seen in andisols.

Question 12.14

The *andic horizon* identifies the presence of well-developed allophanic soils.

The *vitric horizon* identifies soils that have the potential to develop and andic horizon, but are still too young.

The *fulvic horizon* is defined by a high percentage of (oxalate)-dissolvable organic matter. Because oxalate is an acid extractant (pH3), the organic matter is supposed to consist of 'fulvic acids'.

The *melanic horizon* has a very dark colour of the oxalate extract, which reflects dark organic matter, probably charred material which can be extracted by oxalate.

The *phosphate retention* reflects high quantities of reactive surfaces on amorphous alumino silicates and/or ferrihydrite.

Problem 12.1

a) Profile CO13 is strongly organic matter-dominated in the upper three horizons (both total C and Al_p/Al_o). From 51 cm downward, the allophane content increases and Al_p/Al_o decreases strongly. In profile INS 36, organic matter is unimportant and the profile is dominated throughout by allophane.
b) Allophane contents and Al/Si ratios are given in the table on the following page.
c) In CO-13, the second allophane calculation leads to unrealistic Al/Si ratios. These may be due to 1) Al_p values that are too high or too low, and 2) dissolution of amorphous aluminium hydroxide and/or 3) partial desilication of allophane in the topsoil. In INS 36, the results of the two methods are very similar, and the Al/Si ratio is almost constant throughout the profile.
d) *CO-13:* Bh and (Bs) horizons are unlikely in andisols. The Al_p/Al_o and Fe_p/Fe_o ratios of these horizons indicate that organic complexes do not dominate. The increase in clay content and (slight) in C, suggests that the Bh is in fact a buried Ah (2Ah); the Bs is probably a 2Bw.

   INS-36: The correct horizon sequence is: Ah - C1 - 2Ah - 2C - 3Ah -3B.

CO 13	*Ah1*	*Ah2*	*AB*	*Bh*	*Bir*	*BC*	*C1*	*C2*
allophane 1	0.7	0.7	3.6	11.4	17.9	20.0	18.6	19.3
allophane 2	0.8	0.5	2.3	17.1	24.2	25.0	20.3	21.1
Al/Si in (2)	2.0	1.0	0.4	2.7	2.6	2.4	2.1	2.1
INS 36	*A*	*C1*	*B*	*C2*	*2A*	*2B*		
allophane 1	20.7	13.6	25.7	17.9	27.9	34.3		
allophane 2	18.9	11.6	22.5	15.2	23.9	28.4		
Al/Si in (2)	1.6	1.4	1.4	1.4	1.4	1.3		

Problem 12.2
Andisol structure is stable, irrespective of the saturating cation, because aggregation is not caused by the interaction of van der Waals forces and double layer thickness. The charge of andisols does depend on the ionic strength of the solution, but this is hardly relevant in the range of soil solution concentrations.

Problem 12.3
Recrystallization of allophane coatings to phyllosilicates causes birefringent coatings that are very similar to those formed by clay illuviation (Chapter XXX). The formation of allophane coatings is a combination of dissolution, transport, and precipitation, while clay illuviation is mechanical transport. The conditions are completely different.

Problem 12.4

a) At pH 7, OH^- activity equals 10^{-7}; at pH 10.5, this is $10^{-3.5}$.
In 50 ml at pH 10.5, the total amount of OH^- equals $50*10^{-3.5}$ mmol, or $16*10^{-3}$ mmol. At pH 7, the total amount equals $5 * 10^{-6}$ mmol.
The increase in OH^- equals $16 * 10^{-3}$ mmol. This was released by 1 gram of soil, so the total release equals 1.6 cmol OH per kg soil.
b) Assumptions are: no increased dissociation of organic matter; no change in activities.
c) At a surface area of 700 m^2 per gram for allophane, this equals 1.6 cmol per 700000 m^2, or $2.3*10^{-6}$ $cmol.m^{-2}$.

Problem 12.5

a) Release of elements from the fine ash fractions is faster than that from the coarse ash fractions. This should be due to the larger contact surface of the fine ash fraction. This faster release is also expressed in the higher pH of the HCl-ash mixture. If silica release is taken as a reference (dissolution of silica is independent of pH), weathering in 0.001 M HCl is of course much faster than in water (2 pH units difference), but the striking difference between the two is in the release of Al. There is virtually no release of Al in water, which means that it will be accumulated as a hydroxide in the residue. In water, there is a relatively larger release of Na with respect to Ca and Mg. This means that the release in water is probably not from feldspars, but from volcanic glass (feldspar weathering alone would give a constant Na/Ca ratio).

b) The release of Si from this fraction equals $2.35 * 10^{-9}$ $mol.kg^{-1}.sec$. During one year and one kilogram of ash, this would be 60*60*24*365 times as much, or about $74*10^{-3}$ mol. This is 74*28= 2.072 $g.kg^{-1}.yr^{-1}$. (= 0.2% of the total weight).

c) For one hectare to a depth of 10 cm, this would amount to $10^6 * 0.002072$ $kg.yr^{-1}$, or 2072 $kg.yr^{-1}$.

12.8. References

Buurman, P., K. de Boer, and Th. Pape, 1997. Laser diffraction grain-size characteristics of Andisols in perhumid Costa Rica: the aggregate-size of allophane. Geoderma, 78:71-91.

Buurman, P., and A.G. Jongmans, 1987. Amorphous clay coatings in a lowland Oxisol and other andesitic soils of West Java, Indonesia. Pemberitaan Penelitian Tanah dan Pupuk, 7:31-40.

Dahlgren, R., S. Shoji, and M. Nanzyo, 1993. Mineralogical characteristics of volcanic ash soils. In: S. Shoji, M. Nanzyo and R. Dahlgren: *Volcanic ash soils - genesis, properties and utilization*. Developments in Soil Science 21, Elsevier, Amsterdam: 101-143.

Deckers, J.A., F.O.Nachtergaele, and O.C. Spaargaren (Eds.), 1998. *World reference base for soil resources*. ISSS/ISRIC/FAO. Acco, Leuven/Amersfoort, 165 pp.

Henmi, T., and K. Wada, 1976. Morphology and composition of allophane. American Mineralogist, 61:379-390.

Inoue, K., 1990. Active aluminium and iron components in andisols and related soils. Transactions 14th International Congress of Soil Science, Kyoto. VII:153-158.

Jongmans, A.G., F. Van Oort, A. Nieuwenhuyse, A.M. Jaunet, and J.D.J. van Doesburg, 1994a. Inheritance of 2:1 phyllosilicates in Costa Rican Andisols. Soil Science Society of America Journal, 58:494-501.

Jongmans, A.G., F. Van Oort, P. Buurman, and A.M. Jaunet, 1994b. Micromorphology and submicroscopy of isotropic and anisotropic Al/Si coatings in a Quaternary Allier terrace. In: A. Ringrose and G.D. Humphries (eds). *Soil Micromorphology: studies in management and genesis*. Developments in Soil Science 22: 285-291. Elsevier, Amsterdam.

Jongmans, A.G., J. Mulder, K. Groenesteijn, and P. Buurman, 1996. Soil surface coatings at Costa Rican recently active volcanoes. Soil Science Society of America Journal, 60:1871-1880.

Leamy, M.L., G.D. Smith, F. Colmet-Daage, and M. Otowa, 1980. The morphological characteristics of andisols. In: B.K.G. Theng (ed): *Soils with Variable Charge*, 17-34. New Zealand Society of Soil Science, Lower Hutt.

Mizota, C., and L.P. van Reeuwijk, 1989. *Clay mineralogy and chemistry of soils formed in volcanic material in diverse climatic regions*. International Soil Reference and Information Centre, Soil Monograph 2: 1-185. Wageningen.

Nanzyo, M., R. Dahlgren, and S. Shoji, 1993. Chemical characteristics of volcanic ash soils. In: S. Shoji, M. Nanzyo and R. Dahlgren: *Volcanic ash soils - genesis, properties and utilization*. Developments in Soil Science 21, Elsevier, Amsterdam: 145-188.

Nanzyo, M., S. Shoji, and R. Dahlgren, 1993. Physical characteristics of volcanic ash soils. In: S. Shoji, M. Nanzyo and R. Dahlgren: *Volcanic ash soils - genesis, properties and utilization*. Developments in Soil Science 21, Elsevier, Amsterdam: 189-207.

Parfitt, R.L., 1990a. Allophane in New Zealand - a review. Australian Journal of Soil Research 28:343-360.

Parfitt, R.L., 1990b. Soils formed in tephra in different climatic regions. Transactions 14th International Congress of Soil Science, Kyoto. VII:134-139.

Parfitt, R.L., & J.M. Kimble, 1989. Conditions for formation of allophane in soils. Soil Science Society of America Journal 53:971-977.

Parfitt, R.L., R.J. Furkert, and T. Henmi, 1980. Identification and structure of two types of allophane from volcanic ash soils and tephra. Clays and Clay Minerals, 28:328-334.

Shimizu, H., T. Watanabe, T. Henmi, A. Masuda, and H. Saito, 1988. Studies on allophane and imogolite by high-resolution solid-state ^{29}Si- and ^{27}Al-NMR and ESR. Geochemical Journal, 22:23-31.

Shoji, S., R. Dahlgren, and M. Nanzyo, 1993. Genesis of volcanic ash soils. In: S. Shoji, M. Nanzyo and R. Dahlgren: *Volcanic ash soils - genesis, properties and utilization*. Developments in Soil Science 21, Elsevier, Amsterdam: 37-71.

Shoji, S., M. Nanzyo, R.A. Dahlgren, and P. Quantin, 1996. Evaluation and proposed revisions of criteria for andosols in the world reference base for soil resources. Soil Science, 161:604-615.

Van Reeuwijk, L.P., and J.M. De Villiers, 1970. A model system for allophane. Agrochemophysica, 2:77-82.

Wada, K., 1980. Mineralogical characteristics of andisols. In: B.K.G. Theng (ed): *Soils with Variable Charge*, 87-108. New Zealand Society of Soil Science, Lower Hutt.

Warkentin, B.P., and T. Maeda, 1980. Physical and mechanical characteristics of andisols. In: B.K.G. Theng (ed): *Soils with Variable Charge*, 281-302. New Zealand Society of Soil Science, Lower Hutt.

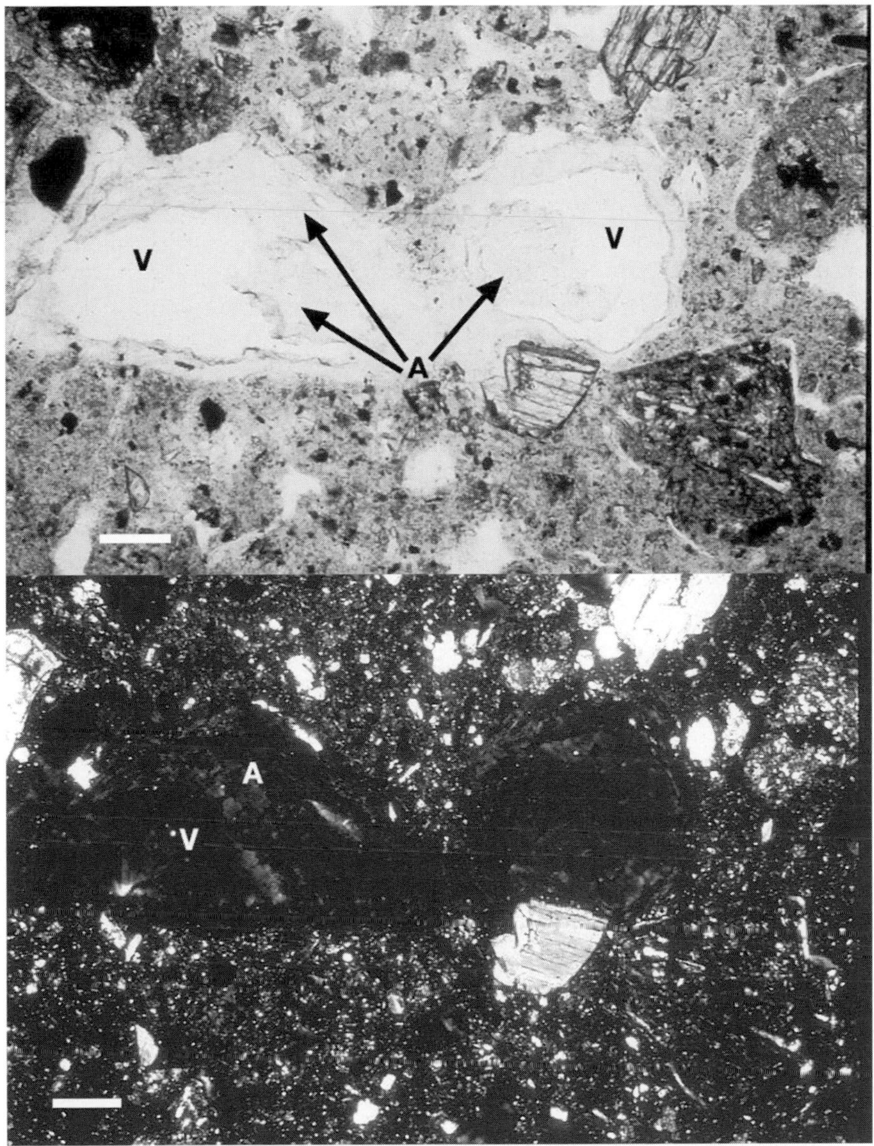

Plate U. Allophane coatings in an Andosol fromGuadeloupe. Top: plain light, bottom: crossed polarisers. Note that allophane is almost colourless in pain light, and isotropic in polarised light. A = allophane; V = void. Scale bar is 215µm. Photographs A.G. Jongmans.

Plate V. Weathering of plagioclase and deposition of allophane. Andosol from Costa Rica. Top; plain light; bottom: crossed polarisers. Light parts in bottom photograph are plagioclase remnants (P). All cracks are filled with allophane (A), which is isotropic under crossed polarisers. Scale bar is 135μm. Photographs A.G. Jongmans

CHAPTER 13

FERRALITISATION

13.1. Introduction

CHARACTERISTICS
We will use the term 'ferralitisation' for the processes associated with strong weathering that lead to the formation of Oxisols (USDA) or Ferralsols (FAO). Soils of these orders have an oxic (USDA) or ferralic B (FAO) horizon that is characterised by the effects of extreme weathering of minerals in all grain-size fractions. Practically all weatherable minerals have been removed from the sand and silt fractions. In the clay fraction, weathering has caused a dominance of kaolinite, gibbsite, and iron minerals, resulting in a low CEC (<16 cmol(+)/kg at pH=7) and a low cation retention (<10 cmol(+)/kg at soil pH). Typical of 'ferralitic' weathering is the removal of silica (desilication) from primary silicates and even from quartz. Desilication leads to residual accumulation of (hydr)oxides of Fe, (Mn), and Al (ferralitisation). Ferralitic weathering also involves a strong depletion of basic cations. As a result of strong weathering of primary minerals, silt contents (2-20 µm) are relatively low (silt/clay ratio in most oxisols <0.15). Desilication and ferralitisation are slow processes; their effects are sufficiently pronounced to cause the presence of Oxisols (Ferralsols) only in old soils of tropical humid climates. Soils with oxic horizons are common also in ustic and dryer climates, but these were formed during a preceding wet climate and are, as such, paleosols.

Question 13.1. *Other terms, used in different publications for ferralitic features are:* ***ferricrete, laterite, laterite, lateritic soils, latosols, and petroferric material.*** *Look these up in the Glossary and provide equivalents used in this text.*

13.2. Desilication

RELATIVE MOBILITY
In soils with excess precipitation over evapotranspiration, the relative mobility of the main soil components (Si, Al, Fe, bases) in solution are determined by their relative solubilities. These, in turn, depend mainly on pH, organic matter, and Eh.

In soil with a high production of organic acids, such as podzols and other soils of cool climates, the solubility of Fe and Al is enhanced by formation of organic complexes that are relatively soluble. Quartz has a low solubility, while only part of the silicate SiO_2

A confusion of terms

Various schools of soil science and of economic geology have used different terms for features associated with ferralitisation. A long list is given by Aleva (1994). We will use the following terminology:

Bauxite - an aluminium ore formed by ferralitisation. Its main constituents are aluminium minerals (gibbsite, boehmite, diaspore, corundum) and iron minerals (goethite, hematite).

Ferralic weathering or ferralitisation - a soil genetic process that leads to removal of bases and silica, thus causing a relative accumulation of Al- and Fe- compounds.

Ferralsols - in FAO classification: soils that have dominant characteristics resulting from ferrallitic weathering.

Oxisols - in USDA classification: soils that have dominant characteristics resulting from ferralitic weathering and/or by the presence of plinthite.

Petroplinthite - an irreversibly hardened soil layer with (absolute) accumulation of iron and aluminium (hydr)oxides.

Plinthite - material common to the gley zone in the subsoil of Oxisols (or other highly weathered soils), that hardens irreversibly upon drying. Gley zones that do not harden should not be called 'plinthite'.

in weathered primary minerals is removed by leaching. As a result, in most temperate humid soils that are rich in quartz, desilication has little effect on the total silica content because quartz is enriched residually.

Question 13.2. *What happens to the Si from weathered silicates that is* <u>*not*</u> *removed by leaching?*

In all, except the most nutrient-poor well-drained soils of the humid tropics, soluble organic acids formed in the A-horizon are decomposed rapidly, and are replaced by the weaker, and non-complexing acid H_2CO_3. This acid cannot lower the pH of the soil below 5. At pH above 5, silica (even from quartz) has a higher solubility than (hydrous) oxides of iron and aluminium. This means that silica is preferentially removed from the system. Such desilication often leads to a residual accumulation of Al and Fe.

Question 13.3. *Why does the solubility of Fe and Al oxides increase with decreasing pH, while that of quartz does not change between pH 3-8? Consider the dissolution reactions of FeOOH,* $Al(OH)_3$ *and* SiO_2 *to respectively* Fe^{3+}, Al^{3+}, *and* H_4SiO_4.

Aluminium and iron compounds are not completely inert, and Fe and Al can be moved during desilication. Fe is especially mobile in waterlogged conditions, while both aluminium and iron are somewhat mobile in acid, highly infertile soils with slow decomposition of organic matter.
In soils with temporary or permanent waterlogging, the mobility of iron (as Fe^{++}) is increased several orders of magnitude relative to that of Al and Si, and Fe may be redistributed in, removed from or added to a soil, depending on its landscape position and water movement.
If the pH drops below 4.5, which may happen in soils in the presence of organic acids, aluminium is much more soluble than silica and iron, and may be leached to some depth in the profile. In case of appreciable complexation of Fe and Al by dissolved organic matter, the dynamics may change completely (see chapter Podzolisation).

Differences in relative mobility are reflected by differences in SiO_2/Al_2O_3, SiO_2/Fe_2O_3, and Al_2O_3/Fe_2O_3 ratios of clay and non-clay fractions in different soil horizons (see Problem 13.2).

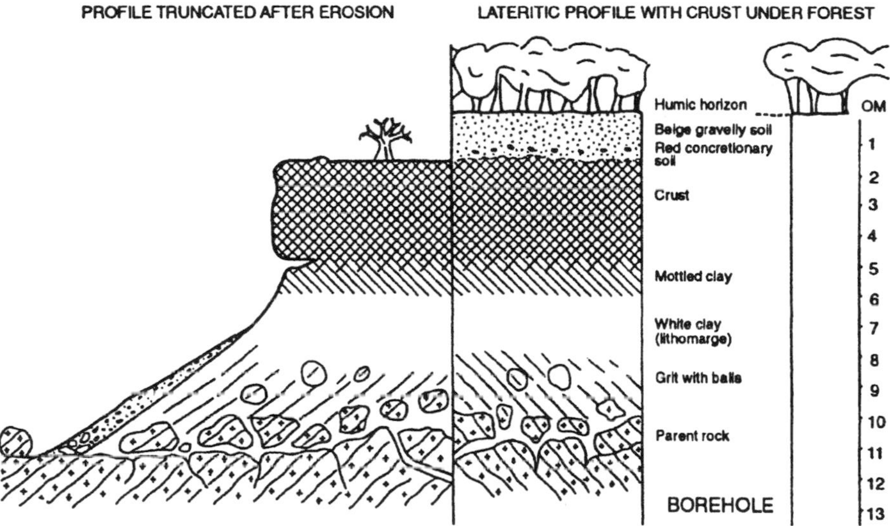

Figure 13.1. A typical ferralitic soil with petroplinthite, and its eroded equivalent, from the Sudan. From Millot, 1970. Reprinted with permission of Springer Verlag GmbH & Co., Heidelberg.

13.3. Formation and profile of a ferralitic soil with (petro)plinthite

Ferralitic soils may have a hardened zone of iron accumulation in the subsurface, and a 'complete' profile is usually depicted with such an accumulation (Figure 13.1). Most ferralitic soils, however, do not contain plinthite or petroplinthite. The profile of Figure

13.1 shows the typical horizon sequence of Ah and B-horizon ('Humic horizon' and 'Beige gravelly soil' in the figure), overlying a cemented iron crust. The crust overlies mottled clay (iron mottles or nodules in a bleached matrix, which overlies white clay. Below the white clay, there is a gradual transition to saprolite and parent rock.

DEVELOPMENT STAGES

Several stages can be distinguished in the formation of a ferralitic soil on a dense felsic rock in a continuously wet climate with high annual rainfall (>2000 mm). Three of these are depicted in Figure 13.2.

- In stage 1, the soil is shallow and has free drainage. Weathering products are easily removed. Primary minerals are fractured. Plagioclases and quartz weather faster than potassium feldspars and micas. Solute concentrations, including those of H_4SiO_4 are very low, so secondary minerals are mainly gibbsite and goethite.

- As the weathering front moves downwards (stage 2), solute concentrations become high enough to permit kaolinite formation. Formation of kaolinite in B- and C-horizons reduces permeability and leaching, and temporary water stagnation in the deeper part of the profile may occur during the wet season. Iron is periodically reduced and redistributed, which leads to mottled clay with absolute iron accumulation (Cgm, plinthite) towards its top. The weathering front is now water-saturated much of the time, and leaching is further impeded, causing still higher solute concentrations. Both kaolinite and smectite clays may form in the saprolite.

- Erosion removes part of the topsoil, and lowering of the erosion level also lowers the periodic water table (stage 3). The top of the mottled zone moves downwards, as does the zone of iron accumulation; the thickness of the iron accumulation zone increases. Above the parent rock, the soil is permanently water-saturated and becomes depleted of iron (pallid zone, Cr). Leaching is still strong in the topsoil, but much less in the subsoil. The zone of rock weathering is predominantly anoxic.

- Lowering of the ground-water table increases drainage in the upper part of the profile, and desilication may cause dissolution of kaolinite and formation of gibbsite.

- In Figure 13.2, the profile is placed on a plateau. This implies that there is no addition of iron from other parts of the landscape. Within the plateau, however, there may be considerable redistribution of iron through lateral groundwater flow: depletion in the middle, and accumulation at the fringes (see also section 13.4).

- Following a change towards a drier climate, the topsoil may be eroded, exposing petroplinthite at the surface.

Question 13.4*. Explain why gibbsite is a major weathering product in stage 1, and why kaolinite and even smectite form in stage 2, as leaching is hampered increasingly. Refer to the silicate stability diagram of Figure 3.1.*

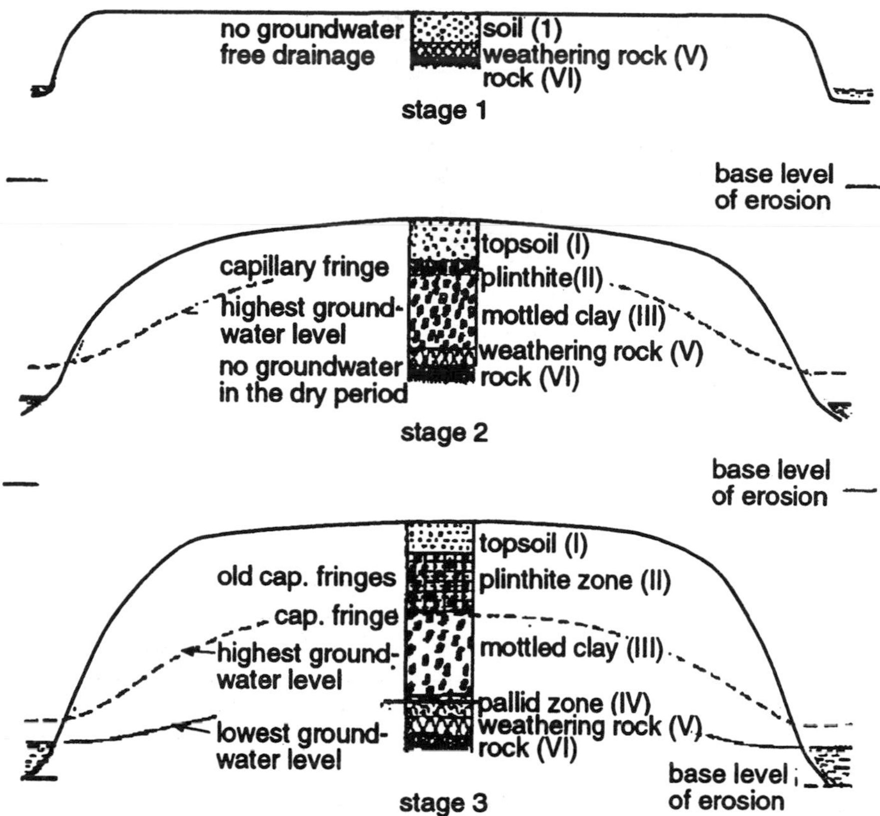

Figure 13.2. Three stages of Oxisol formation. From Mohr et al., 1972.

Question 13.5. *Explain the terms 'relative' and 'absolute' accumulation?*

Question 13.6. *What is the source of the iron that accumulates in the Cg horizon? Why does iron accumulate at the top of the mottled horizon?*

Question 13.7. *What are the causes of iron depletion in the 'pallid zone'?*

The development sequence illustrates that the weathering environment at the rock/saprolite/soil interface changes with time and that great differences exist between the weathering environment of the top and the bottom of the profile. These differences are clearly reflected in the mineralogy of the clay fraction (section 13.5).

On mafic rocks, the weathering is different. The lower amounts of available silica decrease kaolinite formation and favours the formation of gibbsite. In addition, mafic rocks have much more iron than felsic rocks, which results in higher iron contents in the soil and redder colours. Part of the Al from weathered silicates is incorporated in Fe oxides (isomorphic substitution). Higher iron contents cause stronger soil structure, which maintains a high permeability. Leaching is therefore stronger in such soils than in ferralitic soils on felsic rocks, and formation of plinthite is less common.

FERRALITISATION AND KAOLINISATION
In French literature, the process on mafic rocks, which leads to little phyllosilicate clay formation, is called 'ferralitisation', while the process on the much less permeable felsic rocks, which causes formation of abundant kaolinite, is called 'kaolinisation'.

13.4. Plinthite and iron pans

The segregation of iron in the mottled and pallid zones typical of oxisols in low-relief landscapes on felsic rocks can be attributed to normal gley processes (see Chapter 7), with only local redistribution of iron (and manganese) compounds. The mottled zone of ferralitised soils, however, is different from normal gley horizons in that it may harden irreversibly upon exposure to air. Because this material was extensively used for making bricks in, e.g., India and Thailand (Fig. 13.3), it was named *laterite* (later (Latin) = brick), a name that was used in the past for all strongly weathered tropical soils without a textural contrast. The material that can be dug with a spade when moist, but hardens upon exposure is now called *plinthite*. The hardening is presumably due to dehydration of hydrous iron oxides and of small amounts of amorphous silica that act as a cement. The hardened material itself is not called plinthite, but ironstone or petroferric material. Figure 13.2 indicates that the plinthite zone may grow downwards upon lowering of the capillary fringe. Many oxisols have one or several old plinthite horizons overlying the mottled zone that is related to the present groundwater table.
In addition to local redistribution, landscapes with lateral movement of groundwater may have considerable accumulation of iron compounds at sites where the water is aerated, such as valley sides. Upon desiccation following climate change or erosion, these *groundwater laterites* may develop into strongly indurated pans or *petroplinthite* (Figure 13.4).

HORIZON SEQUENCE
The 'complete' ferralitic soil has the following horizons (FAO horizon designations):

- Ah - humic topsoil, to be discussed below; 0.1-0.5 m thick;
- Bs - B-horizon with relative accumulation of aluminium and iron compounds; up to ten metres thick
- Cgm- A cemented iron accumulation; up to several metres thick;
- Cg - A mottled, non-cemented iron redistribution zone; up to 1 or 2 metres thick;
- Cr - A reduced, iron-depleted zone; up to several metres thick
- CR - The saprolite; up to tens of metres thick;
- R - The parent rock

Figure 13.3. A 'laterite' quarry with drying bricks in Burkina Faso. From Aleva, 1994.

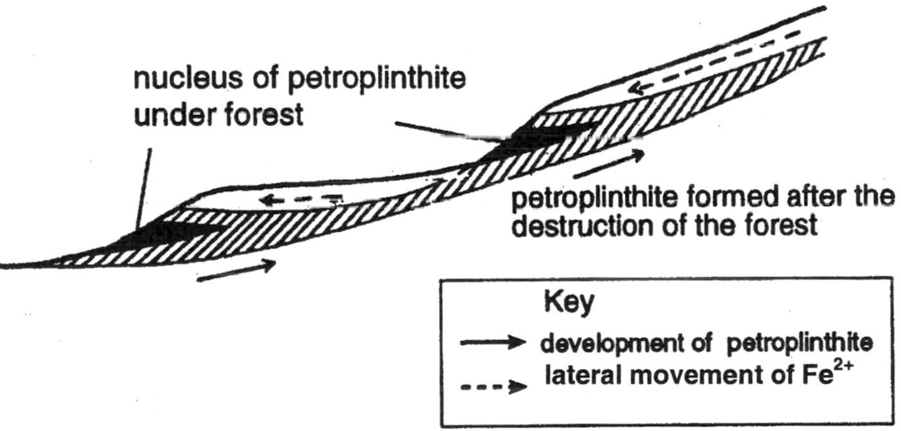

Figure 13.4. Formation of an iron pan (cuirasse) as a result of lateral water flow, deforestation, and erosion. From Duchaufour, 1977.

As indicated, the thickness of each of these horizons can vary considerably. Not all horizons are necessarily present in a given soil.

Question 13.8. *Explain from Figure 13.4: a) the upward growth of the iron accumulation, and b) why destruction of the forest leads to a second iron pan (take account of changes in the landscape).*

Question 13.9. *The sequence of horizons given above is for a ferralitic soil on felsic rocks. What differences do you expect in horizon sequence and thickness in similar soils on mafic rocks? Consider the mineralogy of the rock and the different nature of ferralitic weathering on mafic rocks and on felsic rocks.*

While surface horizons vary with vegetation, land use, and climate, the oxic (B) horizon itself is usually very homogeneous and has gradual to diffuse boundaries between subhorizons. These gradual boundaries are partially due to a very high activity of e.g. termites and ants in tropical rain forests and savannahs (Plate G and H, p. 82). The combination of high biological activity and high contents of free iron and aluminium usually leads to a very stable, porous structure, consisting of silt-size micro aggregates, which may aggregate to crumbs of up to 0.5 cm (Plate J, p. 140). The strong aggregation precludes the movement of clay particles. High biological activity may lead to a stoneline (usually of quartz gravel) in the subsoil, if such fragments are present in the parent rock. The combination of high biological activity and surface erosion may cause loss of fine fractions from the topsoil (see also Chapter 8).

The main factors in aggregation are organic matter, iron minerals, and net charge. Aggregation by organic matter, which causes sand-size aggregates, is only present in the topsoil. Aggregation by iron, which usually causes silt-size aggregates, depends on the form of the iron minerals. Well-crystalline hematite, which is formed in the saprolite and is frequently found as micron-sized grains, causes very little aggregation, while poorly ordered, finer goethite and hematite, e.g., caused by gleying, cause strong aggregates. Free aluminium keeps clay flocculated, but does not cause strong aggregation (Muggler et al., 1999). It appears that aggregation is strongest as the pH approaches the Point of Zero Net Charge (see 13.6).

Pore volume increases with weathering and the large pore volume accounts for a relatively low bulk density of the dry soil, especially on mafic parent materials. Notwithstanding the high biological activity and high porosity, oxisols may contain stable, non-porous inclusions of sand size ('soil nodules', sometimes up to 1 cm diameter) that are more or less inert entities. Such nodules are also called *pseudo sand.* They form either by physical fragmentation and isolation of clayey material, or as faunal excrements.

Question 13.10. *The 'soil nodules' are more or less isolated from the freely draining soil solution. What does this imply for possible mineralogical composition of such nodules?*

CLAY DISPERSION
Because strong weathering causes the disappearance of virtually all primary minerals, and accumulation of clay-sized secondary minerals, ferralitisation leads to clayey soils. Ferralitic soils on felsic parent materials (e.g. granite), however, can be quite sandy because of relatively high quartz contents. The strong aggregation effectively counters dispersion of clay in water. This causes low contents of so-called 'water dispersible clay'. Absolute clay percentages can only be determined after removal of organic matter and iron compounds. Because of this problem, the absolute clay percentage of such soils is frequently estimated from the water retention at pF 4.2:

clay = 2.5 * [water at pF 4.2] (water is expressed as % of dry soil).

This indirect determination is possible because the clay in stable, porous aggregates does interact with water.

Although ferralitisation may occur on rocks that are poor in iron, many oxic horizons are red due to iron oxides, of which over 90% of the iron compounds is well-crystalline. Because of higher iron contents, soils on mafic rocks are usually redder than on felsic rocks.

13.5. Mineralogical profiles

CLAY MINERALS
A sequence of mineral composition with depth of a typical Oxisol (Ferralsol) on basalt is given in Figure 13.5. At the weathering front, smectite is formed in cracks of ferro-magnesian minerals, vermiculite in cracks of biotite.
In addition, weathering of primary minerals may result in local concentrations of silica, aluminium, and cations that are sufficiently high to form halloysite and iron-rich smectites (nontronite). Smectite is only found close to the weathering front. Higher up in the profile it is subject to desilication to kaolinite or iron oxides, and disappears. The smectite zone moves downward with the weathering front. If leaching just above the weathering front is sufficiently rapid to preclude a build-up of solute concentration, smectite formation is restricted to cracks in primary minerals, and virtually all liberated aluminium is recombined with silica to form halloysite or kaolinite. Halloysite is not stable and is gradually converted to kaolinite. Kaolinite, once formed, is usually stable. The solubility of remaining quartz (4-6 mg/L of SiO_2) or of plant opal (phytoliths) (solubility between that of quartz and amorphous silica, 120 mg/L of SiO_2) keeps concentrations of dissolved SiO_2 around 4-8 mg/L in topsoils and 1-4 mg/L in B-horizons of oxisols. This is sufficiently high to prevent desilication of kaolinite to gibbsite. Kaolinite formation is less pronounced in well-drained soils than in those with a stagnant layer. Isomorphic substitution of Fe for Al in kaolinites of Oxisols causes a low crystallinity of this mineral.
Gibbsite is formed only at very low concentrations of dissolved silica (see Fig. 3.1). It is found in cavities of weathering minerals (feldspars), or in the soil matrix when leaching rates are rather high. Gibbsite is not usually found in the topsoil (absence of weatherable primary minerals), except if very low silica concentrations cause

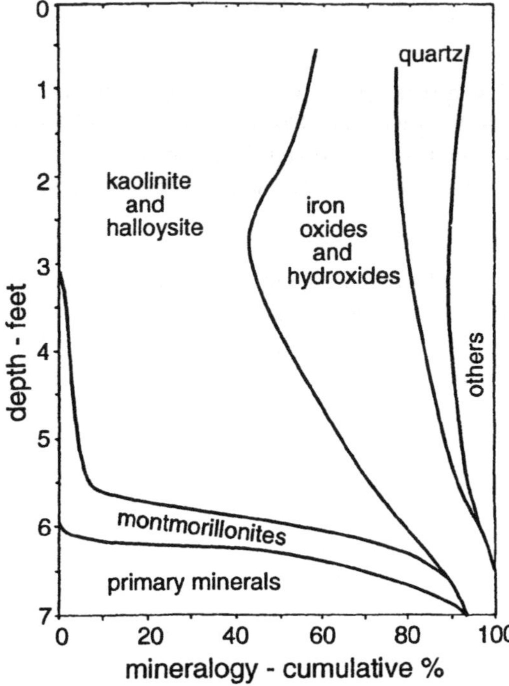

Figure 13.5. Changes of mineralogical composition with depth of an Oxisol on an intermediary volcanic rock. From Singer, 1979. Reprinted with permission of Elsevier Science, Amsterdam.

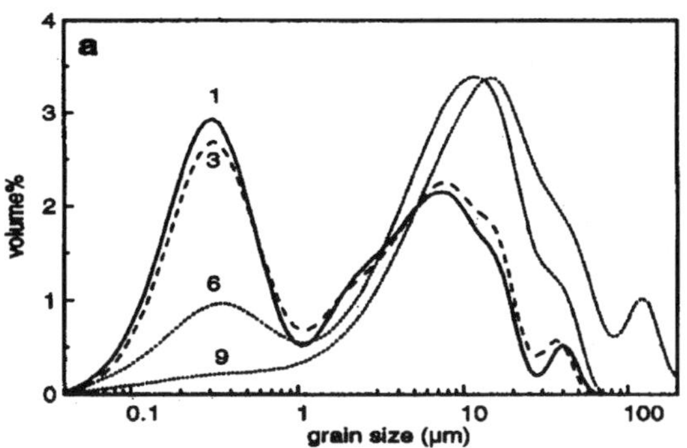

Figure 13.6. Grain size distribution in an old Oxisol from Brazil. (1) topsoil, (9) saprolite. Note formation of clay (< 1 μm) and general decrease of fractions coarser than 10 μm. Laser-diffraction grain-size data. From Muggler, 1998.

dissolution of kaolinite. In highly gibbsitic soils, the mineral is found in thin layers (sheets) or gravel-size aggregates.

IRON MINERALS

Hematite, goethite, and other iron minerals are common in the oxic horizon and in the underlying zones of iron enrichment. In the topsoil, hematite may be converted to goethite through an intermediate iron-organic matter complex.

Iron contents in oxisols on ultramafic rocks, where up to 90 % of the mass of the rock has been removed by leaching, may be as high as 25-50% Fe_2O_3 (Nipe series, Puerto Rico) and even more than 50% on the hematite-quartz rock itabirite.

Given sufficient time, all weatherable minerals (muscovite, biotite, chlorite, olivine, pyroxenes, amphiboles, garnets, phosphates, feldspars, etc) disappear completely from the solum. More stable minerals diminish in size (Figure 13.6). Of the primary minerals listed above, muscovites are most resistant to weathering. It appears that the hydrolysis of potassium in the interlayer is countered by the formation of an aluminium hydroxide interlayer, so that the complete transition of mica to vermiculite is prevented, and a 'soil chlorite' is formed instead. These soil chlorites are very stable at the prevailing pH and aluminium concentrations.

STABLE MINERALS

Even relatively stable minerals, such as quartz (SiO_2), titanium minerals (titanite, TiO_2; ilmenite, $(FeTi)O_2$), spinels ($MgAl_2O_4$, spinel; $FeIICr_2O_4$, chromite; $FeIIFeIII_2O_4$, magnetite), and zircon ($ZrSiO_4$), are gradually dissolved from old ferrallitised soils. Dissolution of some of these minerals may give rise to secondary minerals deeper in the profile.

Secondary Ti oxides (anatase, TiO_2) can be found as cementing agents in the subsoil (Hawaii). Secondary silica accumulations in the subsoil, such as opal or chalcedony, are uncommon in recent oxisols but may occur (see also Chapter 14: duripans). Silica is usually removed to the surface water and finally to the sea.

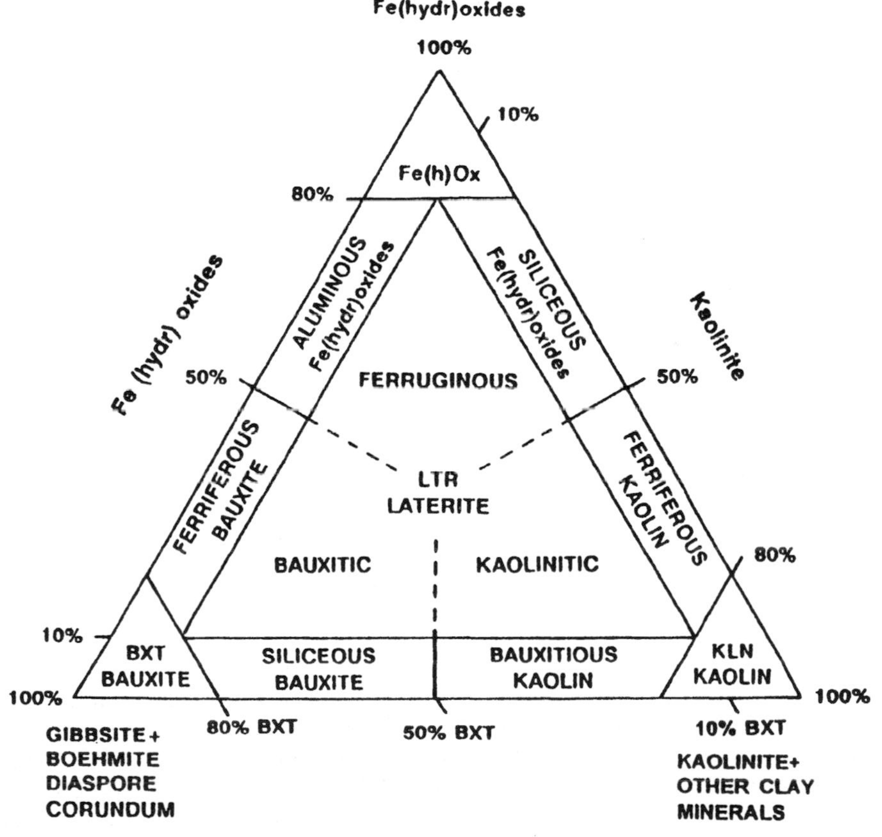

Figure 13.7. Mineralogical classification of ferralitised soils. From Aleva, 1994.

***Question 13.11.** The effects of changing leaching environments with depth in a single profile, and the changes in mineralogical composition with depth resulting from it, are comparable to the relation between rainfall intensity and mineralogy of topsoils as shown in Figures A and B of Problem 3.3. Turn graphs A and B of this assignment in such a way that the vertical scale represents soil depth and the horizontal scale the relative clay mineral content. Which of the figures A and B resembles figure 13.5?*

***Question 13.12.** In chapter 5 you have seen that a chemical profile balance is always based on a constituent that is not mobile (e.g. element) or is not altered (e.g. grain-size fraction), or on a constant volume. Which constituent in ferrallitic soils might be a better index compound than the Ti and Zr, that are commonly used for that purpose?*

The large variation in mineralogy of ferralitised soils is reflected in Figure 13.7. In extreme cases, ferralitization leads to considerable accumulation of economically interesting metals. Bauxites, the Al-rich variety, form on feldspar-rich rocks, such as arkoses and specific igneous and metamorphic rocks. On (ultra)mafic rocks, such as basalts, serpentinites, dunites, etc., ferralitization may give rise to residual accumulation of ores of Fe, Ni, V, and Cr.

***Question 13.13.** Which of the rock types 'mafic', 'felsic', and 'feldspar-rich' would give rise to the end-member mineralogies represented by the three corners of Figure 13.7?*

13.6. Charge, CEC, and base saturation

Surface horizons in strongly weathered soils may have considerable amounts of organic matter, which results in a significant CEC. The main charged components of the underlying mineral horizons are kaolinite clay, iron compounds, and gibbsite. The CEC of kaolinite clays is close to zero at pH values between 3.5 and 4.6, and in the order of 10 cmol(+)/kg at pH 7. Hydrous iron oxides have a much higher Point of Zero Net Charge (see box on next page), and have a positive charge below this pH value. Electrostatic bonds between clay and iron compounds reduce both positive and negative charge of the combination. In soils that are very rich in iron, this may even result in a net positive charge at soil pH values of 4-6.
Soil materials with negligible negative charge and some pH-dependent positive charge are said to have *geric properties* according to the FAO legend, or *acric properties* according to USDA.

***Question 13.14.** Suppose that a soil contains 30% kaolinite (with a constant CEC = 4 $cmol^+/kg$), a specific amount of goethite, and that other components have no charge. How much goethite is present if ZPNC is at pH 5? Suppose that CEC and AEC change linearly with pH.*

***Question 13.15.** Argue that pH_{KCl} can be higher than pH_{H2O} in soils with positive net charge.*

Point of Zero Net Charge

Amphoteric components have a pH value at which their net charge (= sum of positive and negative charge) is zero. This pH value depends on the dissociation constants of the various species of the component. Both Fe(III) and Al components form variously charged surface groups according to (from low to high pH):

Fe^{3+} --> $Fe(OH)^{2+}$ --> $Fe(OH)_2^+$ --> $Fe(OH)_3^o$ --> $Fe(OH)_4^-$

Al^{3+} --> $Al(OH)^{2+}$ --> $Al(OH)_2^+$ --> $Al(OH)_3^o$ --> $Al(OH)_4^-$

The following are ZPNC's and charges at various pH values for common soil minerals (from Parfitt, 1980).

		Charge (cmol/kg) at:	
	PZNC (pH)	pH3.5	pH8
Goethite	8.1	+12	0
Ferrihydrite	6.9	+80	-5
Allophane	6.5	+8	-32
Gibbsite	9.5	+6	0
Kaolinite	-	+1	-4

In a soil, combinations of anion exchange capacity (sesquioxides) and cation exchange capacity (clays, organic matter) may interact so that the soil also has an 'apparent' PZNC. This apparent PZNC can be either due to effective blocking of charges (surface binding) or to equal positive and negative charges without blocking. Because of the large number of components that each have a different pH-charge relation, a soil cannot have one single PZNC (Meijer and Buurman, 1987).

Exchange characteristics of ferralitic soils are usually not captured by the standard CEC at pH 7 (see Appendix 3), because for soil with much variable charge, the CEC at pH7 would deviate considerably from the actual CEC at soil pH, and the ECEC (effective CEC) is often more useful. ECEC is defined as the sum of exchangeable basic cations + aluminium. If >70% of the ECEC is occupied by aluminium, many crops suffer from aluminium toxicity. The base saturation of oxisols is usually low, but at low CEC values, the significance of base saturation for soil classification and soil fertility is questionable.

13.7. The rate of ferralitisation

The rate of ferralitization strongly depends on parent rock, temperature, and leaching. Formation of 1 metre of ferralitic material (oxic horizon) from granite in a perhumid tropical climate is supposed to take 50,000 to 100,000 years. Weathering is much faster on mafic rocks and on porous volcanic deposits (mafic rocks weather 10 times faster than felsic rocks in laboratory leaching experiments). In perudic moisture regimes where weathering is fast, young soils may have ferralitic properties in the clay fraction while the sand fraction still contains weatherable primary minerals.

POLYGENETIC SOILS

The long times necessary for ferralitisation, and the climate fluctuations in the past million years imply that nearly *all* ferralitic soils at the present land surface are polygenetic soils. In most areas, present wet tropical climates have been in effect for less than 10,000 years and were preceded by much dryer periods. Therefore, many ferralitic soils may exhibit traces of an earlier - and frequently different - phase of soil formation.

This effect is especially strong in ferralitic soils on very old surfaces. Ferralitic soils are found on much of the old tectonic shields of South America, Central Africa, Western India, and Australia. In some of these locations the surfaces at which they formed are more than 50 million years old, and many of the changes in climate, erosion, hydrology, etc. are still reflected in such soils. In areas where the climate changed from humid to dry, as in Australia, overprint of arid soil formation are found on ferralitic soils (see Chapter 15).

13.8. Transitions to other processes/soils

Oxisols are commonly associated with podzols, ultisols (acrisols and alisols), inceptisols (cambisols) and quartzipsamments (arenosols). The transition to quartzipsamments is related to parent material rather than due to soil forming processes: ferralitisation has little effect on the mineralogy and chemistry of quartz-dominated sands. Podzolisation may supersede ferralitisation, especially in soils formed on felsic parent materials in perhumid climates. On the other hand, podzols may occur side by wide with oxisols, the podzols occupying the poorly drained and sandier sites.

Associations of oxisols and ultisols may be caused by superposition of ferralitisation (including homogenisation) on soils with clay illuviation, or by true association, with ultisols occupying higher (cooler) parts of a catena and oxisols the lower parts.

In addition, some associations are due to past climatic/landscape history, as indicated in the previous section.

13.9. Problems

Problem 13.1

Figure A depicts the changes with depth of mass fractions of total SiO_2, Al_2O_3, and Fe_2O_3 with reference to a felsic parent rock (100%) for a ferralitic soil. (I) in a moderately humid climate with low organic matter contents in the soil, and (II) a very humid climates with high organic matter contents in the soil. Explain the behaviour of the three components and the differences between situation (I) and (II).

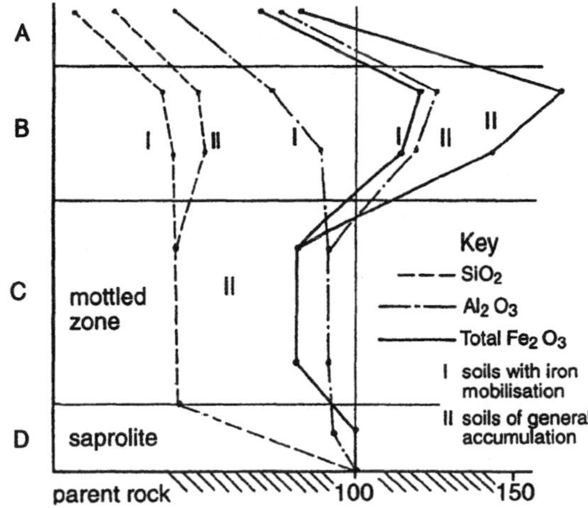

Figure 13A. Geochemical balance of a ferralitic soil. From Duchaufour, 1982.

Problem 13.2

Table 13B gives the molar ratios of silica and sesquioxides in a ferralitic soil on granite. R_2O_3 is the sum (in moles) of Al_2O_3 and Fe_2O_3.

a. Explain the relative movement of silica, iron, and aluminium for both the total soil and the clay fraction. For the total soil, be aware of the mineralogical composition of the granite.

b. Suggest horizon designations for the upper four horizons, on the basis of movement of Al and Fe.

Table 13B. Molar ratios of (total) silica/sesquioxides in soil and clay fraction of an oxisol on granite from South Africa (from Mohr et al, 1972)

SOIL					CLAY			
horizon	SiO_2/Al_2O_3	SiO_2/Fe_2O_3	SiO_2/R_2O_3	Al_2O_3/Fe_2O_3	SiO_2/Al_2O_3	SiO_2/Fe_2O_3	SiO_2/R_2O_3	Al_2O_3/Fe_2O_3
	14.5	56	11.5	3.8	2.7	10.8	2.2	4.0
	12.4	27	8.5	2.2	2.7	10.3	2.1	3.8
	7.1	11.8	4.4	1.7	2.5	8.9	2.0	3.5
	6.1	10.7	3.9	1.7	2.4	6.0	1.7	2.5
	6.7	16.1	4.7	2.4	2.6	7.4	1.9	2.9
C+R	6.3	23	4.9	3.7	2.6	11.6	2.1	4.5
R	8.8	46	7.4	5.3	-	-	-	-

Problem 13.3

Calculate the change in volume if 1 kg of microcline (K-feldspar) is weathered to kaolinite without loss of Al. Suppose that all soluble H_4SiO_4 is lost from the system. For formulas, atomic weight, and bulk densities, see Appendix 2.

Problem 13.4

Figure 13C gives the profile, and figure 13D the relations with depth of a number of chemical compounds in a (fossil) Oxisol from Western Australia.

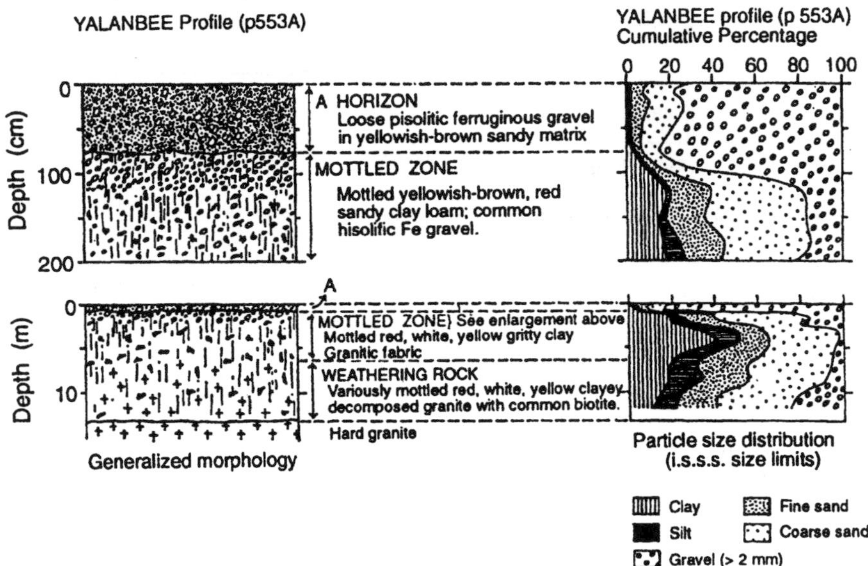

Figure 13C. Profile morphology of a (fossil) Oxisol from Australia. Upper and lower drawings are of the same profile at different scales. Source: see Figure 13D.

a. The profile was formed on granite. Compare the profile with Figure 13.1. Which horizons are missing?
b. Study the silica content for the whole profile and compare it with that of the parent rock (x in lower drawing). What is the reason that silica decreases and iron increases sharply at depth 1 metre? Can you find a reason why the Ti content remains constant?
c. Is any element in figure D suitable for calculation of the profile balance?
d. Can you explain why Si is higher in the fine earth than in the total soil while the opposite is true for Ti?
e. Na and Ca are strongly depleted with regard to the parent rock. Can you explain why Ca increases towards the top of the profile (so does K).
f. Is the iron enrichment relative (only local redistribution) or absolute one?

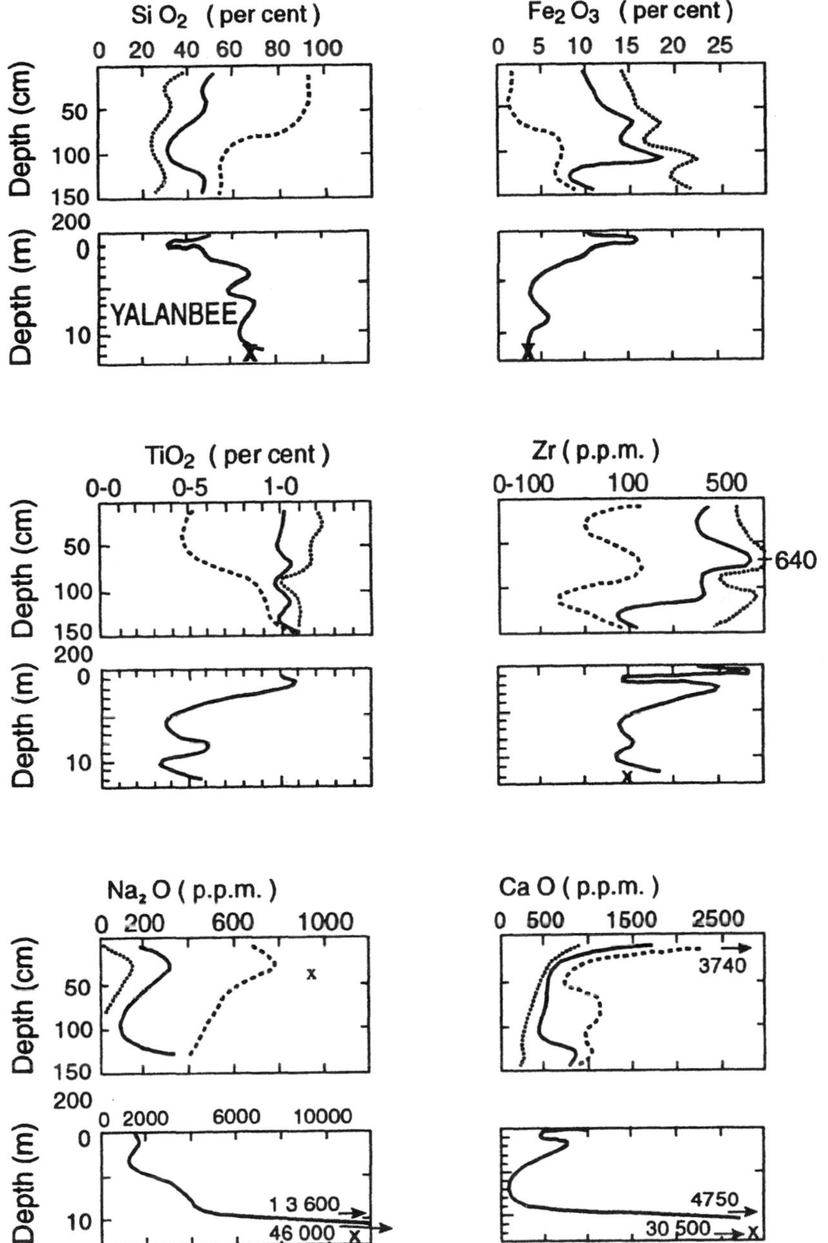

Figure 13D. Element contents in the Oxisol of Figure 13C. Full lines: whole soil; ---: fraction < 2 mm; ....: fraction > 2 mm. From Gilkes et al., 1973. Reprinted with permission of Blackwell Science Ltd., Oxford.

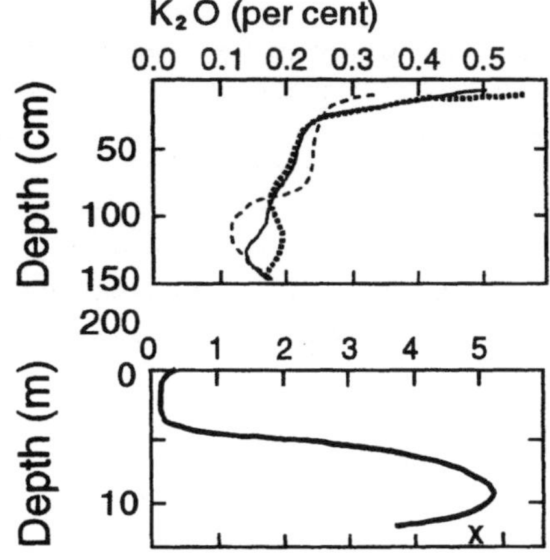

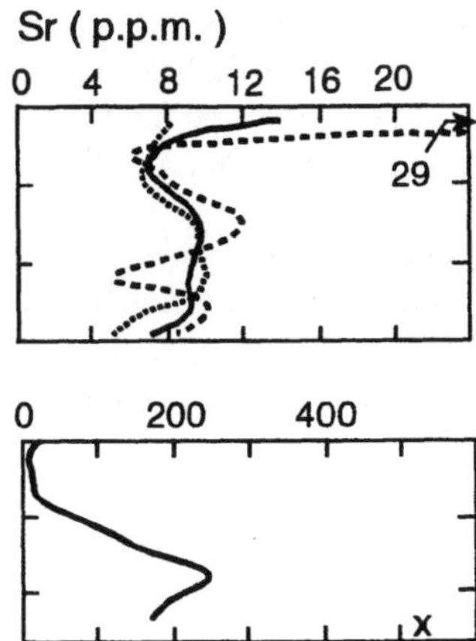

Figure 13D. Continued.

13.10. Answers

Question 13.1

Ground water laterite is equivalent to petroplinthite. Ferricrete is eroded and redeposited petroplinthite. Lateritic soils and latosols are old terms for oxisols. Petroferric material is equivalent to petroplinthite.

Question 13.2

The Si that is not removed by leaching is incorporated in clay minerals.

Question 13.3

The solubility of Fe and Al is determined by the dissolution of (oxy)hydroxides, and decreases with increasing pH (e.g.: $Al(OH)_3 + 3H^+ \rightarrow Al^{3+} + 3\ H_2O$). The solubility of SiO_2 ($SiO_2(s) + 2\ H_2O \rightarrow H_4SiO_4\ (aq)$ is independent of pH up to pH 9 (Figure 14.5).

Question 13.4

In stage 1, drainage is very good and therefore all silica that is liberated by weathering is removed from the profile. At later stages, with impeded percolation, higher Si concentrations can build up in pore waters deeper in the profile. The stability diagram indicates that gibbsite is only stable at very low Si concentrations, and smectite at the highest concentrations. Kaolinite can be formed at intermediate Si concentrations.

Question 13.5

Relative accumulation is caused by removal of other components. Absolute accumulation is caused in a profile by addition from outside the profile and in a horizon by redistribution within the profile.

Question 13.6

The iron in the Cg horizon is mainly due to movement of iron from the reduced zone towards the Cg, and redistribution within the Cg itself. It accumulates at the top of the mottled horizon because this is where iron comes most readily in contact with oxygen.

Question 13.7

Iron depletion in the pallid zone takes place through reduction and transport. Main transport is probably lateral and very slow, but upward transport towards the mottled zone also occurs. Transport in such non-porous clayey substrates is very slow.

Question 13.8

a) Formation of iron pans blocks water transport, so that the ground water level occurs higher in the profile. This means that also the reduction/oxidation boundary lies higher in the profile and that the iron accumulation zone tends to grow upwards.
b) The drawing suggests that deforestation resulted in erosion of the soil upwards of the original iron pan. This means that a new nucleus is formed upstream, where the reduced iron comes in contact with atmospheric oxygen.

Question 13.9

Ferralitic soils on mafic rocks do not have a clearly expressed mottled zone because kaolinite formation and the accompanying reduced permeability are less pronounced or absent. Instead they have a thin zone (several centimetres) of weathered rock fragments still rich in silica. Less formation of silicate clays on mafic than on felsic rocks is due to more eluviation of silica (higher porosity), lower aluminium contents, and incorporation of part of the aluminium in the lattice of iron oxide minerals.

Question 13.10

Soil nodules can have a mineralogy that is different (less weathered) than the rest of the same horizon, because they are hardly effected by percolating water

Question 13.11

If you turn Figures A and B so that the X-axis (annual precipitation) becomes the depth axis, with the highest precipitation at the top of the profile, you have a graph similar to Figure 13.5. Figure B is closest (no illite in the parent rock).

Question 13.12

Most elements are mobile in ferralitised soils, including Ti and Zr. Least mobile is Al, so this element would be the best index.

Question 13.13

Kaolinite is dominant is normal felsic rocks, so the right hand corner represents the felsic rocks. Iron is most common in mafic rocks (top), and gibbsite most common on feldspar-rich rocks

Question 13.14

1 kg of this soil has 300 g kaolinite, with a CEC of 1.2 cmol.
The AEC of goethite at pH 5 is about 8 $cmol^+.kg^{-1}$ (supposed linear relation between pH and charge of goethite). Therefore, a content of 1.2/8kg of goethite, which equals 150 gram or 15% would result in zero net charge.

Question 13.15

If the soil has positive net charge, addition of an electrolyte will liberate more OH^- than H^+ from adsorption complex and oxide surfaces. Therefore, pH in electrolyte will be higher.

Problem 13.1

Situation I.
Significant loss of silica (>50%) is already visible in the weathering zone (D), while losses of Al and Fe are very small. The lost components must have been removed laterally.
In the pallid zone (C), which is permanently waterlogged, Al remains stable (Al is not reduced!), but there is a slight loss of Fe, which is removed in reduced form, either laterally or upwards. In the mottled zone (plinthite, B) there is a net accumulation of Fe (oxidation), but not of Al.
In the topsoil (A+B horizons), there is loss of all three components, indicating that there may be a)transport of iron to the subsoil, and b) decomposition and removal of clay minerals.
Situation II
The differences with situation I are found in the layers A and B. Iron accumulation in B is much stronger than in situation I, while also Al and Si are higher in layer B. The concurrent higher values for Al and Si in layer B suggest that they occur as kaolinite clays, with some excess Al in the form of gibbsite. Because the profile has more humus, it is likely that more acidic circumstances have caused the transport of Al (and perhaps also of Fe). On the other hand, the Al and Si content in the topsoil remain higher than in situation I, which may indicate that there is less destruction of clay.

Problem 13.2

Note that the ratios only give *relative* mobility.
The R (parent rock) gives the ratios of the original material, so all changes should be regarded with respect to these values. In the following we will use metal ratios instead of oxide ratios, for easier reading.
Whole soil:
The Si/Al ratio increases upwards. Because this is contrary to desilication, it must be an accumulation of residual quartz towards the top of the profile. There appears to be some accumulation of Al at the bottom of the profile and some removal from the top.
The Si/Fe ratio has lowest values in the middle of the profile, indicating accumulation of Fe. The Al/Fe ratio suggests that there is an absolute accumulation of Fe in the middle of the profile.

Clay fraction:
The Si/Al ratios are fairly stable throughout. The values suggest some quartz in the clay fraction in addition to kaolinite. The middle part of the profile is slightly more aluminous.
Also in the clay fraction, we note an increase in Fe in the middle of the profile, both with respect to Si and to Al.

Problem 13.3

The solid phase part of the weathering reaction is:
$2KAlSi_3O_8 \rightarrow Al_2Si_2O_5(OH)_4$ (the other reaction products are soluble).
2 moles of feldspar result in one mole of kaolinite. 1 kg of feldspar equals

1000 / (39 + 27 + 3*28 + 8*16) Mol = 3.6 Mol.

This results in 1.8 Mol of kaolinite, or

1.8 * (2 * 27 + 2 * 28 + 5 * 16 + 4 * 17) = 464 g kaolinite.

1 kg of feldspar has a volume of 1 / 2.6 L = 0.38 L
464 g kaolinite has a volume of 0.464 / 2.65 L = 0.18 L,
so the volume loss is 0.2 L, or more than 50% of the original volume.

Problem 13.4

a) Although the profile description contains an A-horizon, it is obvious that all horizons above the plinthite layer are absent and that the present A was developed in the plinthite or reworked plinthite (gravel instead of cemented horizon).
b) The loss of silica is relatively low because the top of the profile is missing. The decrease in silica at 1 m is due to an addition of Fe. Ti can only remain constant if some is added with the Fe (actually, many iron pans and silcretes (see chapter 14) are enriched in Ti).
c) If an element is suitable for a profile balance, it should not be redistributed or lost from the profile. In the case of strong loss of weathering products, it should show a (relative) accumulation in the weathered layers. Of the elements shown, Si is lost; Fe is redistributed; Ti must be redistributed (see b); Zr appears to show a relative accumulation that is due to removal of other components - it might be suitable; Na, K, Ca, and Sr are lost.
d) Si is higher in the fine earth because the total soil is dominated by iron gravel, which has little silica. Ti is enriched in the iron gravel.
e) K and Ca are accumulated in the top of the profile because of the nutrient recycling (pumping) by plants. A large part of the Ca and K in surface horizons of old, deeply weathered soils, is normally derived from cyclic salts in rainwater and from dust transported over long distances (Chadwick et al., 1999). Na is little used by plants and is leached.
f) The high iron contents in a weathering mantle of granite suggest absolute accumulation.

13.11. References

Aleva, G.J.J., 1994. *Laterites - Concepts, geology, geomorphology and chemistry.* International Soil Reference and Information Centre, Wageningen, 1-169.

Bowden, J.W., A.M. Posner, and J.P. Quirk, 1980. Adsorption and charging phenomena in variable charge soils. In B.K.G. Theng (ed): *Soils with Variable Charge*, 147-166. New Zealand Society of Soil Science, Lower Hutt.

Chadwick, O.A., L.A. Derry, D.M. Vitousek, B.J. Huebert and L.O.Hedin, 1999. Changing sources of nutrients during four million years of ecosystem development. Nature, 397:491-497.

Duchaufour, P., 1977. *Pedology - pedogenesis and classification.* Allen & Unwin, London.

El-Swaify, S.A., 1980. Physical and mechanical properties of oxisols. In B.K.G. Theng (ed): *Soils with Variable Charge*, 303-324. New Zealand Society of Soil Science, Lower Hutt.

Eswaran, H., & R. Tavernier, 1980. Classification and genesis of oxisols. In B.K.G. Theng (ed): *Soils with Variable Charge*, 427-442. New Zealand Society of Soil Science, Lower Hutt.

Gilkes, R.J., G. Scholz & G.M. Dimmock, 1973. Lateritic deep weathering of granite. Journal of Soil Science, 24(4):523-536.

Herbillon, A.J., 1980. Mineralogy of oxisols and oxic materials. In B.K.G. Theng (ed): *Soils with Variable Charge*, 109-126. New Zealand Society of Soil Science, Lower Hutt.

Meijer, E.L., and P. Buurman, 1987. Salt effect in a multi-component variable charge system: curve of Zero Salt Effect, registered in a pH-stat. Journal of Soil Science, 38:239-244.

Mohr, E.C.J., F.A. van Baren & J. van Schuylenborgh, 1972. *Tropical Soils, a comprehensive study of their genesis*. Mouton, The Hague.

Muggler, C.C., 1998. Polygenetic Oxisols on Tertiary surfaces, Minas Gerais, Brazil - Soil genesis and landscape development. PhD Thesis, Wageningen University, 185 pp.

Muggler, C.C., C. van Griethuysen, P. Buurman, and Th. Pape, 1999. Aggregation, organic matter, and iron oxide morphology in Oxisols from Minas Gerais, Brazil. Soil Science, 164:759-770.

Paramananthan, S., & H. Eswaran, 1980. Morphological properties of oxisols. In B.K.G. Theng (ed): *Soils with Variable Charge*, 35-44. New Zealand Society of Soil Science, Lower Hutt.

Parfitt, R.L., 1980. Chemical properties of variable charge soils. In B.K.G. Theng (ed): *Soils with Variable Charge*, 167-194. New Zealand Society of Soil Science, Lower Hutt.

Van Wambeke, A., H. Eswaran, A.J. Herbillon & J. Comerma, 1983. Oxisols. In: L.P. Wilding, N.E. Smeck & G.F. Hall (eds): *Pedogenesis and Soil Taxonomy II.* The Soil Orders, 325-354. Developments in Soil Science 11B, Elsevier, Amsterdam.

CHAPTER 14

DENSE AND CEMENTED HORIZONS:

FRAGIPAN, DURIPAN, AND TEPETATE

14.1. Introduction

Pans are dense or cemented pedogenic soil horizons that obstruct root penetration and movement of air and water. Hard or dense layers of geological origin are not considered pans. Examples of non-cemented, dense pans are plough pans, found at the bottom of the Ap-horizon and fragipans. Examples of cemented pans are humus pans (Chapter 11), petroferric horizons (Chapter 13), thin iron pans (placic horizons) at the contact between hydromorphic surface horizons and aerated subsoils (Chapter 7), iron concretion horizons at the basis of a puddled layer in wet rice cultivation, (petro)calcic and (petro)gypsic horizons (Chapter 9), B horizons of hydromorphic podzols, and duripans. Dense pans slake in water, while cemented pans do not. Most pans have a bulk density above 1.6 kg/dm^3. In the following, fragipan, duripan and tepetate will be discussed. Both fragipans and duripans are diagnostic horizons in the USDA soil classification, and appear as phases in the FAO Legend.

Question 14.1. *What are the soil forming processes responsible for the formation of, respectively, a petroferric horizon, a thin ironpan, a dense humus pan?*

14.2. The Fragipan

DEFINITION
A fragipan (Fig 14.4) is a subsurface horizon that is compact, but not or only weakly cemented. It is brittle when moist and hard when dry. Soil material is 'brittle' if it ruptures suddenly when pressure is applied, rather than undergo plastic deformation. The fragipan is loamy, sometimes sandy, and often underlies a cambic, spodic, argillic, or albic horizon. It has a very low content of organic matter, and a high bulk density relative to the horizons above it. A dry fragment slakes or fractures when placed in water. A fragipan is usually mottled, is slowly or very slowly permeable to water, and has few or many bleached, roughly vertical planes that are faces of coarse or very coarse polyhedrons or prisms (SSS, 1990).

Question 14.2. *What would be the chemical nature of mottles in fragipans, and why?*

Fragipans are always parallel to the soil surface, usually at a depth of 40-80 cm. They are absent in calcareous materials, and form preferably in sediments with a high fine sand and silt content. The original vegetation is usually forest. Fragipans are found in cold climates as well as in the tropics. In profile descriptions, fragipan character of a horizon is indicated with the suffix *x*.

FRAGIPAN GENESIS

Several causes of fragipan genesis have been put forward:

1. Compaction due to slight shrinking and swelling followed by filling of cracks.

Many fragipans have polygonal cracks that are attributed to slight shrinking and swelling due to alternate wetting and drying. Cracking is limited to polygonal contraction cracks, with distances that decrease with increasing clay content of the material. Further fragmentation as observed in Vertisols does not occur. The open cracks are filled with material from above. Upon rewetting, the polygons are closed, but as the space has already been filled, considerable pressure is exerted on the soil bounded by the polygon. This pressure cannot be neutralized by upward displacement, because slip faces are missing. As a result, the material between the cracks is compacted, and pore space decreases.

Question 14.3. *Why would fragipans not be found in clayey soils, but rather in coarser textured soils?*

2. Accumulation of amorphous aluminium silicates.

In some soils, aluminium silicates appear to have precipitated above a lithological discontinuity. Removal of water to a certain depth by the roots of forest vegetation could have led to dehydration of these precipitates and formation of weak cement (see also the paragraph 14.4 on tepetates).

Question 14.4. *In Chapter 12 you have seen that amorphous silicates (allophane) harden irreversibly upon dehydration. Given that fact, argue why the presence of allophane cannot cause the 'hard when dry, brittle when moist' character of fragipans.*

3. Translocation of clay and arrangement of silt.

Many fragipans have oriented clay. Illuviation of clay on grain contacts and in pores is a possible cause of higher bulk density and binding of particles by sedimented (or crystallised amorphous) material. Arrangement of silt-size material in spaces between much coarser grains, caused by water percolation, increases density (Figures 14.1,2). The *densipan* of New Zealand (Wells and Northey, 1985) appears to be a fragipan with rearrangement of silt in the pores between sand grains.

4. Influence of overburden.

Fragipans in moraine material were explained by the pressure of overlying ice mass, or by compaction resulting from repeated freezing and thawing of water-saturated material.

Many fragipans have probably formed as the result of two or more of these processes, of which 1 and 3 are considered most important. In surface water gley soils and podzols, breakdown of clay may favour process 2.

Question 14.5. *Which of the four processes could cause fragipan formation in the tropics, and which in formerly glaciated areas?*

Figure 14.1 and 14.2 illustrate the effect of arrangement of silt and of clay particles, respectively, in a fragipan of two different textures. The profile is a fossil (sub)tropical podzol of Oligocene age (±35 M years), with several superposed processes. Some profile characteristics are given in Figure 14.3. The fragipan character of the horizons is indicated with the suffix 'x'. In the upper three horizons, the fragipan character is due to clay coatings that form bridges between grains (Figure 14.1). The Bhx2 horizon has a higher silt content, which is of sedimentary origin. In this horizon, the fragipan character is due to arrangement of silt particles rather than to clay bridges (Figure 14.2). Rutile (TiO_2) contents in the clay indicate mobility of Ti. The fine texture of the Bhx2 horizon appears to have aided organic matter accumulation.

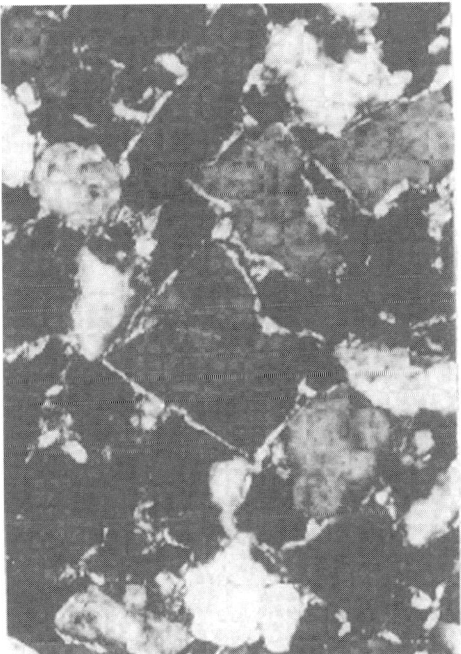

Figure 14.1. Dense layer of sand, cemented by bridges of clay coatings. *Ex* horizon of Figure 14.3. Width 0.5 mm. From Buurman and Jongmans, 1976.

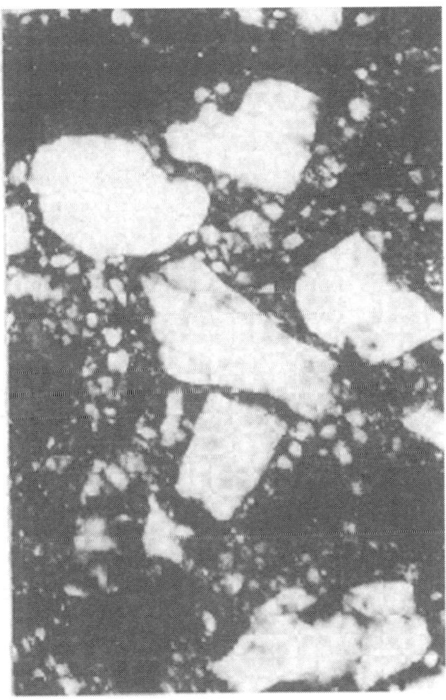

Figure 14.2. Dense arrangement of particles in a loamy fragipan. *Bhx2* horizon of Figure 14.3. Width 0.5 mm. From Buurman and Jongmans, 1976.

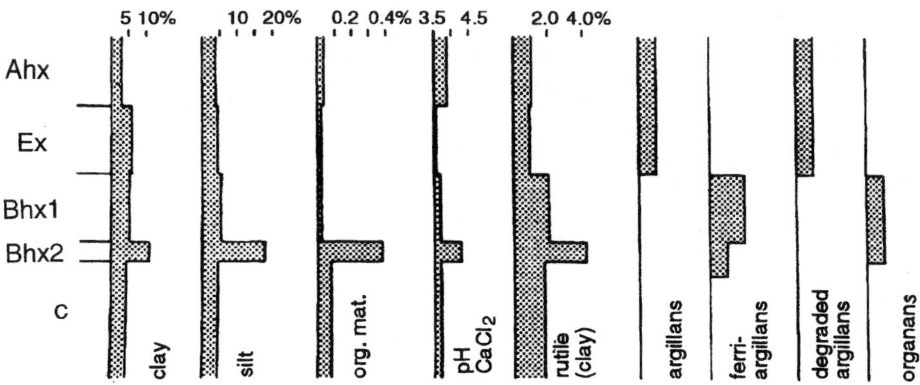

Figure 14.3. Characteristics of a fossil soil with a fragipan. From Buurman and Jongmans, 1976.

Dense arrangement in a New Zealand *densipan* in a podzol is illustrated in Plates W-Y (p. 342-343). In alfisols (luvisols, lixisols), fragipans may occur in the argillic horizon, and sometimes below it. In ultisols (alisols, acrisols) and podzols, fragipans appear in horizons that have undergone breakdown of clay (by ferrolysis, see Chapter 7). Here, the breakdown of clay may have resulted in some amorphous cementing material.

Question 14.6. *The profile of Figure 14.3 has undergone several stages of soil formation that can be inferred from its micromorphological characteristics. 'Argillans' stand for clay illuviation coatings that are depleted of iron; 'ferri-argillans' for clay illuviation coatings with iron; 'organans' are organic matter coatings. What are the processes that have acted upon this profile, and what was their sequence? Which of the processes is related to fragipan formation?*

Fragipans are considered important as soil classification features, because they strongly hamper root penetration and hydraulic conductivity. This is clearly illustrated by Figure 14.4, which depicts variations in saturated hydraulic conductivity and in bulk density with soil depth in two Canadian podzols in glacial till with a fragipan. The Arago soil has about 20% clay, and the Ste. Agathe soil around 5%.

Question 14.7. *The horizon designations in Figure 14.4 suggest that the two profiles have a different genesis. What are the processes that played a role in the Arago and Ste-Agathe profiles? Do these processes have a relation to fragipan formation?*

Because the fragipan is caused primarily by a rearrangement of soil particles, it does not have special mineralogical characteristics, although its presence may cause changes in the mineralogy of the overlying layers.

Question 14.8. *What new soil forming processes in the overlying horizons could be induced by a fragipan?*

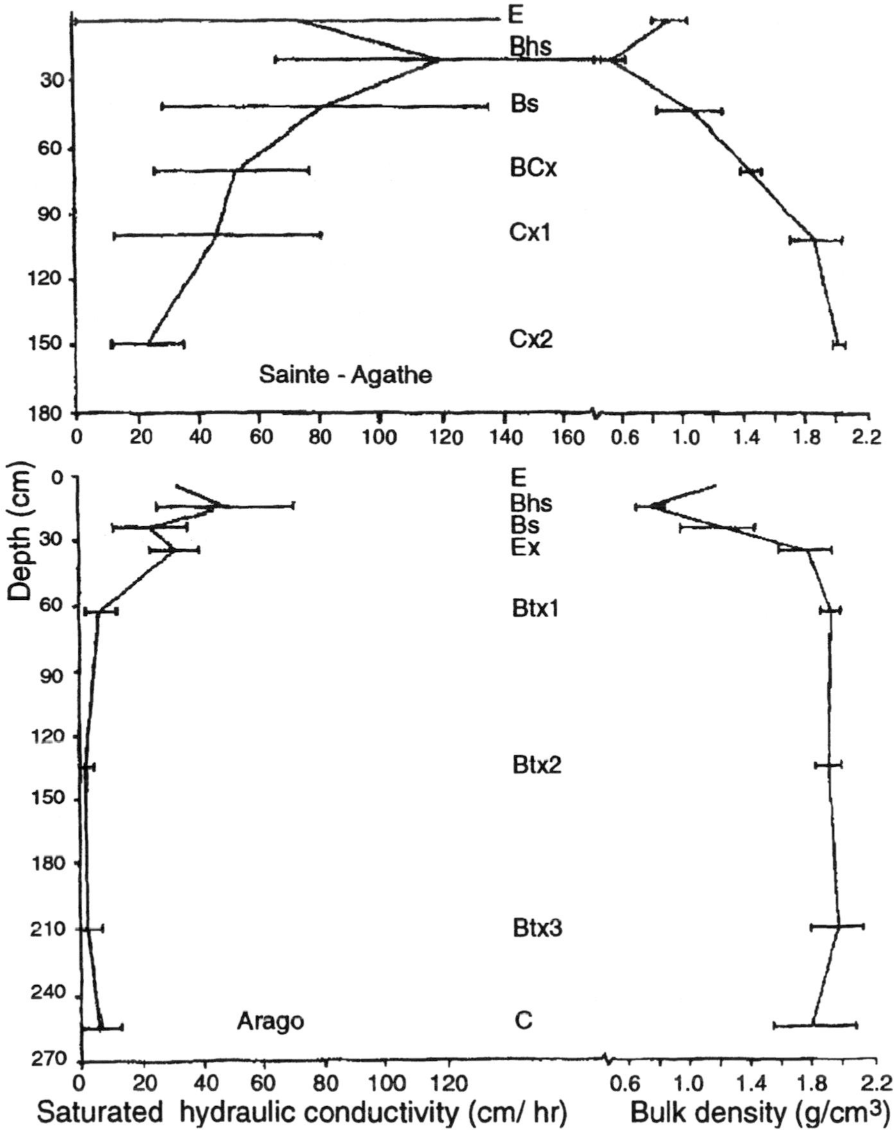

Figure 14.4. Saturated hydraulic conductivity and bulk density values in two Canadian podzols with a fragipan. From Mehuys and De Kimpe, 1976. Reprinted with permission of Elsevier Science, Amsterdam.

14.3. The duripan

The duripan is a subsurface horizon that is cemented by silica to the degree that fragments of the air-dry horizon do not slake during prolonged soaking in water or in HCl (SSS, 1990). Duripans are also known as silcretes, e.g. in Australia, where they are widespread. In profile descriptions, duripans are indicated with the suffix *qm* (*q* for silica accumulation, *m* for cementation).

Duripans are usually found as part of, or in association with calcic horizons. They are restricted to ustic and dryer climates, and are found in Ustolls, Xerolls (Calcic Chernozems and Kastanozems), and Aridisols (Calcisols).
Duripans are formed by precipitation of amorphous silica (plate Z, p.344) at some depth in the profile. The source of silica may vary. In arid soils, weathering of primary silicate minerals is usually restricted. However, if pH is very high (≥10) as in alkaline soils, solubility of silica and silicates may increase considerably (Figure 14.5).
At normal pH values, silica sources are usually easily weatherable silica-rich materials, such as volcanic glass (Chadwick et al., 1989), or even loess (Blank and Fosberg, 1991).

Question 14.9. *Give two reasons why weathering of primary silicate minerals is very slow in arid regions.*

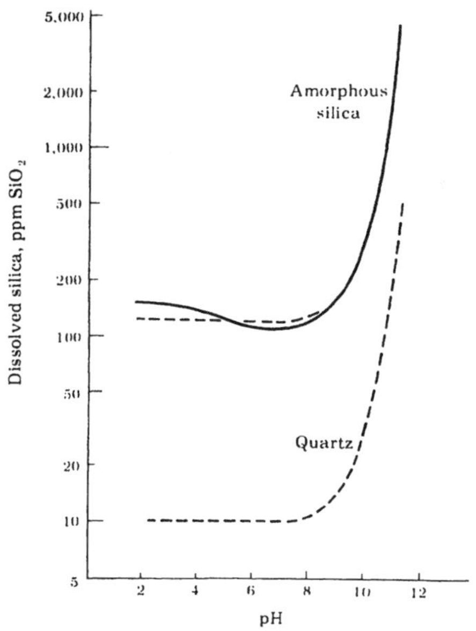

Figure 14.5. Solubility of silica in dependence of pH. Solid line, measured data; dashed line calculated from thermodynamic data. From Krauskopf, 1967.

Question 14.10. *What are the causes of high pH in soils of arid regions? (see Chapter 9.3).*

At pH values below 8, precipitation of amorphous silica takes place when the concentration in solution exceeds about 120 ppm SiO_2 (= 2mmol/l SiO_2 = 192 ppm H_4SiO_4). Such concentrations can only be reached by evaporation (see also 14.4). This implies that duripans can only be formed in climates with a distinct dry period. On the other hand, the formation of duripans requires sufficient moisture for weathering of primary minerals and for transport of dissolved silica, but the climate should not be so wet that dissolved silica is leached. The duripan forms at the maximum depth of wetting and is therefore encountered at shallower depth in more arid environments.

NATURE OF SILICA PRECIPITATES

At high concentrations of silica in solution, opal precipitates. This is a water-rich, amorphous silica compound (opal-A). Opal with some crystallographic organisation is called opal-CT (CT stands for cristobalite and tridymite-like structure). Eventually, opal crystallises to chalcedony, a water-free, cryptocrystalline form of quartz. Direct precipitation of quartz only occurs at low concentrations of dissolved silica. Most duripans have various forms of silica accumulation, indicating changing circumstances during formation. In arid regions, gypsum and palygorskite are found in association with duripans.

In duripans, silica precipitates in finer soil pores, where it forms a spongy fabric, while larger pores remain open. Unlike calcite and gypsum cementation, silica cementation is not usually related to groundwater. Weak duripans can still be penetrated with a hand auger (strongly hydrated opal, or little cement). In more humid climates (ustic-udic), the duripan has properties transitional to those of the fragipan.

SILCRETES

The Australian *silcretes* are probably fossil duripans with a more complex history. They occur on a large variety of rocks of widely different ages (Precambrian-Tertiary) and are associated with extremely old landscapes (40-50 million years) of former probably humid tropical climates. Silcretes are found on pallid and mottled zones of ferralitic soils (Chapter 13) and in floodplain positions. In old, exposed, silcretes, all silica is crystallised to chalcedony, in young, buried ones opal is dominant. Old silcretes may show strong evidence of weathering of primary and secondary minerals; many have high contents (up to 20%) of secondary TiO_2 as anatase (Langford-Smith, 1978; Milnes et al., 1991), which indicates that they formed in connection with a high-weathering environment. Because most silcretes occur in old landscapes, they have a complex history that reflects changing climatic and hydrological regimes: massive silcretes may have formed through evaporation of groundwater. Silcrete formation still has many secrets.

Question 14.11. *What could be the source of silica in silcretes associated with ferralsols?*

In poorly developed duripans, cementation is restricted to isolated nodules (durinodes) of opal and some opal material on pebble bottoms (opal *pendants*). Opaline silica is a more effective cement than $CaCO_3$. While as little as 6% silica cement is sufficient for slight cementation, petrocalcic horizons need about 40% authigenic calcite for significant cementation. Strongly developed, continuous duripans of old land surfaces (Mid-Late Pleistocene) have a laminated top, consisting of imbricated (tile-like overlying, slightly inclined) silicified plates, 1-15 cm thick, that may be interlayered with calcite. All transitions from weak to very strongly expressed duripans and from duripans to petrocalcic horizons can be encountered. Silica accumulations may be covered by coatings of iron and manganese compounds (stagnating water), or illuviated clay.

Question 14.12. *What does the presence of opal pendants (compare calcite pendants, Chapter 9.2) tell you about the mechanism of Si accumulation?*

14.4. The tepetate

Tepetate or *cangahua* are names for a group of cementations and compactions in volcanic deposits. Such pans occur predominantly in climates with pronounced wet and dry seasons. The names are from Latin America, and there is a variety of other local names. Tepetates include cementations by calcite, precipitated clays, allophane, and silica, and dense structures that are included in the fragipan (Zebrowski et al., 1997). We will use *tepetate* for pans that are cemented by weathering products typical for volcanic deposits.

Weathering of volcanic deposits sets free large amounts of silica and aluminium, which can be transported down the profile with percolating rain water or groundwater. Drying out of zones with accumulated weathering products may lead to the formation of allophane/imogolite coatings (Jongmans et al., 2000) that fill pores and cement coarse grains (Figure 14.6).

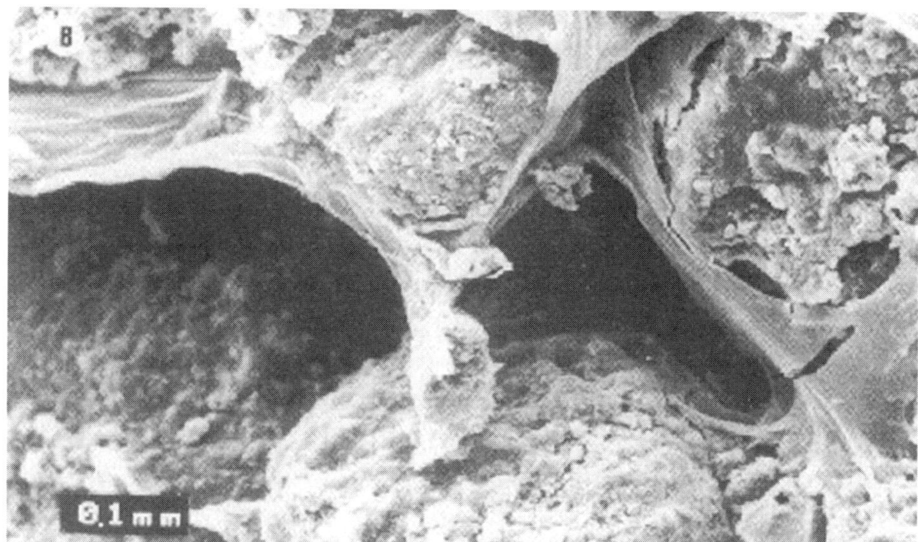

Figure 14.6. Scanning electron micrograph of allophane/imogolite coatings from indurated volcanic ash in Costa Rica. From Jongmans et al., 2000.

Alternating dry and wet periods may favour crystallisation of amorphous aluminium silicates to phyllosilicate clays, so that many tepetates have clay coatings that are due to recrystallisation rather than to clay illuviation (see also Chapter 12). Phyllosilicate clay-cemented tepetates (2:1 minerals, interstratified 2:1 minerals, or halloysite) have been described by Hidalgo et al. (1997) and by Quantin and Zebrowski (1997).
Many secondary accumulations have Si/Al ratios of up to 5. These high ratios are probably caused by leaching of Al and Si at slightly acid to neutral pH, when silica is more mobile than aluminium. At such ratios, most of the silica cannot be accommodated in the allophane or imogolite structure, and silica precipitates separately as opal. High-silica tepetates are (transitions to) real duripans. Opal-cemented tepetates have been described, among others, by Poetsch and Arikas (1997) and Mora and Flores (1997).

14.5. Problems

Problem 14.1
In the Netherlands, fragipans are encountered in sediments of the last glacial period (coversands, loess). All these fragipans have polygonal networks of cracks. What could have caused these cracks? What does this imply for the age of the fragipan?

Problem 14.2
Do you expect fragipans in loamy soils of the perhumid tropics? Explain your answer.

Table *14A.* Selected properties of two soils with a duripan. From Soil Survey Staff, 1975; horizon designations recoded.

horizon	depth cm	sand	silt	clay	$CaCO_3$	Org. C	b.d. $g.cm^{-3}$	sum bases	CEC pH7
		mass fraction %						cmol(+)/kg	
pedon 63									
Ah1	0- 3	86	12	2	3	0.1	-	13.6	5.6
Ah2	3-10	80	15	5	5	0.1	-	18.0	7.3
Ah3	10-41	75	14	11	7	0.2	-	23.1	11.6
Cqm1	41-58	87	7	6	14	0.2	-	20.6	12.1
Cqm2	58-76	81	13	6	9	0.0	-	26.9	14.3
C	76 94	87	9	4	3	0.0	-	22.4	12.2
pedon 98									
Ah1	0- 5	19	65	16	-	2.2	1.34	3.5	18.6
Ah2	5-10	17	61	22	-	1.1	1.33	5.3	19.7
Ah3	10-20	18	57	25	-	0.8	1.37	6.4	20.3
Bt1	20-30	19	54	28	-	0.6	1.32	8.0	22.3
Bt2	30-53	14	26	60	-	0.5	1.70	39.9	49.3
2Bt3	53-63	35	24	41	-	0.5	1.43	7.6	47.3
2Cqm1	63-69	82	14	4	-	0.3	1.55	9.6	20.0
2Cqm2	69-105	85	11	5	-	0.2	1.44	7.9	22.1
2Cqm3	105-125	72	19	9	-	0.1	1.45	7.5	19.1

Problem 14.3

Table 14A(page 337) gives analysis of two soils with duripans. Texture analysis of the duripan is carried out after crushing of the samples.

a) Which analyses reflect the presence of the duripan?
b) Which profile is from the driest climate?
c) Calculate base saturation for all horizons. What is the cause of discrepancies between sum of bases and CEC in pedon 63? (This is not relevant to the duripan, but it is to arid soils).

Problem 14.4

Study the analyses of a profile with a fragipan in Table 14B.

a) Which horizon(s) constitute(s) the fragipan?
b) What other soil forming process(es) have influenced this profile?
c) Give a correct horizon designation.

Table ***14B***. Selected analysis of a profile (pedon 26) with a fragipan. From Soil Survey Staff, 1975.

hor.	depth	sand	silt	clay	Org. C	b.d.	sum bases	CEC pH7	Al	Fe
	cm	mass fraction %				g/cm^3	cmol(+)/kg		%pyroph. extr.	
	4- 0	nd	nd	nd	23.0	nd	27.9	73.1	0.1	0.67
	0- 4	59	32	9	2.2	nd	11.6	15.4	0.1	0.20
	4- 18	57	30	12	2.3	nd	25.4	27.5	0.8	1.32
	18- 33	58	32	10	1.5	1.28	18.5	20.4	0.4	0.71
	33- 51	58	32	10	0.8	1.37	12.6	13.8	nd	nd
	51- 71	67	27	6	0.1	1.92	3.8	4.6	nd	nd
	71-105	66	26	8	0.1	1.85	3.2	4.9	nd	nd

14.6. Answers

Question 14.1

Both petroferric horizons and thin iron pans are due to reduction and oxidation of iron in combination with transport by ground water. A dense humus pan is due to (hydromorphic) podzolisation.

Question 14.2

The mottles in the fragipan are usually iron mottles. Their presence is due to (periodic) stagnation of water on the fragipan.

Question 14.3

In clayey soils, pressure is released by failure along slip faces (see Chapter 10) so that compaction does not occur. Dry clay is also hard when dry, but not brittle when moist. It is therefore excluded from the definition of the fragipan.

Question 14.4
Dehydration of amorphous substances is irreversible when the soil dries out completely. So a pan formed this way would remain hard and not return to a brittle state when remoistened.

Question 14.5
Under a seasonal climate in the tropics, processes 1-3 may operate. In areas with seasonal temperate climates that were glaciated in the past, all four processes may have played a role, but the influence of overburden is limited to very small areas.

Question 14.6
The profile clearly shows the effects of clay illuviation (argillans) and podzolisation (organans). The presence of the argillans in the top of the profile indicates that the eluvial horizon was eroded. The removal of iron from the clay coatings in the upper two horizons is due to either periodic wetness above the fragipan or, more likely, to podzolisation. Clay illuviation takes place before podzolisation, because it is impeded by low pH. Organic matter accumulation in the form of organans is mainly above the fragipan: the finer texture of the Bhx2 may act as a sieve. Clay illuviation is clearly responsible for the fragipan in the upper horizons.

Question 14.7
The horizon codes indicate podzolisation in both profiles. In the Arago profile, the podzol developed in the E horizon of a profile with clay illuviation. Illuviated clay may have caused the sharper upper boundary of the fragipan. Because the podzols are formed on glacial till, a loamy sediment, clay illuviation will not have much influence on fragipan formation. This fragipan is probably due to periodic freezing.

Question 14.8
Because of its density, the fragipan may cause stagnation of water and therefore also stagnation of dissolved and suspended substances. In addition, water stagnating may cause gleying and, eventually, ferrolysis.

Question 14.9
Weathering of primary minerals in arid soils is slow because of lack of water (no leaching of weathering products) and high pH.

Question 14.10
Alkalinity in arid soils is due to highly soluble carbonates, such as Na_2CO_3.

Question 14.11
Ferralitization causes strong liberation of silica, which can be transported with groundwater or accumulated deep in the profile (ferralitization needs a precipitation excess). Duripan formation could either follow as a result of climate change (drying out of subsoil) or when the silica is transported by flowing groundwater to a drier zone, as is sometimes supposed for Australian silcretes.

Question 14.12

Pendants of opal suggest that silica precipitated from water dripping down (compare calcite pendants, Chapter 9.2).

Problem 14.1

Polygonal cracks in the subsoil of Dutch soils cannot be formed under the present humid climate. Cracking is very common in soils of Polar Regions: at temperatures below -4°C, the volume of ice decreases, causing cracking of the soil. Polygon networks are common in tundra soils. Strong evidence of frost action implies that the soils should have formed during or before the last glacial period: Late Weichselian.

Problem 14.2

Fragipan formation requires a seasonal climate and/or illuviation and orientation of fine material. In the perhumid tropics (perhumid = permanently wet, so not seasonal), neither the seasonal climate, nor the prerequisites for clay illuviation are present. Fragipan formation is therefore restricted to movement of fine silt into fine pores of sandy soils.

Problem 14.3

a) Because there is no specific analysis for silica cements (Appendix 3), a duripan cannot be identified chemically. In the second profile, the duripan is indicated by high bulk density in the Bt2 horizon.

b) Soils from drier climates have $CaCO_3$ accumulation closer to the surface. Pedon 63 is therefore from a drier climate than pedon 98.

c) The sum of 'exchangeable' bases is higher than the CEC in the calcareous horizons of pedon 63. Because this is theoretically not possible, excess Ca should be due to dissolution of $CaCO_3$ during extraction of bases.

Problem 14.4

a) the fragipan is indicated by very high bulk densities, which start at 51 cm depth.

b) the slight increase in clay between 4 and 51 cm may indicate some transport of clay (clay illuviation). The relatively high pyrophosphate-extracted Al and Fe, and relatively high organic C at 4-18 cm depth suggest accumulation of humus with Al and Fe in a podzol-B horizon.

c) if we take both clay illuviation and podzolisation in account, the horizon coding would be:

O - Ah/E - Bth1 - Bth2 - Bt(h)3 - Cx1 - Cx2

14.7. References

Blank, R.R., and M.A. Fosberg, 1991. Duripans of Idaho, U.S.A.: in situ alteration of eolian dust (loess) to an opal-A/X-ray amorphous phase. Geoderma, 48:131-149.

Buurman, P. & A.G. Jongmans, 1975. The Neerrepen soil: an Early Oligocene podzol with a fragipan and gypsum concretions from Belgian and Dutch Limburg.

Pédologie, 25:105-117.

Chadwick, O.A., D.M. Hendricks and W.D. Nettleton, 1989. Silicification of Holocene soils in northern Monitor Valley, Nevada. Soil Science Society of America Journal, 53:158-164.

Hidalgo, C., P. Quantin, and F. Elsass, 1997. Caracterización mineralógica de los tepetates tipo fragipán del valle de México. In Zebrowski, C., P. Quantin, and G. Trujillo (Eds.), 1997. *Suelos volcánico endurecidos*. III Simposio Internacional (Quito, diciembre de 1996), p. 65-72. ORSTOM.

Jongmans, A.G., L. Denaix, F. van Oort, and A. Nieuwenhuyse, 2000. Induration of C horizons by allophane and imogolite in Costa Rican volcanic soils. Soil Science Society of America Journal, 64:254-262.

Krauskopf, K.B. 1967. *Introduction to geochemistry*. McGraw Hill, New York, 721 pp.

Langford-Smith, T., 1978. *Silcrete in Australia*. Department of Geography, University of New England.

Maliva, R.G., and R. Siever, 1988. Mechanisms and controls of silicification of fossils in limestones. Journal of Geology, 96:387-398.

Mehuys, G.R., and C.R. De Kimpe, 1976. Saturated hydraulic conductivity in pedogenetic characterization of podzols with fragipans in Quebec. Geoderma, 15:371-380.

Milnes, A.R., M.J. Wright and M. Thiry, 1991. Silica accumulations in saprolites and soils in South Australia. In: W.D. Nettleton (ed.): *Occurrence, characteristics, and genesis of carbonate, gypsum, and silica accumulations in soils*, pp. 121-149. Soil Science Society of America Special Publication No. 26. Madison, Wisconsin.

Mora, L.N., and D. Flores R., 1997. Pedogénesis de capas endurecidas de suelos volcánicos del altiplano de Nariño (Colombia). In Zebrowski, C., P. Quantin, and G. Trujillo (Eds.), 1997. *Suelos volcánico endurecidos*. III Simposio Internacional (Quito, diciembre de 1996), p. 48-55. ORSTOM.

Poetsch, T., and K. Arikas, 1997. The micromorphological appearance of free silica in some soils of volcanic origin in central Mexico. In Zebrowski, C., P. Quantin, and G. Trujillo (Eds.), 1997. *Suelos volcánico endurecidos*. III Simposio Internacional (Quito, diciembre de 1996), p. 56-64. ORSTOM.

Quantin, P., and C. Zebrowski, 1997. Caractérisation et formation de la *cangahua* en Équateur. In Zebrowski, C., P. Quantin, and G. Trujillo (Eds.), 1997. *Suelos volcánico endurecidos*. III Simposio Internacional (Quito, diciembre de 1996), p. 29-47. ORSTOM.

Soil Survey Staff, 1975. *Soil Taxonomy*. Agriculture Handbook 436. Soil Conservation Service, U.S. Dept. of Agriculture, Washington.

Soil Survey Staff, 1990. *Keys to Soil Taxonomy*. Soil Management Support Services Technical Monograph 19. Virginia Polytechnic Institute and State University.

Zebrowski, C., P. Quantin, and G. Trujillo (Eds.), 1997. *Suelos volcánico endurecidos*. III Simposio Internacional (Quito, diciembre de 1996). ORSTOM, 512 pp.

Plate W. Podzol in marine sands with a densipan (arrows) in the E horizon, New Zealand. The densipan is formed where water percolates. Photograph P. Buurman.

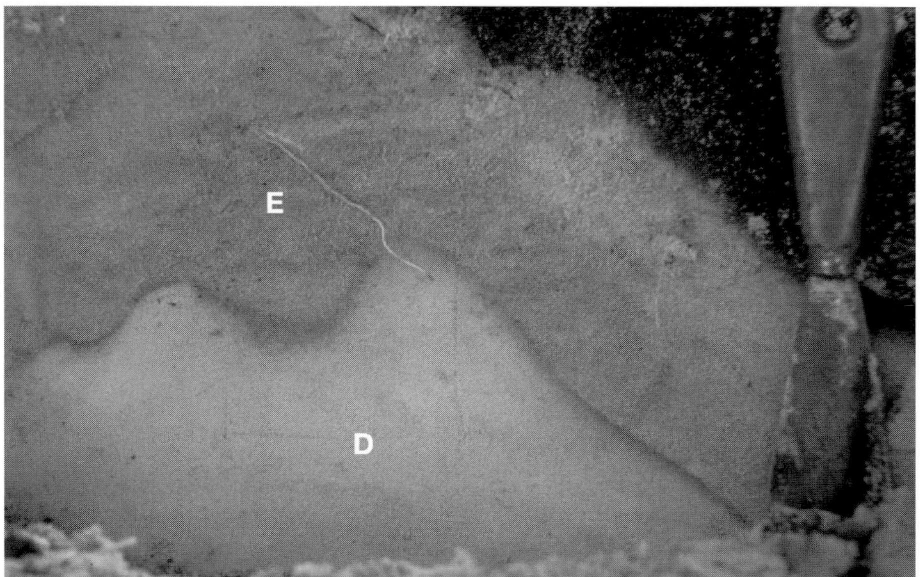

Plate X. Detail of a densipan (D). Podzol in marine sands, New Zealand. Because of its closer packing, the densipan appears whiter than the uncompacted E horizon. Photograph P. Buurman.

Plate Y. Dense packing of sand and silt in a densipan. Scanning electron micrograph of an E-horizon of a podzol on marine sands, New Zealand. Scale units are 10μm. Photograph A.G. Jongmans.

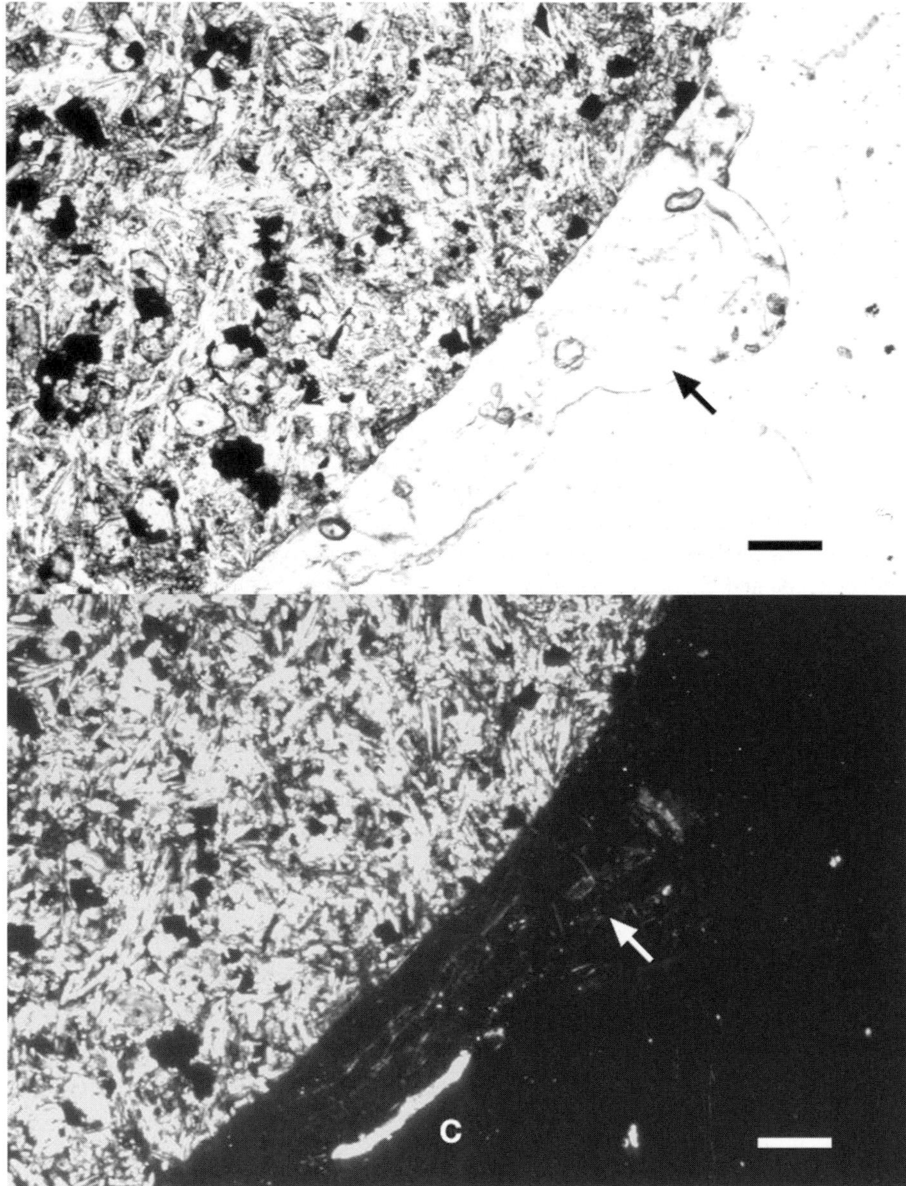

Plate Z. Opal pendant (arrow) at the bottom of a basaltic pebble from France. Top: plain light; bottom: crossed polarisers. Note isotropic nature of opal. The birefringent part is calcite (C). Scale bar is 215 μm. Photographs A.G. Jongmans.

CHAPTER 15

ANALYSING GENETICALLY COMPLEX SITUATIONS

15.1. Introduction

Many soils have been influenced by several different soil forming processes acting at different intensities or at different times. Even slight shifts over time in soil forming factors may cause a shift in the dominant process. In fact, most soils have a complex genesis due to both internal and external causes. Major shifts in climate - an external factor - may completely change soil formation at a specific location. Such shifts are evident in the Holocene of Western Europe (e.g. different climates during the Boreal, Atlantic, and Subboreal), but much stronger over a longer period of time: glacials and interglacials, and changes from tropical Tertiary climates to Quaternary cooler ones. Internal factors are caused by soil formation itself. For example, accumulation of clay in an illuvial horizon may cause stagnation of water, and thereby induce gley processes and ferrolysis, which overprint illuviation.

In northern Europe and America, most parent materials are relatively young, and most soils are of Holocene age. In areas not influenced by glaciations, both younger sediment covers and glacial erosion are absent, and remnants of older soils are found at the surface, now in disequilibrium with the present climate. In geologically and climatically more stable parts of the world (Eastern South America, Australia, central and South Africa, parts of India) the present geomorphic surfaces are extremely old, and soil formation may have acted on - and changed - the same material continuously during more than 50 million years. Such soils may reflect many phases of soil formation.

PERSISTENT FEATURES

Although many former phases of soil formation can be traced by careful study of profiles, some characteristics are very stable in time, while others disappear completely. Petroplinthites are extremely stable and, due to their resistance to erosion may dominate a landscape long after their formation has ceased. Salinization, on the other hand, though extremely important in recent soils, may leave no trace other than, perhaps, changes in soil structure or associated clay mineralogy when the climate becomes wetter. Only persistent features are reliable in the reconstruction of the genetic history of a soil or a landscape. Reconstruction of soil genesis *must* take account of changes in landscape morphology. At the profile scale, especially soil micromorphology is very helpful in distinguishing the sequence in which features have formed.

15.2. Unravelling soil genesis in complex situations

For the problems in section 15.3, you will depend on data gathered by others for specific purposes not always related to soil formation. To interpret such data for pedogenetic goals is often problematic. It is useful to consider several levels of information: 1) geomorphic and climatic information, 2) profile descriptions, and 3) analytical data. We have already mentioned the importance of micromorphological data, but because such data are difficult to interpret without prior experience, they are not used here.

Geomorphic information. The history of a soil profile can only be understood well with good knowledge of its environment. Climate, hydrological position, parent material, etc. give valuable information on the processes that may have acted upon the soil. Especially landscape information may give clues to a polycyclic history and about the importance of solute- or mass transport at a large scale.

Profile descriptions. Profile descriptions carry the interpretation given by the person who studied and sampled the profile. Horizon designations are keys to major soil forming processes. For this book, original horizon codes have sometimes been translated into FAO or USDA terminology to become intelligible. There may be a discrepancy between original horizon codes and processes that can be deduced from chemical and mineralogical evidence.

Soil analyses. Because it is extremely expensive - and rarely useful - to perform many different analyses on a specific soil profile, the investigators usually have made a choice of analyses that reflect the question they had in mind. This means that, in such cases, the choice of analyses already contains information on expected soil forming processes. First check for what purpose a specific analysis was carried out, and then interpret the results. Many analyses are performed expressly for soil classification purposes and provide only limited information about soil formation. The purpose of a number of analyses is outlined in Appendix 3. If you want to interpret a soil in terms of genesis, you have to know exactly by what properties processes are identified and which analyses are used to quantify such properties. The previous chapters give the required information.

Question 15.1. *What are the properties typical for 1) andosols, 2) podzols, and 3) acid sulphate soils? Which analyses would you select to identify each of these groups of soils?*

In the following, we present a number of complex situations with different levels of primary information to deduce the genetic processes that lead to the specific properties.

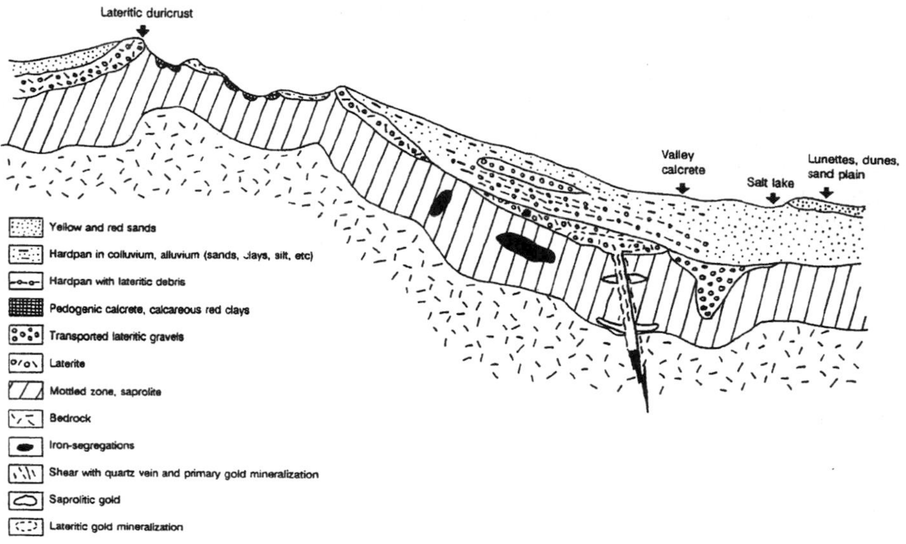

Figure 15.1. Generalised cross section through a landscape near Yilgarn, Western Australia. From Aleva, 1994.

15.3. Problems

Problem 15.1. Complex polygenetic landscape with limited soil information

The landscape of Figure 15.1 has a complex history and part of the soils is polygenetic. The text and the legend in this figure help to understand the landscape history but do not explain the soil genesis. The duricrust is iron-cemented. The horizontal scale is several kilometres. Analyse the situation as follows:

Questions:

a) What could be meant by the term *lateritic duricrust*?
b) Distinguish major soil forming processes that can explain Fig. 15.1. First try to understand each of the legend units. Don't forget to look at the description of the main landscape regimes. What is the layer above the mottled horizon in the right-hand part?
c) Reconstruct former and present ground water regime for various parts of the sequence.
d) Explain the formation of 1) the profile in the left hand part of the sequence; 2) the location of the valley calcrete and salt lake. What does the reconstruction tell about present and past climate?

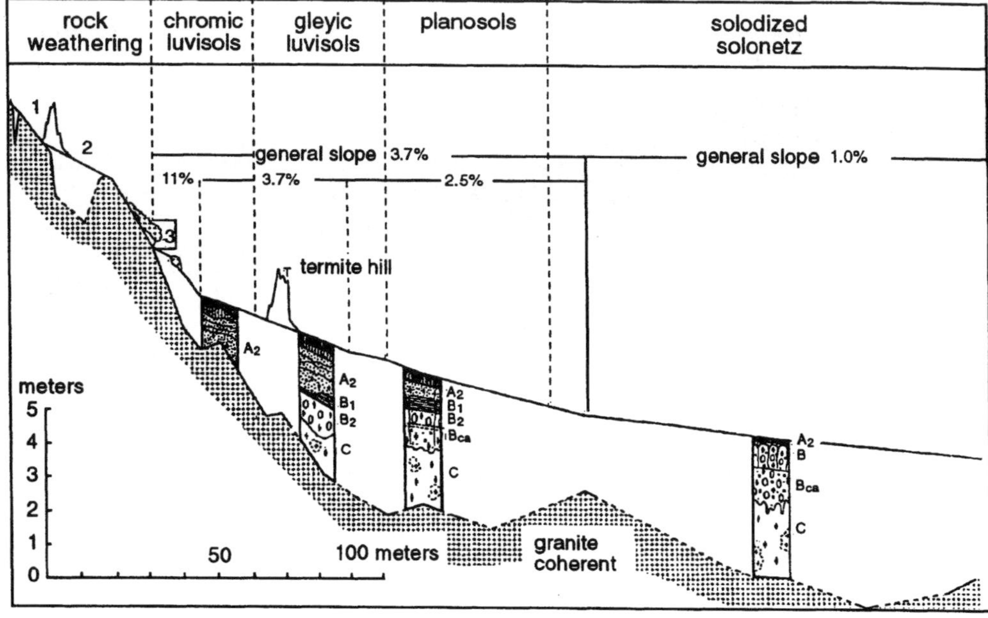

Figure 15.2. Soil catena of the Kossélili area, Chad. After Bocquier, 1973.

Problem 15.2. Catena in a dry tropical climate

Figure 15.2 gives a cross section of a catena near Kossélili, Chad (from Bocquier, 1973). The present climate is tropical, with a mean annual temperature of 19°C and an annual precipitation of 850mm, concentrated in a wet season of four months. The vegetation varies from tree savannah at the left to grassland at the right. The parent material, biotite granite, is found everywhere, within 5 metres depth. Very little soil remains on the granite rock, but on the pediment is a sequence of soils that ranges from Luvisols, at left, to Solonetz and Vertisols (not shown) at right.

Questions:

a) Which soil forming processes dominate in the various parts of the catena? (Remember that A2 = E, ca = calcite accumulation)
b) Suppose that the granite basement is impermeable. What is the main drainage direction in various parts of the sequence?
c) In which part of the catena is a net removal of weathering products?

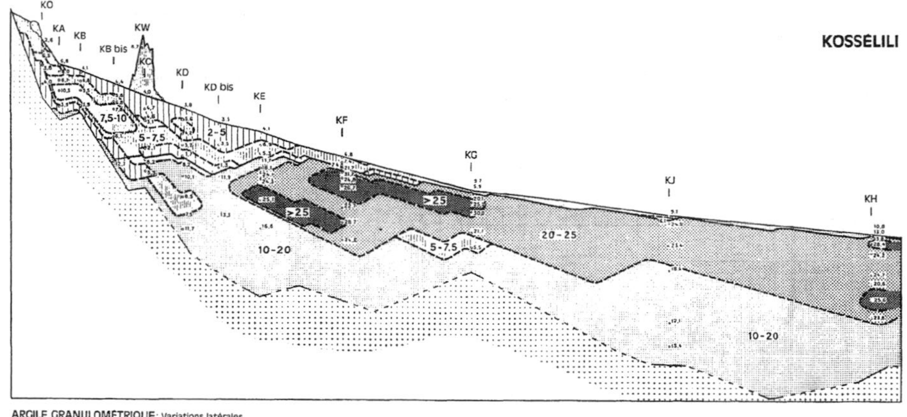

Figure 15.3. Clay contents in the Kossélili catena. From Bocquier, 1973

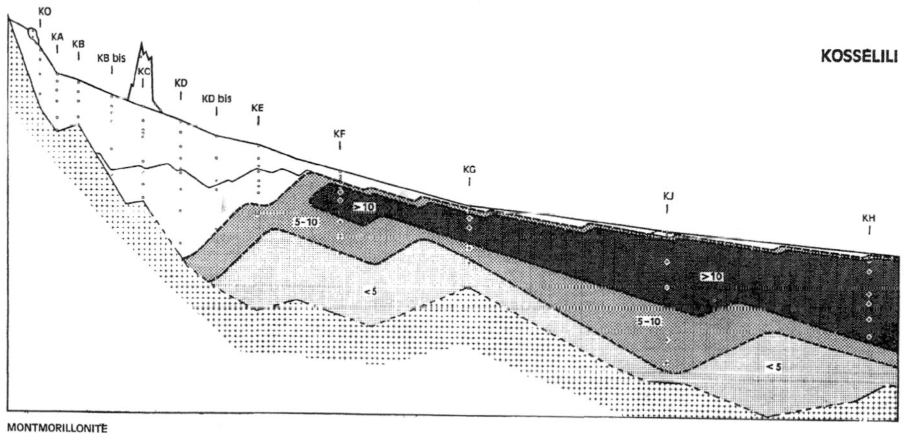

Figure 15.4. Smectite contents in the Kossélili catena. From Bocquier, 1973.

Figures 15.3 and 15.4 give clay contents and smectite contents in the catena.

d) List at least four processes that could have caused the differences in clay contents with depth in various part of the catena.
e) What is the cause of smectite formation in the lower parts of the catena?

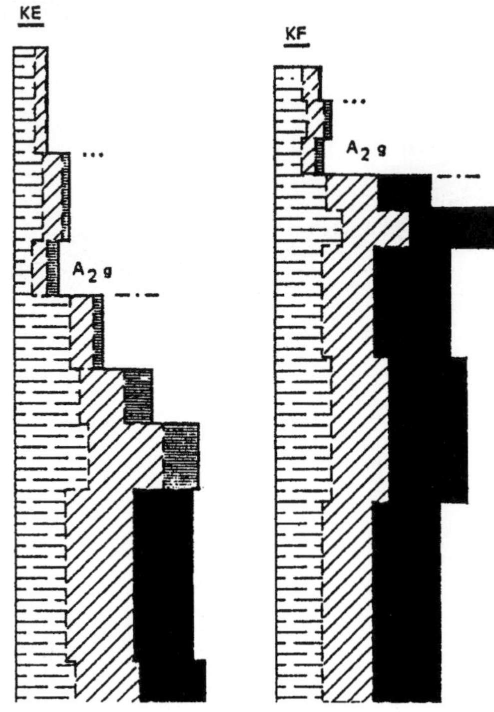

Figure 15.5. Clay mineral assemblage in profile KE and KF of Figure 15.4. From left to right: biotite, kaolinite, biotite-smectite, smectite (black). From Bocquier, 1973.

f) What clay minerals do you expect in the higher part of the catena? Figure 15.5 gives the clay mineral assemblage of profiles KE and KF in the catena.

Figure 15.5 gives the clay mineral assemblage of profile KE and KF in the catena.

g) Explain the mineralogical transition below the A2g (Eg) horizon. Is there any evidence of ferrolysis?

Table 15.1 gives the composition of water from a well close to the granite basement, and Figure 15.6 gives the soil pH in the catena.

h) Calculate the HCO_3^- concentration in the water of the well.

i) What happens to the pH of this water if it evaporates?

j) Explain the pH values in the catena.

Table 15.1. Water composition of a well at Kossélili ($mmol_c.L^{-1}$)

Ca^{2+}	20	Cl^-	1
Mg^{2+}	10	SO_4^{2-}	2
K^+	10	HCO_3^-	
Na^+	7		

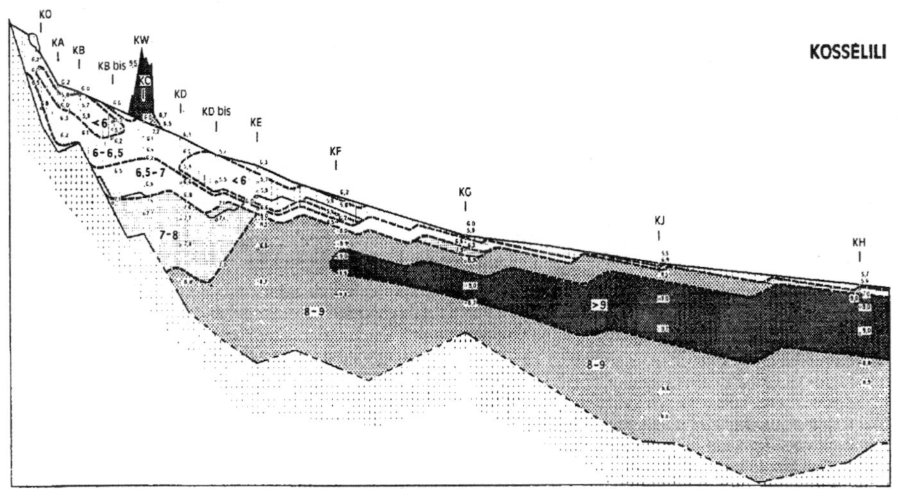

Figure 15.6. pH values in the Kossélili catena. Black: pH > 9. From Bocquier, 1973.

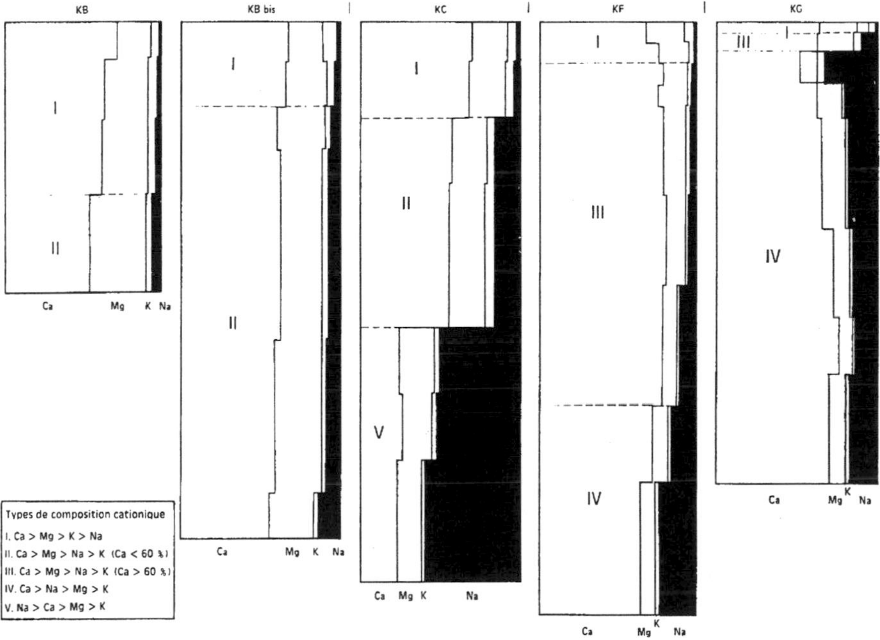

Figure 15.7. Composition of the exchange complex of some Kossélili profiles. From left to right in each column: Ca^{2+}, Mg^{2+}, K^{+}, Na^{+}. From Bocquier, 1973.

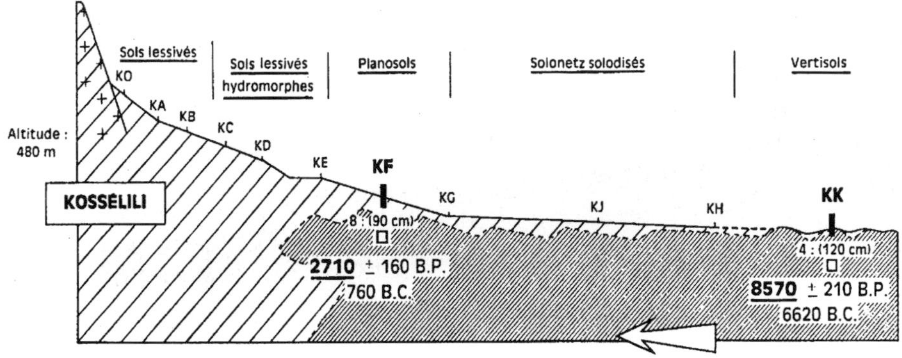

Figure 15.8. [14]C ages of calcite in the Kossélili catena. The fine hatching indicates the presence of $CaCO_3$. From Bocquier, 1973.

Figure 15.7 gives the composition of the exchange complex of five profiles from this catena. The locations can be found in Figure 15.6.

k) Explain the variations in Na^+ adsorption.

Figure 15.8 gives radiometric (^{14}C) ages of carbonates in the Kossélili catena.

l) Assume that $CaCO_3$ precipitated of dissolved $Ca(HCO_3)_2$ formed during weathering of granite upstream. i) What does this imply about the relationship between the ^{14}C age of the carbonate-C and the age of the calcic horizon? ii) Would you expect a similar ^{14}C-age distribution in the secondary carbonates if the upstream parent rock were a Cretaceous limestone?
m) What differences do you expect between ^{14}C ages of carbonate-C and of organic matter-C?
n) What do the apparent ages suggest with respect to the history of the calcic horizon?

Problem 15.3. Landscape-soil relationships and polycyclic soils in the tropics

Soils of the perhumid tropics may show various transitions - both in time and in space - between Ferralsols, Acrisols and Podzols. The result is a superposition of characteristics belonging to several soil-forming processes.

Questions:

a) What are the dominant soil forming processes in Ferralsols, Acrisols, and Podzols?
b) List the conditions necessary for each of these processes. Can these processes occur together in one soil?

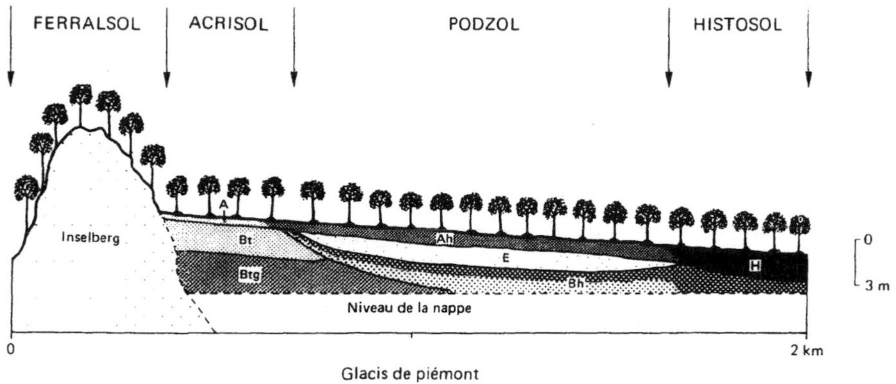

Figure 15.9. Cross-section of an Oxisol-Podzol transition in Kalimantan. From Brabant, 1987.

Figure 15.9 gives a soil transition in a recent Oxisol - Histosol sequence in the lowlands of Kalimantan, Indonesia. On the left is a granite inselberg. The sediments to the right of the inselberg are loamy granite weathering products. The podzol-peat transition is a common sequence in sandy, quartz-dominated coastal deposits with a high ground water table.

c. Explain the occurrence of the four soil groups in terms of differences in soil genesis related to their location.

Transitions between Acrisols and Podzols are also found on other loamy materials in the same area. Sometimes the podzols are found on the flanks of plateau-like units, sometimes they are in the central part of the plateau. The Ferralsol - podzol transition of Figure 15.10 has the same cause as the Acrisol - Podzol transition of Figure 15.9.

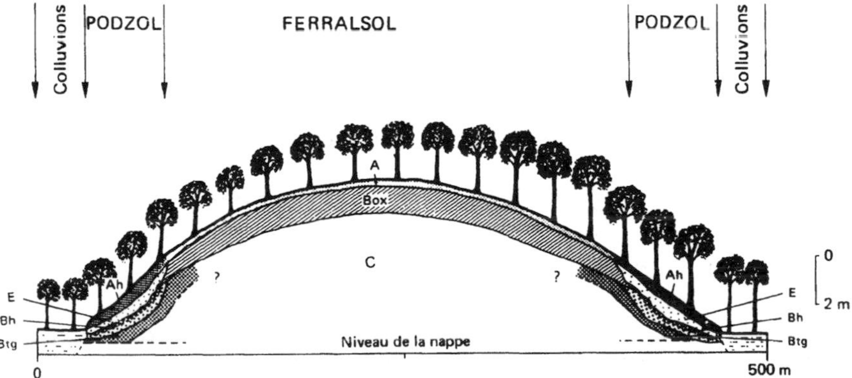

Figure 15.10. Cross section of an Oxisol - Podzol transition in Kalimantan. From Brabant, 1987.

d. Acrisols are not identified In Figure 15.10, but the process that formed such soils can be recognised from the horizon codes. What can you deduce about the sequence of ferralitization, podzolisation, and clay illuviation in this Figure? Is there an alternative interpretation for the texture differentiation (Btg horizon)?

Podzolisation in the centre of plateaus is illustrated in the following: In the coastal area of the Guyanas, where loamy beach ridge-like deposits overly coastal clays, Oxisols are 'eaten up' by a sequence of processes that start at the centre of slightly convex plateaus. The various stages of the removal of the Oxic horizon are depicted in Figure 15.11.

e. Suggest hypotheses that can explain the stages 2-5 of Figure 15.11 in terms of soil genetic processes and their effects. Propose laboratory analyses to test your hypothesis.

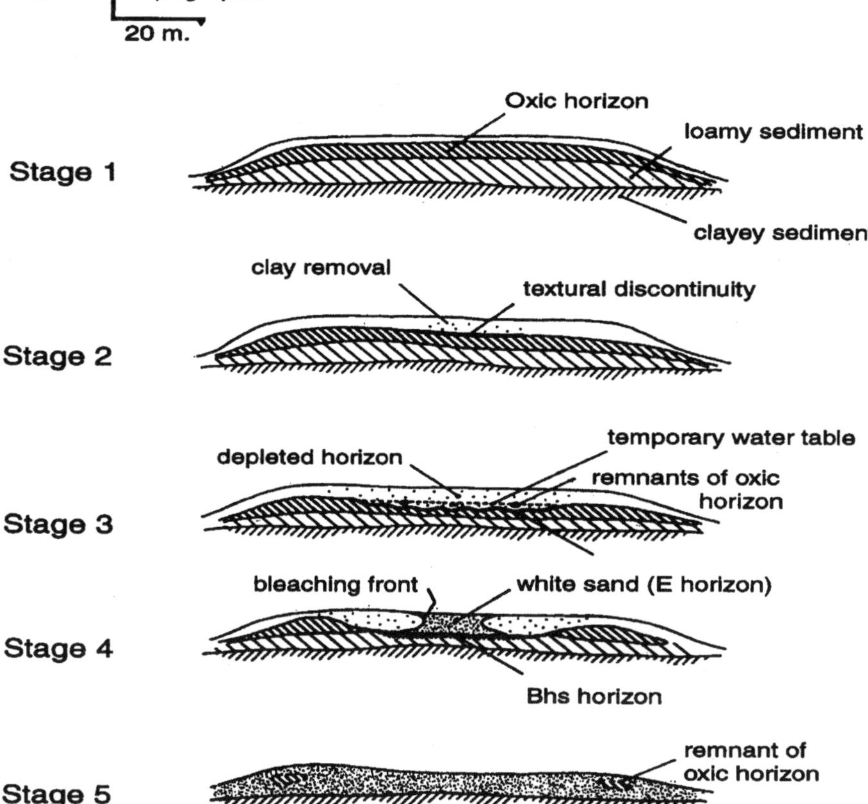

Figure 15.11. Stages of podzolisation of an Oxisol in the coastal area of French Guyana. From Lucas et al., 1987.

Problem 15.4: Analyses only

Of the following three profiles (pages 335 and 336), only the analyses are presented. Look carefully at the trend of each analysis down the profile.

a) Note changes, trends, and discontinuities that can be attributes to sedimentary and to soil forming processes
b) Attribute horizon codes (see Appendix 1).
c) Suggest a sequence of processes (if applicable) and related circumstances of soil formation.

Pedon 5. (From SSS, 1975)
The profile is from a flat former lake bottom, without erosion.
In addition to some standard analyses, also electrical conductivity ($mS.dm^{-1}$) and ionic composition (in $cmol_c.L^{-1}$) of the saturated soil extract are given.
- Cl^- in the soil extract is negligible.
- The remaining anions are HCO_3^- and CO_3^{2-}.
- The extractable SO_4^{2-} content for the third horizon is probably wrong.

Hints for pedon 5: What indications do you use to distinguish sedimentary and pedogenic causes of differences in sand, silt, and clay content? How does organic matter content change with depth in an unlayered soil? Is clay illuviation possible under the present circumstances? What is the relation between high pH and exchangeable Na^+?

Hor.	Depth cm	sand %	silt %	clay %	Org C %	$CaCO_3$ %	pH H_2O	Exchangeable (meq/100g)				EC	Extractable (meq/l)		
								Ca	Mg	Na	K		Ca +Mg	Na	SO_4
	0-15	55	30	15	2.7		7.2	13.7	4.8	0.6	1.0	0.8	2.9	9.6	7.4
	-18	56	34	11	2.1		7.8	9.6	4.6	1.8	0.5	1.3	3.1	24.4	25.9
	-24	46	34	20	1.3		8.7	10.0	8.1	4.9	0.7	2.2	1.6	18.2	71.6
	-30	46	28	26	1.1		9.0	11.0	10.8	10.0	0.7	3.6	2.9	63.0	59.9
	-43	49	26	25	0.6	2	9.2	11.6	12.5	8.1	0.7	4.8	3.2	108	106
	-61	26	37	36	0.1	30	9.5		12.9	6.0	0.5	4.3	3.9	91.5	87.8
	-91	16	58	26	0.3	27	9.3		10.7	4.8	0.6	4.0	3.4	88.0	81.7
	-125	6	70	24	0.3	20	9.2		12.9	6.0	0.5	3.0	2.7	72.8	66.3
	-150	6	70	24	0.3	15	9.1		12.3	5.4	0.5	2.4	2.8	59.7	

Pedon Kruzof. From Rourke et al., 1984.

The upper 20cm of the mineral soil have dark colours. The layer of 30-33 cm is cemented. Base saturation is very low throughout. The soil is about 8000 year old. The horizons from 5 cm downwards are thixotropic. The lower horizon is light coloured and consists mainly of partly weathered rock fragments.

Hints for pedon Kruzof: Do textural analyses and C-profile suggest layering? Which expected properties are suggested by the choice of pH and extraction methods? Is this corroborated by water contents and bulk density?

	Depth cm	Sand %	Silt %	Clay %	Org C %	pH H_2O	pH NaF	Oxalate Si %	Fe %	Al %	Pyrophos. Fe %	Al %	BD pF 2.5*	Water content% PF 2.5	pF 4.2
	25-0				50.0	3.6	5.9						0.17	231	
	0-5	41	52	7	5.1	4.1	6.8		0.1	0.3	0.1	0.1	0.86	52	10
	-8				6.9	4.0	9.1								26
	-20	51	39	10	18.1	4.2	9.8	0.8	0.7	2.6	0.6	2.2	0.37	210	118
	-30				17.3	4.5	10.9	3.7	0.3	8.7	0.2	2.9	0.30	258	160
	-33				11.7	5.0	10.8	5.0	11.2	8.3	3.6	2.0	0.37	216	153
	-56				4.3	5.5	11.0	5.4					0.24	338	190
	-91				3.1	5.6	10.8								162
	-185	30	44	26	2.8	5.6	10.8	5.3	2.3	6.7	0.1	0.5	0.27	272	159
	-230				0.8	6.5	10.2	2.9	0.9	2.7	0.1	0.2			20

*. BD pF 2.5 = bulk density calculated on the volume the soil has at pF 2.5.

Pedon Echaw *(next page)* (From Rourke et al., 1984)
This soil has developed in a beach ridge. The layer with increased clay contents has clay coatings around sand grains. There are no micromorphological data. Precipitation is 1370 mm/year and mean annual temperature is 17 °C.

Hints for pedon Echaw: What process causes the formation of clay coatings? Does textural information also point in this direction? What kind of soil formation do you normally find on soils of this texture? Do the analyses also indicate this (expected) soil formation? What is the sequence of the soil forming processes? Does this explain the absence of covers on the clay coatings in the 55-73 cm zone? Are the pH values relatively low or high for the most recent soil formation? Is the most recent soil formation strongly expressed?

Hor.	Depth cm	San d %	silt %	clay %	Org. C %	pH H_2O	Pyrophosphate extract		
							C %	Fe %	Al %
	0-23	94	4	2	0.73	4.4	2.3	0.1	0.1
	23-37	91	6	3	0.23	4.6	2.9	0.2	0.3
	37-55	90	5	5	0.17	4.6	0.1	0.2	0.3
	55-73	83	4	13	0.24	4.6	0.1	0.4	0.4
	73-91	88	3	9	0.18	4.7	0.1	0.4	0.5
	91-105	96	2	2	0.36	5.2	0.3	0.3	0.3
	105-126	97	2	1	0.34	5.4	1.7	0.2	0.2
	126-146	98	2	0	0.68	5.4	0.7	0.1	0.3
	146-169	98	1	1	0.68	5.3	0.4	0.0	0.3
	169-190	99	1		0.39	5.3	0.3	0.0	0.2

Problem 15.5. Profile drawings only

Figure 15.12 gives three profile drawings of loess soils from New Zealand. The drawings indicate horizon development and structure development; the vertical channels

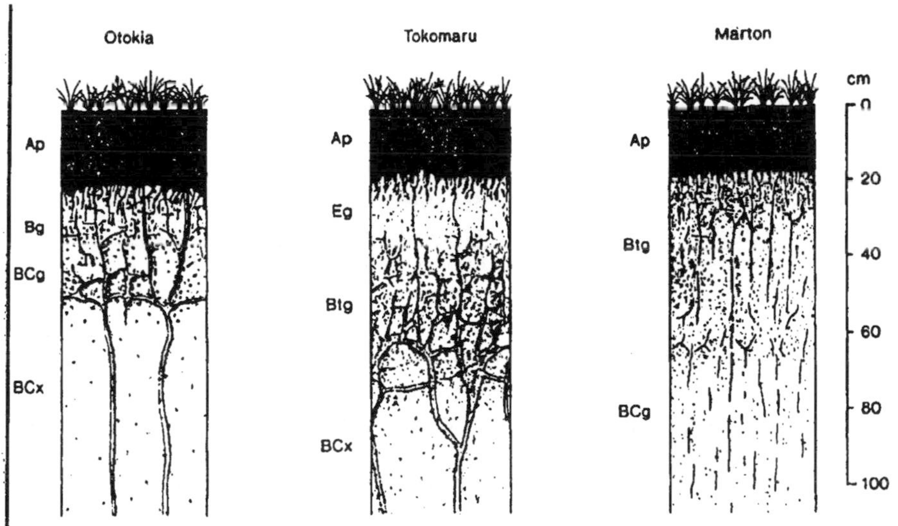

Figure 15.12. Three loess profiles from New Zealand. From Clayden et al., 1992.

in the subsoils of Otokia and Tokomaru are bleached; on a horizontal cross section, they show a reticulate (network) pattern. The Tokumaru profile has Al-interlayered clays in the upper three horizons and vermiculite in the BCx and 2Cg2 (not shown) horizons. The Marton profile has Al-interlayered clays throughout and lacks vermiculite completely in the Btg horizons. Assume that the horizon codes are correct.

Questions:

a) Formulate a hypothesis about the dominant soil forming processes in the three profiles.
b) What causes the differences between the three profiles?
c) What research must be done to test these hypotheses?

15.4. Answers

Question 15.1

Typical for andosols are: presence of amorphous silicates and Al-organic complexes. Indicative of these compounds are: high water retention; high oxalate-extractable Al, Si; high phosphate retention; high pH-NaF (suggested analyses). Podzols have relatively high amounts of metal-organic matter complexes in the illuvial horizon. This can be assessed by pyrophosphate extraction. Most podzol-B horizons do not have high oxalate-extractable silica contents, and water retention is much lower than in andosols. Typical for acid sulphate soils are; pyrite contents (when not oxidised), sulphate and exchangeable Al contents (when oxidised).

Problem 15.1

a) A lateritic duricrust is a petroplinthite layer.

b) The cross section indicates the presence of the remains of a ferralitic soil (mottled zone, laterite gravel, lateritic duricrust). Calcite accumulation is found in valleys on top of the mottled zone, and lower down the slope. Salt accumulation is found in a depression. The right-hand side of the figure consists of transported remnants of a petroferric horizon overlying the mottled zone. In the mottled zone, there is local accumulation of iron.

c) There is a fossil and a present water regime. During the formation of the Ferralsol, drainage must have been good, but the petroferric horizon points to a fossil groundwater level. Erosion has removed most of this horizon towards the valley at right. At present, there appears to be water stagnation on the mottled zone, and at various levels in the transported material (hardpan formation due to iron segregation). The salt lake must have a high groundwater level. The valley calcrete has slightly better drainage and possibly lateral flow.

d) The Ferralsol must have formed under a humid, warm climate. It is very deep, so its formation will probably have taken a long time. Erosion removed most of the soil, down to the petroferric horizon, or even lower. At a later stage, the climate must have changed, because weathering products are now accumulated instead of leached from the

profile. Least soluble salts, such as calcite, are accumulated in sites with slightly impeded drainage, while soluble salts accumulate where groundwater evaporates. Iron accumulations in the mottled zone point to a (fossil?) preferent ground water flow and aeration.
The whole sequence points to a strong climate change: tropical humid to arid, but the timeframe is unknown.

Problem 15.2

a) The soil names indicate the following processes:
- Clay illuviation in the higher part of the catena, together with weathering of primary minerals. The Gleyic Luvisols indicate the effect of groundwater at shallow depth. Also the presence of planosols indicates stagnating water, but in this case it must be stagnating surface water. The presence of a solodised solonetz indicates sodium accumulation. The planosols and the Solonetz have a calcic horizon. In the Planosol this is not accompanied by sodium, so this profile still has relatively good drainage and net removal of salt. Vertisols at the extreme right hand side of the catena suggest smectite clay formation, which may be related to removal of weathering products from the higher part of the catena.
b) If the subsoil is impermeable, the drainage is mainly lateral.
c) Net removal of weathering products occur on the weathered rock and in the Luvisols. If the soils have a calcic horizon, weathering products are not removed completely.
d) Surface removal of clay may have occurred in the higher parts of the landscape (possibly with concomitant sedimentation in the lower part). Clay formation occurred where evaporation of the soil solution led to super-saturation with respect to clay minerals (in the lower part of the landscape); illuviation is obviously found in the Luvisol; clay destruction by ferrolysis may have happened in the Planosol, where surface water stagnates (see also g), but the cross section suggests that the textural contrast may be due to clay precipitation in the subsoil rather than to ferrolysis in the topsoil (see also g).
e) The cause of montmorillonite formation is the accumulation of weathering products in the lower parts, and seasonal concentration of the soil solution through evaporation.
f) The higher parts of the catena probably have minerals that are due to transformation of biotite (mainly vermiculite) to kaolinite. If the clay fraction contained kaolinite alone, the soil would not be a Luvisol (CEC_{clay} would be too low).
g) The diagrams reflect both the total clay content and the clay mineralogy. The lower clay content in the topsoil is probably due to surface removal. The montmorillonite should be due to new formation from the soil solution (lateral accumulation through ground water flow), and is therefore restricted to the subsoil. Ferrolysis should be accompanied by aluminium interlayering of montmorillonite, which is not evident (see also d).
h) The HCO_3^- concentration is the amount that is necessary to balance the positive and negative charge: 44 $mmol_c.L^{-1}$.
i) The water has a very high residual alkalinity (bicarbonates of Na and K), and would cause both strong alkalinization and sodication.
j) The pH values in the upper part of the profile reflect slightly desaturated soils without

free calcite. In the soils with calcite, but without alkalinization, pH is buffered by the dissolution of calcite. The pH values above 9 reflect the combined effect of sodication and alkalinization.

k) In profiles KB and KBbis, there is hardly influence of sodication, because Na is removed from the soil and there is no evaporation from ground water. In profile KC, there is some sodication in the subsoil, due to Na-rich groundwater percolating through the lower part of the profile. The profile is too shallow to remove all Na. Profile KF is much deeper, and the ground water is also deeper than in KC. Soluble salts are therefore removed to greater depth. The presence of a calcic horizon causes more Ca saturation on the adsorption complex. Profile KG appears to have shallow ground water and therefore there is some increase in Na saturation close to the surface. Nevertheless, lateral and vertical removal is strong enough to remove most Na.

l) i) Because the parent rock does not contain carbonate, all carbonate-C is derived from CO_2. This means that the ^{14}C age of the carbonate may reflect the moment of weathering of Ca from the granite.

ii) If the upstream rock were limestone, half of the HCO_3^- in solution would be derived from this rock. Such carbonate does not contain ^{14}C (after 50.000 years, ^{14}C contents are virtually zero), so that it would have an apparent age that is much too high and not related to soil forming processes. The addition of ^{14}C-depleted carbonate from old rocks is called the 'hard-water effect' ('*hard*' water contains much bicarbonate).

m) ^{14}C ages of calcite should be interpreted differently from those of organic matter, because calcite is not broken down biologically. Therefore calcite-C age may reflect a real age of precipitation (apart from effects of mixing, redissolution and reprecipitation), while organic matter ages in old soils are always a reflection of MRT.

n) the apparent ages suggest that the calcic horizon started to form in the lowest part of the catena and grew laterally, from right to left.

Problem 15.3

a) The dominant soil forming processes are: Ferralsols: strong weathering, removal of weathering products, inclusive of silica, residual accumulation of iron and aluminium compounds. Acrisols: clay eluviation and illuviation, fairly strong weathering and removal of weathering products. Podzols: eluviation and illuviation of organic matter and sesquioxides; acidification, strong depletion of weathering products, but not of silica.

b) Ferralsols: carbonic acid-dominated weathering, warm climate, no strong accumulation of humus. Acrisols: depleted adsorption complex, not acid enough to liberate Al^{3+}. Podzols: strong production of unsaturated organic acids; low pH in topsoil, low buffer capacity of parent material. The circumstances are mutually exclusive, but can occur sequentially in one profile.

c) The four processes that can be recognised in the cross section are: ferralitization (Ferralsol), clay illuviation (Acrisol), podzolisation (Podzol), and accumulation of organic matter under poorly drained conditions (Histosol). The drawing suggests that there are two landscapes that are not linked with respect to soil genesis: the inselberg and the pediment to the right. On the inselberg, ferralitization has been the dominant process. This precludes clay illuviation. The fact that the well-drained part of the younger piedmont deposits shows clay illuviation (Acrisol) suggests that this soil formation took place under a seasonal climate. The 'acric' character of the soil may be due to the pre-weathered sediment derived from the inselberg. In the lowest part of the landscape, the Histosols point to water saturation during most of the year. The authors indicate a Bh-horizon under the Histosol. This may be due either to organic matter eluviation from the Histosol, or to lateral humus accumulation from the adjacent podzol. Podzolisation occurred as a result of bad drainage (normally, granite-derived, clayey parent material would have too much buffer capacity to allow podzolisation, and organic matter breakdown under well-drained circumstances would not cause podzolisation under a tropical climate). Podzolisation may also have caused breakdown of the Acrisol.

d) The indication of a Btg below the Box in Figure 15.10 suggests that clay illuviation preceded ferralitization (remember that the reverse is unlikely!). The Bt horizon would normally be destroyed by the intense biological homogenisation that occurs in ferralitised soils. Podzolisation on the fringes clearly post-dates both ferralitization and clay illuviation. It appears to due to periodically high groundwater tables.

e) Clay removal in stage 2 can be due to either 1) surface removal, or 2) ferrolysis. It is not possible to decide which process is responsible in this case, although the temporary water table in stage 3 suggests that there is some iron pan formation at the contact, which favours ferrolysis. Ferrolysis effectively destroys clay minerals, and may also lead to a reduction in pH and removal (through lateral flow) of iron, which reduces the buffer capacity. Once the soil has become more acid, organic matter is broken down more slowly. This induces podzolisation. Podzolisation results in further breakdown of clay through complexation of aluminium. A B-horizon is formed where iron and aluminium of the oxic horizon precipitate organic matter. Through continued attack by organic acids, the oxic horizon is completely eaten away. A few remnants, isolated from percolating water, may remain. Such islands may remain stable for a considerable time.

Problem 15.4

Pedon 5

The textural analysis points to 1) a horizon with more clay, from 43-61 cm, and 2) and increase in clay in the zone of 18-43 cm. Because sand/silt ratios change dramatically, the boundary at 43 cm is probably a lithological discontinuity. Although a loss of fine particles in the topsoil might be caused by erosion (see Chapter 8), the position of the profile precludes this, so that clay illuviation is a clear possibility. The present range in pH is high for clay illuviation, so that it probably preceded the accumulation of calcium carbonate and salts, although clay movement caused by high sodium saturation cannot be excluded (Chapter 9).

Calcium carbonate contents indicate the presence of a calcic horizon (and sufficient water percolation to completely remove calcite from the upper horizons).

pH suggests some alkalinization. Alkalinization should be accompanied by high exchangeable Na^+, which is indeed found between 24 and 43 cm. The high EC suggests the presence of soluble salts. The analysis of the soil extract supports this: the main salt is Na_2SO_4. Alkalinity is only slight, which is in accordance with pH values.

If the profile has an argillic horizon with high Na saturation, this is equivalent to a natric horizon. In addition, there is accumulation of soluble salt, a calcic horizon and a lithological discontinuity. Horizon coding should be:
Ah - E - Bt1 - Btnz1 - Btnz2 - Btnzk - 2Btnzk - 2Cnzk - 2C.
The distribution of sodium sulphate with respect to calcite suggests that salt accumulation is due to evaporation from ground water. (Indeed, the profile description indicates that the 2C horizon is reduced).

Pedon Kruzof

Although granular analysis is incomplete, the data suggest that the parent material may be layered. If weathering were the cause of textural differences, we would expect an increase of the sand/silt ratio with increasing clay content.
Organic carbon shows an accumulation between 8 and 33 cm. The maximum organic carbon accumulation does coincide with maximum Al_p, but not with the Fe_p maximum. This suggests that not all iron is transported and accumulated as organic matter complex. In the horizon from 8-20 cm, most of the extractable aluminium is organic matter- bound, but below 20 cm, most of the Al must be silicate bound. This is supported by Si_o contents.
The cementation in the layer 30-33 cm, in combination with its high non-organic iron content, suggests that this is a thin iron pan (placic horizon).
All horizons below 8 cm depth have very high water contents. Together with low bulk density, high Si_o, and thixotropy, this suggests the presence of allophane. The lowest horizon does not have this high water content.
The combination of properties suggests 1) andosol formation, with 2) slight podzolisation in the topsoil, which is probably caused by the thick forest floor (25 cm thick!). The combination suggests soil formation in volcanic ash under a cold climate. In addition, there is some iron segregation due to stagnating water (surface-water gley), in the form of a thin iron pan. Because textural and water conductivity data are lacking, we cannot establish the cause of water stagnation. The lowest horizon is only slightly affected by soil formation.
A possible horizon coding would be: O - Ah1 - Ah2 - Bh - Bw1 - Bwsm - Bw3 - Bw4 - 2Bw5 - 2C. The number of (ash) layers, however, cannot be established with the available data. The lower organic matter contents in layers 0-5 and 5-8 cm suggest that this is a young ash layer.

Pedon Echaw

The soil is predominantly sandy. There is formation of an Ah horizon (although the pyrophosphate C contents are at variance with the total C contents). The soil has some clay illuviation in the 55-73 cm layer. In addition, there appears to be illuviation of organic matter together with aluminium and iron in the layers below 105 cm. The bottom two layers lack iron, which suggests that iron has been removed by reduction and lateral flow. Considering the landscape position, this is not illogical.
Normally, we would expect superposition of podzolisation on clay illuviation, but this would cause at least some breakdown of illuviated clay and (remnants of) organic matter coatings. As micromorphological evidence is lacking, we cannot judge whether clay

coatings are weathered or whether remnants of organic matter coatings occur in the55-75 cm horizon. The simplest reconstruction, however, which is that of podzolisation after clay illuviation, requires both; otherwise we would have to assume that the organic matter were accumulated through lateral transport. A logical horizon coding would be: Ah1 - Ah2 - E1 - Bt - E2 - EB - Bh1 - Bh2 - BC1 - BC2.
This is an uncommon situation. Podzolisation of a clay-depleted layer is frequently found, especially in acid planosols and podzoluvisols. Such soils have a Bh horizon above the Bt horizon. The sandy character of the parent material (low buffer capacity) and the high precipitation may have caused podzolisation to reach deeper than the clay illuviation.

Problem 15.5
Pedon Otokia has a fragipan and stagnation of water above it. The bleached channels in the fragipan are preferent water ducts with effects of surface water gleying. There is no bleached E horizon, so ferrolysis is still negligible.
Pedon Tokomaru has a clay illuviation horizon and a strongly bleached horizon above it. Together with the bleached channels in the underlying fragipan, this strongly suggests the effect of ferrolysis. The clay mineral distribution corroborates this. Clay illuviation in the, originally well-drained, profile caused periodic water stagnation on the Bt horizon, which in its turn led to ferrolysis. The fragipan is probably older (late Pleistocene), but this cannot be inferred from the drawings. The bulk of the BCx horizon shows no effects of ferrolysis in its clay mineralogy, but such effects would be found along the bleached channels.
Pedon Marton does not have a fragipan or any other horizon of strong water stagnation. Its horizon sequence suggests, however, that water passes through the soil very slowly and causes local anaeroby during the wet season: there is some gleying throughout. The effect of water stagnation is found back in the Al-interlayered clays, which are also found throughout the profile.

15.5. References

Aleva, G.J.J., 1984. *Laterites - Concepts, geology, morphology and chemistry*. ISRIC Wageningen, 153 pp.

Bocquier, G., 1973. *Génèse et évolution de deux toposéquences de sols tropicaux du Tchad*. Memoires ORSTOM No. 62. ORSTOM, Paris, 350 pp.

Brabant, P., La répartition des podzols a Kalimantan. In: D. Righi and A. Chauvel (eds.): *Podzols et Podzolisation*, pp.13-24. Institut National de la Recherche Agronomique, Paris.

Brewer, R., 1976. *Fabric and Mineral Analysis of Soils*. Krieger Publ. Co., Huntington, New York.

Bullock, P., N. Fedoroff, A. Jongerius, G. Stoops, T. Tursina and U. Babel, 1985. *Handbook for Soil Thin Section Description*. Wayne Research Publications, Wolverhampton, U.K., 152 pp.

Clayden, B., A.E. Hewitt, R. Lee, and J.P.C. Watt, 1992. The properties of pseudogleys in New Zealand loess. In: J.M. Kimble (ed.): Eighth International Soil Correlation

Meeting (VIII ISCOM): *Characterization, classification, and utilization of wet soils.* pp. 60-65. USDA Soil Conservation Service.

Lucas, Y., R. Boulet, A. Chauvel and L. Veillon, 1987. Systèmes soil ferralitiques - podzols en région Amazonienne. In: D. Righi and A. Chauvel (eds.): *Podzols et Podzolisation*, pp.53-65. Institut National de la Recherche Agronomique, Paris.

Rourke, R.V., F.T. Miller, C.S. Holzhey, C.C. Treffin, and R.D. Yeck (eds.), 1984. *A brief review of Spodosol taxonomic placement as influenced by morphology and chemical criteria.* ICOMOD Circular No. 1.

SSS (Soil Survey Staff), 1975. *Soil Taxonomy - a basic system of soil classification for making and interpreting soil surveys.* US Dept. of Agriculture Handbook No. 436. Washington

APPENDIX 1

FAO HORIZON CODES

In the following, only a short description of the horizons is given. For complete definitions we refer to FAO (1990).

Major horizons

- H Organic horizon due to (periodic) water saturation, e.g. in peats.
- O Well-drained organic surface horizon (forest floor)
  In this book, the following subdivisions of the O horizon are used:
  - L Litter layer, undecomposed fresh litter
  - F Fragmentation (fermentation) layer: partially decomposed, but still recognisable litter
  - H Humified layer, without identifiable plant remains.
- A Mineral surface horizon characterised by accumulation of organic matter.
- E Eluvial horizon characterised by loss of silicate clay, iron, and/or aluminium.
- B Horizon characterised by weathering, illuviation, or structure formation. Little rock structure.
- C Layer that is relatively little affected by soil formation. A C-horizon is penetrable to roots.
- R Solid rock underlying the soil.

Transitional horizons

If horizons have characteristics transitional between two major horizons, the codes of the major horizons can be combined: AB, CR. If a horizon is a mixture of characteristics of two major horizons, e.g. by destruction of an earlier horizon, or by tonguing, the sign '&' is used between the major horizon codes: B&R, E&B.

Additional codes

- b buried soil horizon, i.e. horizon covered by a sediment.
- c presence of concretions of nodules (always together with a code that specifies the kind of nodules).
- f permanently frozen soil.
- **g** gleying, characterised by mottling of iron and/or manganese compounds.
- **h** accumulation of humus, both in topsoils (Ah) and in illuvial horizons (Bh).

j mottles of jarosite (in acid sulphate soils).
k accumulation of calcium and magnesium carbonates.
m cementation or induration (usually with a code identifying the cement).
n accumulation of sodium on the adsorption complex.
o residual accumulation of sesquioxides (B-horizons of Ferralsols; *not* plinthite of petroplinthite).
p disturbance by ploughing.
q accumulation of silica (duripan).
r strong reduction: grey colours; usually in combination with C.
s illuvial accumulation of sesquioxides (e.g. in podzols).
t illuvial accumulation of silicate clays.
v presence of plinthite (not hardened).
w weathering: development of colour, structure, removal of carbonates, etc.
x fragipan character.
y accumulation of gypsum.
z accumulation of salts that are more soluble than gypsum.

In case of combinations of suffixes, those that are printed in bold character precede the others: Btn, Btb, Bkm. The suffix *b* is always used last. There is no limit to the number of suffixes that can be combined into a horizon code.

Stratified materials

If the soil has sedimentary stratification, i.e. is developed in a stratified geological unit, the second, third and following geological layers are given number prefixes: Ah-B2-2C1-3C2. The number for the top sediment is always omitted. Different sedimentary layers may coincide with buried horizons: Ah-Bw-2Ah-2C-3Ah… as, e.g., in volcanic soils.

Reference

FAO, 1990. Guidelines for soil profile description. Rome.

APPENDIX 2

FORMULAS AND ATOMIC WEIGHTS

Atomic mass (g/mole)

H = 1	C = 12	N = 14	O = 16	Na= 23
Mg= 24	Al= 27	Si= 28	P = 31	Cl= 35
K = 39	Ca= 40	Fe= 56	S = 32	Ti = 48

Formulas and specific density (kg/m^3) of selected minerals

mineral	formula	specific density
albite	$NaAlSi_3O_8$	2620
amphibole	$(Mg,FeII,Ca)_7(Si,Al)_8O_{22}(OH)_2$	3000-3400
anatase	TiO_2	3900
anorthite	$Ca[Al_2Si_2O_8]$	2760
biotite	$K_2(Mg,FeII)_4(FeIII,Al)_2[Si_6Al_2O_{20}](OH,F)_4$	3000
calcite	$CaCO_3$	2720
garnet	$(Mg,FeII,Mn,Ca)_3Al_2Si_3O_{12}$	3580-4320
gibbsite	$Al(OH)_3$	2400
goethite	FeOOH	4300
gypsum	$CaSO_4.2H_2O$	2300
halite	NaCl	2160
hematite	$Fe(III)_2O_3$	5500
illite	$KAl_4[Si_7AlO_{20}](OH)_4$	2750
kaolinite	$Al_4[Si_4O_{10}](OH)_8$	2650
lepidocrocite	FeOOH	4090
magnetite	$Fe(II)Fe(III)O_4$	4520
microcline	$K[AlSi_3O_8]$	2600
muscovite	$K_2Al_4[Si_6Al_2O_{20}](OH,F)_4$	2800
olivine	$(Mg,Fe)_2[SiO_4]$	3220-4390
pyrite	FeS_2	4950
pyroxenes-ortho	$(Mg,Fe)[SiO_3]$	3210-3960
pyroxenes-clino	$(Ca,Mg,Fe,Al)_2Si_2O_6$	3200-3600
quartz	SiO_2	2650
serpentine	$Mg_3[Si_2O_5](OH)_4$	2550
smectite	$(M^+)_{0.7}(Al,Mg,Fe)_4(Si,Al)_8O_{20}(OH)_4.nH_2O$	2500
vermiculite	$(M^{++})_{0.7}(Mg,Fe,Al)_6[(Al,Si)_8O_{20}](OH)_4.8H_2O$	2300

Chemical formulae of selected organic compounds (brackets[ ] denote aromatic ring).

acetic acid	CH_3CO_2H
aspartic acid	$HO_2CCH_2CH(NH_2)CO_2H$
citric acid	$HOC(CH_2CO_2H)_2CO_2H$
p-coumaric acid	$[C_6H_4OH]CH=CH\text{-}CO_2H$
galacturonic acid	$HOC(COH_2)_4CO_2H$
glucuronic acid	$HCO_{1/2}OH(COH_2)_3CO_{1/2}HCO_2H$
p-hydroxybenzoic acid	$[C_6H_4OH]CO_2H$
lactic acid	$CH_3CH(OH)CO_2H$
malonic acid	$CH_2(CO_2H)_2$
oxalic acid	HO_2CCO_2H
succinic acid	$HO_2C(CH_2)_2CO_2H$
vanillic acid	$HO_2C[C_6H_3OH]OCH_3$

Equilibrium constants for a number of reactions.

Data from Bolt and Bruggenwert (1976), Martell and Smith (1977) and Nordstrom (1982). H_2O is in the liquid phase; all other species not indicated as solid (s) or gaseous (g) are aqueous (aq). Log K values at 25°C and 1 bar pressure.

mineral/species	reaction	log K
CO_2	$CO_2 + H_2O \leftrightarrow H_2CO_3$	-1.46
	$H_2CO_3 + H^+ \leftrightarrow HCO_3^-$	-6.35
	$HCO_3^- + H^+ \leftrightarrow CO_3^{2-}$	-10.33
calcite	$CaCO_3 \leftrightarrow Ca^{2+} + CO_3^{2-}$	-8.35
	$CaCO_3 + H^+ \leftrightarrow Ca^{2+} + HCO_3^-$	-2.0
gypsum	$CaSO_4.2H_2O \leftrightarrow Ca^{2+} + SO_4^{2-} + H_2O$	-4.6
iron - goethite	$FeOOH + H_2O \leftrightarrow Fe^{3+} + 3OH^-$	-44.0
- amorphous	$Fe(OH)_3 \leftrightarrow Fe^{3+} + 3\ OH^-$	-39.1
gibbsite	$Al(OH)_3 \leftrightarrow Al^{3+} + 3\ OH^-$	-34.3
	$Al^{3+} + H_2O \leftrightarrow Al(OH)^{2+} + H^+$	-5.02
quartz	$SiO_2 + 2\ H_2O \leftrightarrow H_4SiO_4$	-4.00
opal	$SiO_2 + 2\ H_2O \leftrightarrow H_4SiO_4$	-2.70
silicic acid	$H_4SiO_4 \leftrightarrow H^+ + H_3SiO_4^-$	-9.77
microcline	$KAlSi_3O_8(s) + 4\ H^+ + 4\ H_2O \leftrightarrow K^+ + Al^{3+} + 3\ H_4SiO_4$	+1.3
Mg-beidellite	$6Mg_{0.167}Al_{2.33}Si_{3.67}O_{10}(OH)_2(s) + 44\ H^+ + 16\ H_2O \leftrightarrow Mg^{2+} + 14\ Al^{3+} + 22\ H_4SiO_4$	+36.60
kaolinite	$Al_2Si_2O_5(OH)_4(s) + 6\ H^+ \leftrightarrow 2\ Al^{3+} + H_2O + 2\ H_4SiO_4$	+7.63
jurbanite	$AlOHSO_4.5H_2O(s) + H^+ \leftrightarrow Al^{3+} + SO_4^{2-} + H_2O$	-3.8
salicylic acid	$H^+ + L^- \leftrightarrow HL$	+13
	$Al^{3+} + L^- \leftrightarrow AlL^{2+}$	+12.9
	$Al^{3+} + 2L^- \leftrightarrow AlL_2^+$	+23.2
water	$O_2 + 4H^+ + 4\ e^- \leftrightarrow 2\ H_2O$	+83.0
hydrogen:	$2H^+ + 2e^- \leftrightarrow H_2$	0

APPENDIX 3

TYPICAL ANALYSES USED IN PEDOGENESIS

Chemical and physical analyses

Commonly used abbreviations in analytical tables:	
Al_{KCl}	Aluminium extracted by 1M KCl, unbuffered.
Al_d, Fe_d, Mn_d	Aluminium, iron and manganese extracted by sodium dithionite (strong reduction).
Al_o, Fe_o, Si_o	Aluminium, iron and silica extracted by acid oxalate (pH 3)
Al_p, Fe_p, C_p	Aluminium, iron, and carbon extracted by sodium pyrophosphate (pH 10)
Bases	Exchangeable bases (Ca, Mg, Na, K), either as sum or specified
Base saturation	Exchangeable bases as percentage of the CEC (usually at pH 7), or of the ECEC.
BD	bulk density, in $kg.dm^{-3}$ or $Mg.m^{-3}$.
CEC	Cation exchange capacity, usually with specified pH, also expressed per kg clay, without correction for organic matter contribution.
EC, EC_e	Electrical conductivity, electrical conductivity of the soil extract, in mS
ECEC	Effective CEC, or sum of bases and exchangeable Al (equivalent to CEC at pH_{KCl} of the soil), also expressed per kg clay.
ESP	Exchangeable sodium percentage, percentage of CEC occupied by Na^+.
pH_{H2O}, pH_{KCl}	pH of a 1:5 soil/water or soil/1M KCl mixture
Total chemical analysis	Expressed as oxide percentages of fraction <2 mm

Soil analyses are usually carried out on the fraction < 2 mm (= fine earth). This is based on the assumption that the coarse fraction is inert in terms of soil properties important for plant growth. Ignoring the coarse fraction may give a misleading picture, both for soil genesis (stone lines, profile element balances) and plant nutrition (Agnelli et al., 2001). At least the percentage of coarser material should be reported. Analytical results are expressed with respect to oven dry (105°C) total fine earth, including organic matter. Exceptions are mentioned in the text.

Chemical analyses of soluble and solid phases are widely used in soil genetic studies. Solid phase analyses are more commonly available, because these are also used in soil classification. Buurman (1996) lists more than 70 analyses, and properties that are calculated from analyses that are used in soil classification and soil genesis. In the following we have made a choice of the parameters that are frequently used in this text.

Allophane
Allophane forms when weathering causes high concentrations of Al^{3+} and H_4SiO_4 in solution. Its general formula is $Al_2O_3.SiO_2.nH_2O$, but the Al/Si atomic ratio may vary between 2 and 1. Allophane is fully extracted by an oxalic acid/sodium oxalate solution of pH 3. There are various ways to estimate allophane content from Si_o, Al_o, and Al_p. The simplest way is to assume 14% Si_o in allophane.

Forms of Al and Fe
We recognise four main forms of aluminium and iron in the solid phase:
- silicate-bound Al and Fe
- 'free' forms
  - crystalline Al and Fe (hydr)oxides (goethite, hematite, gibbsite)
  - 'amorphous' or poorly crystalline Al and Fe (hydr)oxides (e.g. ferrihydrite)
- organically bound Al and Fe

The various forms are distinguished by selective extraction. The following figure indicates the various determinations:

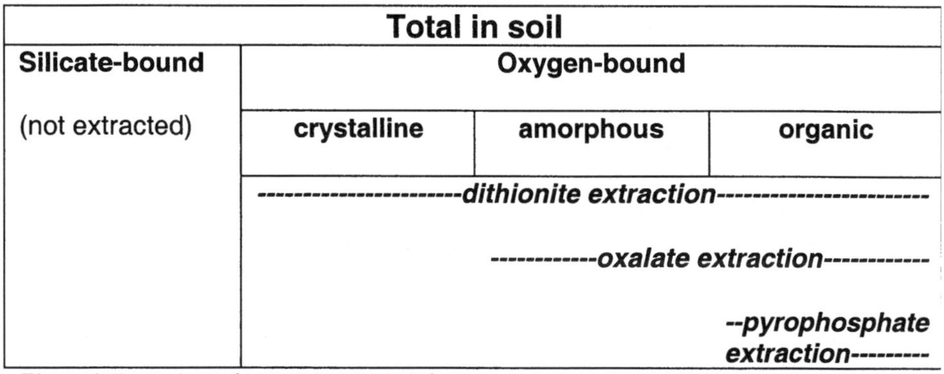

Total in soil			
Silicate-bound	**Oxygen-bound**		
(not extracted)	**crystalline**	**amorphous**	**organic**
	---------------------***dithionite extraction***----------------------		
		-----------***oxalate extraction***-----------	
			--***pyrophosphate extraction***---------

Figure 1. Extractions of iron and aluminium fractions.

Dithionite extracts all non-silicate-bound (generally O-bound) iron. Extraction of certain crystalline Fe-oxides, such as magnetite or ilmenite may be incomplete. Dithionite is also used to extract aluminium, e.g. to determine Al substitution in iron oxides, but it is not exactly known which aluminium forms are extracted. *Free iron* refers to dithionite-extractable iron.
Oxalate extracts all 'amorphous' and organic aluminium and iron, but also part of the crystalline gibbsite ($Al(OH)_3$) and perhaps poorly crystalline halloysite. As mentioned under *allophane*, it also extracts non-crystalline aluminium silicates.

Pyrophosphate at pH=10 only extracts organically bound Al and Fe. The high pH strongly decreases the solubility of (hydr)oxides. Frequently, carbon is also measured in this extract, to quantify all metal-bound C, which is dissolved by pyrophosphate. Because of its high pH the extractant probably extracts more than the metal-bound C alone.

Further details about extractable Al are obtained by extraction/exchange with $CuCl_2$ (Juo and Kamprath, 1979), $LaCl_3$ (Bloom et al., 1979), and KCl (Lin and Coleman, 1961). Copper extracts strongly bound fractions, but usually less than pyrophosphate; Lanthanium extracts lightly bound fractions, and KCl only removes exchangeable aluminium.
The extractions are not sequential, and 'amorphous' and 'crystalline' Fe and Al are obtained by subtraction: Amorphous Fe = Fe_o - Fe_p; Crystalline Fe = Fe_d - Fe_o. Values are usually reported as % metal, and not as % metal oxide.

Carbon fractions
Organic carbon is usually reported as percentage with respect to the oven dry soil, so not with respect to the mineral oven-dry soil. If *organic matter* is reported instead of *organic carbon*, the value is usually based on oxidizable material, using a fixed factor to convert amount of oxidant used to amount of organic matter. Because the organic carbon content of organic matter varies (45-60% C in organic matter), there is no straightforward correlation between the two.
Pyrophosphate carbon (C_p) is always reported as carbon and with respect to the total oven-dry soil.

CEC, ECEC, and Bases
Although the Cation Exchange Capacity was introduced as a soil fertility characteristic, it also gives information that is useful in soil genetic studies. The measured CEC is a combination of CEC of organic matter (pH dependent), clay (pH independent) and (hydr)oxides and amorphous silicates (pH dependent), and the Anion exchange capacity of Al- and Fe- (hydr)oxides. In soils, the AEC usually blocks part of the CEC.
Although CEC is commonly determined at a fixed pH, buffered CEC values (pH 7, pH 8.2) are not a realistic for acidic soils. For such soils, the 'effective' or ECEC, or the CEC at soil pH is used, which more closely reflect the CEC under soil conditions. When CEC (preferably corrected for the contribution of organic matter) is recalculated for the clay fraction, it gives an indication of the clay mineralogy.

Exchangeable bases may give information on soil processes such as sodication, salinization and desalinisation or alkalinisation. Base saturation decreases as a larger fraction of the weathering products is leached. Upon decreasing pH, basic cations are increasingly replaced by Al^{3+} and H^{+}.

Bulk density

Bulk density of common rocks ranges from 2.6 to 2.9 $kg.dm^{-3}$. Bulk density of soils depends on pore volume and organic matter content and usually ranges between 1.0 and 1.6 $kg.dm^{-3}$. Because allophane binds large amounts of water in micropores, soils with amorphous silicates have a very large pore volume. Such soils usually have bulk densities below 0.9 $kg.dm^{-3}$, and sometimes as low as 0.2 $kg.dm^{-3}$. Also peat soils (high organic matter contents and pore volume) have low bulk densities (sometimes less than 0.1 $kg.dm^{-1}$).

Phosphate retention

Soils with highly reactive aluminium and iron hydroxides strongly bind phosphate, and phosphate retention is used to characterise such soils. High phosphate retention is typical for allophanic Andisols, but it also occurs in podzol-Bs horizons and Oxisols. Phosphate retention, as defined by the method, indicates the speed of phosphate fixation, not the capacity.

pH

pH is the outcome of soil formation and in its turn influences the stability of many primary and secondary minerals, as well as many biological properties. Usually pH is determined both in water and in 1M KCl at soil:solution ratios of 1:2.5 or 1:5. The difference is a measure for the amount of exchangeable H^{+} at the adsorption complex that is removed by the electrolyte. If pH_{KCl} is higher than pH_{water}, this indicates net positive charge of the adsorption complex.

To indicate the presence of allophane, the pH of 1 gram of soil in 50 ml 1M NaF is sometimes determined. F^{-} form a strong complex with Al^{3+}, and so replaces Al-bound OH^{-} ions. A large content of amorphous Al in the soil gives a high pH_{NaF} (usually above 10). Soils high in amorphous sesquioxides (podzol-B) also give a high pH_{NaF}.

Water retention

Water retention of soils is related to a) aggregate pores, b) intra-particle pores, and c) inter-particle pores. Water contents at high negative pressure (pF 4.2) is indicative of soil mineralogy. In soils with crystalline clays, the water at pF 4.2 is held at the clay surfaces and is proportional to the clay content (0.25 to 0.3 x clay content). In soils with allophane, water at high tension is held in very small pores, and can exceed 100% even at pF 4.2.

Soil features in thin section

Micromorphology is the microscopic study of soil in its natural arrangement. Undisturbed, oriented samples, taken from the field, are impregnated with a resin, cut, and polished to a 10-50μm thick transparent slice, a thin section, that can be studied with a light-microscope. The following features are commonly studied:

- Shape, arrangement, composition, and form of coarse particles
- Arrangement of coarse and fine particles into soil structure
- Abundance, kind, and shape of voids
- Weathering features of primary mineral particles and newly formed mineral phases
- Rearrangement patterns caused by pressure, illuviation, etc.
- Illuviation patterns (clay, matrix material, organic matter, iron)
- Removal and accumulation patterns, and forms of translocated minerals (iron, manganese, calcite, gypsum, salts, jarosite)
- Form, location, and decay of organic components
- All evidence of biological activity (burrows, excrements, remnants, etc.)
- Recognisable vegetation remnants, such as phytoliths and pollen

Micromorphology and submicroscopy

Micromorphology

In this text, we have approached soil-forming processes mainly from a chemical-physical-mineralogical point of view. We have linked characteristics of the bulk of soil material to horizons that are diagnostic to specific soil forming processes. This is a simplistic approach, because the bulk of a soil only rarely reflects present soil formation. Neither chemical analysis of bulk samples, nor of pore waters or percolating waters can give a complete insight into the history of the soil.

Most soil forming processes leave morphological (visible) traces at the microscopic (mm to μm) or submicroscopic (μm to nm) level. Micromorphology, the study of microscopic features, concentrates on arrangement of soil particles at various levels (soil - aggregate - intra-aggregate), the weathering of particles, and evidence of translocation of material, such as organic matter, clay, matric material, iron and manganese, etc., often reveals effects of different, subsequent phases of soil formation.

In most soils, relicts of former processes are preserved in aggregates, where they are chemically or physically protected against decay, or otherwise excluded from later change. Frequently, the arrangement of features gives a clue as to their sequence of formation. Concentrating on such fossilised features, and only using those visible characteristics that can be interpreted in terms of specific soil forming processes, eliminating the scores of features that can be observed but not properly interpreted, much of a soil's history can be reconstructed. Some of the properties that can be studied by micromorphology are listed in the box above.

Submicroscopy

Sub-microscopy, the study of features smaller than 1μm, is the realm of electron microscopes. With the Scanning Electron Microscope (SEM), which gives a three-dimensional image, arrangement of particles, coatings, crystal shapes, etc. can be studied at magnifications between 10 and 10.000 times. Because of its large focal depth, it gives very sharp pictures and may augment microscopic evidence on specific particles or groups of particles (e.g. Figure 4.12). In addition, it may provide information on the chemical composition of the studied features, through X-ray Dispersive Analysis.

Transmission Electron Microscopy (TEM) makes use of dried suspensions or ultra-thin sections of soil material. Its large magnification (100 - >100.000 times) allows the distinction and identification of weathering products and their arrangement. It is used for the study of products of weathering and new formation, and their arrangement, e.g. of clays, of which the basal distances can be seen and measured. It also has possibilities for elemental analysis, especially to create 'maps' of the abundance of a specific element. Figure 12.4 is an example of a TEM image.

The foregoing is not an introduction into micromorphology or submicroscopy, but here and there, evidence obtained through these techniques is presented. A full description of micromorphological features and their classification is found in Bullock et al. (1985), while principles of interpretation can be found in Brewer (1976).

References

Agnelli, A., F.C. Ugolini, G. Corti, and G. Pietramellara, 2001. Microbial biomass-C and basal respiration of fine earth and highly altered rock fragments of two forest soils. Soil Biology and Biochemistry, 33:613-620.

Bloom, P.R., M.B. McBride, and R.M. Weaver, 1979. Aluminium organic matter in acid soils: salt-extractable aluminium. Soil Science Society of America Proceedings, 43:813-815.

Brewer, R., 1976. *Fabric and Mineral analysis of soils*. Krieger Publ. Co., Huntinton, New York.

Bullock, P., N. Fedoroff, A. Jongerius, G. Stoops, T. Tursina, and U. Babel, 1985. *Handbook for Soil Thin Section Description*. Wayne Research Publications, Wolverhampton, U.K., 152 pp.

Buurman, P., 1996. Use of soil analyses and derived properties. pp. 291-314 in: P. Buurman, B. van Lagen and E.J. Velthorst (Eds.): *Manual for Soil and Water Analysis*. Backhuys Publishers, Leiden, 314 pp.

Juo, A.S., and E.J. Kamprath, 1979. Copper chloride as an extractant for estimating the potentially reactive aluminium pool in acid soils. Soil Science Society of America Proceedings, 43:35-38

Lin, C., and M.T. Coleman, 1960. The measurement of exchangeable aluminum in soils. Soil Science Society of America Proceedings, 24:444-446.

GLOSSARY

acid neutralising capacity
the total amount of H^+ buffer capacity in the soil: weatherable minerals, exchangeable cations, protonation of organic matter, etc.

acric properties
low CEC of the clay fraction (<16 $cmol.kg^{-1}$ clay)

acrisols
soils with an argic B horizon with CEC <16 cmol/kg clay, <50% base saturation and a moist soil regime

aec see anion exchange capacity

albic horizon (USDA), albic E horizon (FAO)
a horizon from which clay and free iron oxides have been removed or where the oxides have become segregated to the extent that the colour of the horizon is determined primarily by the colour of the uncoated sand or silt particles. It has a light colour and occurs above a spodic, argillic, or natric horizon, a fragipan, or an impermeable layer causing a perched water table

alfisols (USDA)
mineral soils that have an ochric epipedon overlying an argillic horizon; with moderate to high base saturation, and which are moist for at least three months during the growing season.

aliphatic acid
a carboxylic acid in which the acidic group is bound to a chain of C atoms, and not to an aromatic (ring) structure

alisols (FAO)
soils with an argillic horizon, CEC >24 cmol/kg clay, base saturation <50% and >35% exchangeable Al on the adsorption complex.

allophane
a non-crystalline 1:1 co-precipitate of silica and alumina, which contains water, exchangeable ions, and frequently iron and organic matter. It is a major solid phase in volcanic soils of humid climates.

amphibole
a single-chain silicate mineral which may contain Mg, Fe, Al, Ca, Na, etc.

amorphous
non crystalline, or without crystal structure (X-ray amorphous). Some non-crystalline substances are partially ordered (as shown by, e.g., broad X-ray diffraction peaks). Such components (allophane, imogolite, opal) are sometimes called *short-range ordered minerals*.

amphoteric
the ability of a metal ion to form both positively and negatively charged hydroxylated forms.

anoxic
absence of molecular oxygen in the environment
anatase
a TiO_2 mineral highly resistant to weathering. It may form in the soil
andesite
a dark coloured volcanic rock consisting of plagioclase and one or more of hornblende, pyroxene and biotite
andisols (USDA), andosols (FAO)
soils with an exchange complex dominated by amorphous alumino-silicates (allophane, imogolite), or aluminium-organic complexes. They have a mollic, umbric, or ochric epipedon overlying a cambic horizon with a low bulk density (<900 kg/m^3). The sand fraction is composed mostly of vitric pyroclastic material
anion exchange capacity
the capacity of sesquioxides and allophane to bind anions. This capacity usually decreases with increasing pH
anthraquic properties
properties due to long-term surface irrigation, usually expressed as a horizon rich in Mn and Fe concretions at the bottom of the irrigated layer
anthrosols (FAO)
soils in which human activity has resulted in profound modifications of the original soil characteristics
appauvrissement (F)
lateral removal of fine particles from the surface soil; see also *elutriation*
arenosols (FAO)
coarse-textured soils in unconsolidated materials, not saline or wet, without clear diagnostic horizons except an ochric epipedon or albic material
argillan
a thin coating of fine clay on grains or walls of pores, visible in thin section under the microscope; usually the result of translocation of clay (clay illuviation)
argillic horizon (USDA) or argic B horizon
a mineral subsurface horizon that is characterised by more clay than the overlying horizon, sometimes due to illuviated layer silicate clay minerals.
aridic soil moisture regime (USDA)
a moisture regime in which the soil is dry most of the time, and never moist for as long as 90 consecutive days
aridisols (USDA)
soils of arid regions that are either salty, or where water is held most of the time at a matrix potential lower than -1500kPa. They usually have an ochric epipedon over an argillic, natric, (petro)calcic, (petro)gypsic, salic or cambic horizon
arkose
a feldspar-rich sedimentary rock, usually derived from granite or gneiss
assimilation factor
the relative amount of C of microbial food that is used to build new biomass.
authigenic
formed or generated in place, a term used for some soil minerals

autotrophic organisms
organisms that can make their biomass directly from inorganic substances

azonal soils
soils that are too young to have obtained characteristics that belong to soils of a specific climate zone. See also *zonal* and *intrazonal* soils

base saturation
the percentage of the cation exchange capacity at pH 7 that is saturated with the cations of Na, K, Ca, and Mg

basic rock
see mafic rock

beidellite
an aluminium-rich clay mineral of the smectite group

biotite
a dark-coloured mineral of the mica group

bioturbation
mixing of soil material by animals, e.g. worms, termites, etc.

birefringence
the property of non-isotropic minerals and, e.g., oriented clay, to split a beam of ordinary light into two unequally refracted, polarised beams, so that the material lights up against a black background if viewed through crossed polarisers

boehmite
an aluminium hydroxide (γ- AlOOH)

bog
swamp, a tract of land without drainage; the environment in which peat forms

brucite
an $Mg(OH)_2$ mineral; trioctahedral phyllosilicates (talc, serpentine, pyrophyllite, chlorite) contain octahedral layers of brucite

bulk density
the mass of dry soil (105°C) per unit bulk volume, expressed as kg/m^3 or g/cm^3. For soils that shrink and swell, the volume at pF 2,5 (field moist) is taken as the reference volume

calcic horizon (USDA, FAO)
a horizon of accumulation of secondary calcium carbonate with at least 5% more calcium carbonate than an underlying C horizon.

calcisols (FAO)
soils of semi-arid areas having a shallow calcic or gypsic horizon, and ochric epipedon and a cambic or argillic horizon. Without signs of wetness, salinity, or abrupt textural change.

cambic horizon (USDA), cambic B horizon (FAO)
a mineral subsurface horizon in the position of a B that does not qualify as argillic, oxic, natric, or spodic. It lacks the dark colours, organic matter, or structure of a histic, mollic, or umbric horizon and it is not cemented or hardened. It is characterised by alteration and/or removal of mineral material as indicated by mottling or grey colours, removal of carbonates, or development of structure.

cambisols (FAO)
soils with a cambic horizon and no other diagnostic horizon except an ochric or an umbric A horizon or a calcic or gypsic horizon

carbon/nitrogen ratio
the mass ratio of total carbon to total nitrogen in organic matter or soil

catena
a laterally adjacent sequence of soils that may differ in age, parent material or drainage (see also toposequence)

cation exchange capacity
the ability of solid material (clay, organic matter) to bind and exchange positively charged ions. This capacity increases with increasing pH for allophane and organic matter and is virtually independent of pH in layer silicate clay minerals

cec
see cation exchange capacity

chalcedony
a cryptocrystalline form of quartz, found in silcretes, flint, etc.

chelation
a usually bidentate binding between an organic component and a di- or trivalent metal ion; hence: chelate, cheluviation. A chelate is usually soluble in water when unsaturated with metal ions; it may precipitate when saturated.

cheluviation
the removal of metal ions by eluviation as chelates

chernozems (FAO)
soil with a thick mollic A horizon and a calcic or gypsic horizon, without salinity, wetness, or signs of bleaching

chlorite
a phyllosilicate mineral with an extra brucite or gibbsite layer between adjacent T-O-T groups. Geogenic chlorites usually have brucite; soil chlorites are predominantly aluminous

chronosequence
a group of soils formed in the same parent material, but differing in age

clay minerals
naturally occurring crystalline phyllosilicates, usually smaller than 0.002 mm, e.g. illite, smectite, chlorite, kaolinite, halloysite. Minerals that are not phyllosilicates, such as gibbsite, quartz, feldspars, anatase, may also be found in the clay fraction.

COLE
Coefficient of Linear Extension. The one-dimensional change in volume of a rod of soil upon wetting.

congruent dissolution
dissolution of a mineral, without leaving behind an insoluble residue

corundum
an aluminium oxide (α - Al_2O_3)

cristobalite
an SiO_2 polymorph common in felsic volcanic rocks. Crystallisation of opal results in a product with structural similarity to cristobalite and tridymite

cryic temperature regime
soils with a mean annual temperature between 0 and 8°C
cryptocrystalline
crystalline, but so fine-grained that the individual components cannot be distinguished at magnifications up to 500x, as with a light microscope
cutan
general name for thin coatings of, e.g. clay, organic matter, silt, iron compounds, etc, that can be found on pore walls or grains
cyclic salt
salt aerosol that is spread by the wind and caused by evaporation of sea spray
desilication
removal of SiO_2, derived from silicate as well as from silica minerals, from a soil by dissolution and drainage of dissolved SiO_2 (H_4SiO_4)
diaspore
an aluminium oxyhydrate (α - AlOOH)
dioctahedral
refers to the structure of a layered mineral (phyllosilicate) in which only two-thirds of the possible octahedrally co-ordinated positions are occupied by cations (e.g.,by Al, Fe). Kaolinite, halloysite, illite, and most smectites are dioctahedral. See also 'trioctahedral'.
dispersion
the breaking-up of compound particles, such as aggregates, into individual particles; e.g., dispersion of clay in water
dissimilatory bacteria
bacteria that use organic matter as an energy source but not to produce biomass
DOC/DOM
dissolved organic carbon/matter
duripan (USDA)
a hardened horizon that is at least partially cemented by silica and/or alumino-silicates which are soluble in strong alkali
effective precipitation
the amount of precipitation that reaches the soil surface and penetrates into the soil
electrical conductivity (EC)
a measure of material or liquid to conduct electricity. In water, conductivity increases with decreasing pH and with increasing salt content
elutriation
the gradual removal of clay from the soil by dispersion and overland flow
eluviation
downward removal of soil material in suspension or solution from part or whole of the soil profile (e.g. E horizon). The opposite process is illuviation
entisols (USDA)
young mineral soils with little or no soil formation. Apart from an ochric, anthropic, or histic epipedon, they have no well-developed diagnostic subsurface horizons.
epipedon
a diagnostic horizon (USDA) at the soil surface

etch pits
dissolution phenomena at the mineral surface, expressed as (sub)microscopically visible, crystallographically determined voids

exchange acidity
the sum of acidity produced by exchanged hydrogen and aluminium in a 1M KCl extract

exchangeable sodium percentage (ESP)
the percentage of the cation exchange capacity that is occupied by sodium

felsic rocks
igneous rocks with more than 66% SiO_2. Common minerals are quartz, potassium feldspar, and muscovite

ferralitic
a term describing highly weathered tropical soils with a SiO_2/Al_2O_3 molar ratio of less than 1.3, a low CEC, and a clay fraction dominated by kaolinite and sesquioxides. The dominant genetic process is a loss of silica and bases. Hence 'ferrallitic weathering', 'ferralsol'

ferralsols (FAO)
soils with an oxic B horizon

ferricrete
a conglomerate, composed of iron oxide-cemented gravel. Eroded material of a petroplinthite

ferrihydrite
a poorly crystalline ferric hydroxide, rich in adsorbed water

ferrolysis
the breakdown of clay minerals in relation to periodic wetting of the soil in surface water gley soils

ferromagnesian
a term referring to minerals with relatively high Fe and Mg contents, and to rocks rich in such minerals.

fine earth
a sieved soil fraction, consisting of particles smaller than 2 mm.

first order reaction
a chemical process the rate of which is linearly related to concentration of one of the reactants

fluvisols (FAO)
soils formed from recent alluvium, without diagnostic horizons other than ochric, umbric, or histic epipedon, and usually with remnants of stratification

fragipan (USDA)
a loamy subsurface horizon, low in organic matter, with high bulk density and slowly permeable to water. It appears cemented when dry but slakes in water.

frigid temperature regime
soils with a mean annual temperature lower than 8°C, and a difference between mean summer and mean winter temperature of more than 5°C

fulvic acid
humus fraction that remains in solution when an NaOH extract is acidified with HCl

geric properties
FAO term, equivalent to *acric properties* of USDA

gibbsite
a mineral of the composition $Al(OH)_3$. The octahedral layer in dioctahedral phyllosilicates has a similar structure and composition, hence 'gibbsite layer'.

gilgai
a microrelief sometimes produced by swelling clays. It consists of a succession of micro-basins and micro-elevations at a scale of decimetres to metres

gley
soil morphological phenomenon resulting from alternating water saturation and drainage (anoxic and oxic circumstances) in the soil. It is characterised by a combination of grey colours and dark (red, brown, black) mottles or concretions of iron and manganese oxides

gleysols (FAO)
very wet soils formed in unconsolidated materials, with no other diagnostic horizons than a histic or ochric A, a cambic B, a calcic, or a gypsic horizon. They are not saline and do not have vertic properties

goethite
α - FeOOH, a common mineral in well-drained soils. It has the same formula as lepidocrocite, but a different crystal structure (dimorphism)

green rust
a ferro-ferri hydroxide that is probably the cause of greenish reduction colours; general composition: $FeII_6FeIII_2(OH)_{18}$

greyzems (FAO)
soils with a black mollic epipedon and bleached coatings on ped surfaces. They are not saline and do not have natric horizons

gypsic horizon
a non- or weakly cemented horizon enriched with secondary sulphates with at least 5% more gypsum than an underlying C horizon. It occurs in arid regions.

halloysite
a flat or tubewise rolled, or concentrically grown, 1:1 phyllosilicate of the kandite family with chemical formula $Al_2O_3.2SiO_2.2H_2O$.

hematite
α - Fe_2O_3. A purple to red iron mineral typical of tropical red soils. In ore deposits hematite is black.

heterotrophic organisms
organisms that build their biomass at least in part from organic components

histic epipedon (USDA) or histic H horizon (FAO)
an organic-rich surface horizon that is saturated with water for some part of the year.

histosols (USDA, FAO)
soils with a histic epipedon, consisting mainly of organic matter and usually saturated with water

humic acid
a mixture of dark-coloured organic substances extracted from soil with dilute alkali and precipitated by acidification

humin
a heterogeneous fraction of organic matter in the soil that does not dissolve in dilute alkali. It consists of partly decomposed plant remains, charcoal, strongly complexed organic substances, non-polar humic material, and strongly clay-bound organics. Humin has a high ash content

hydrolysis
a chemical process in which a mineral dissolves by a reaction with ionised water (H_3O^+, OH^-)

hydromorphism
morphologic features resulting from reduction and oxidation due to intermittent or permanent presence of water. Gley features are the result of hydromorphism.

hydrophilic
having an affinity for water.

hydrophobic
repellent to water

hydrosequence
a group of soils in one landscape that differ by hydrology alone

hydroxy-interlayering
the formation of an extra (Al-)octahedral sheet between the plates of a 2:1 clay mineral, leading to chlorite-like structures. Hydroxy-interlayering leads to the formation of 'soil-chlorites'.

hygroscopic
having the property of readily absorbing water

hyperthermic temperature regime
soils with a mean annual temperature above 22°C, and a difference between summer and winter temperature of more than 5°C at 50 cm depth

igneous
said of a rock or mineral that solidified from a molten material

illite
a 2:1 layer silicate (phyllosilicate) mineral of the clay fraction that forms upon weathering of muscovite

illuviation
the accumulation through precipitation or immobilisation of dissolved or suspended material. Illuvial horizons are usually B horizons. The removal of material is called 'eluviation'

ilmenite
an iron-titanium oxide ($FeTiO_3$) that is very resistant to weathering. It is common in volcanic rocks

imogolite
a thread-shaped para-crystalline (with some order) hydrated 2:1 aluminium silicate occurring in volcanic soils

inceptisols (USDA)
mineral soils of humid areas with umbric, ochric, or anthropic epipedons, but no diagnostic subsurface horizons other than cambic, fragipan or duripan

incongruent dissolution
dissolution of a mineral, leaving an insoluble residue (e.g. a residue of iron hydroxides upon dissolution of pyroxene or olivine)

interstitial water
water held at the contacts between soil particles and in small pores

interstratified clay mineral
a layer silicate in which different types of unit layers alternate in a regular or irregular manner. Interstratified minerals are indicated by the names of the constituents, e.g. smectite-illite interstratification.

intrazonal soils
soils in which the effect of climate is hardly visible because one of the other soil forming factors (e.g. water logging, extremely poor patent material) dominates. See also *zonal* and *azonal* soils

ironpan, thin
see placic horizon

iso- temperature regimes
the prefix iso- in temperature regimes denotes a difference of less than 5°C between mean summer and winter temperatures

isomorphic substitution
partial or complete replacement of one ion in a crystal lattice by another of similar size, without major changes in the structure of the lattice. Substitution of an ion by another of different charge results in charge excess or deficit, which is compensated by adsorption of anions, resp. cations.

isovolumetric weathering
weathering during which the total volume of the parent material does not change (e.g. loss of mineral matter is compensated by increase in pore volume). This is usually restricted to the saprolite. Isovolumetric weathering is used in calculation of weathering losses

jarosite
$KFe_3(SO_4)_2(OH)_6$, a yellow mineral that is typical of oxidised pyritic sediments at pH<4

kandic horizon (USDA)
a strongly weathered B horizon characterised by a clear textural contrast at its top, and low CEC (<16 cmol/kg clay) and exchangeable bases (<12 cmol/kg clay) throughout

kandites
the group of 1:1 clay minerals to which kaolinite and halloysite belong

kaolinite
a non-swelling 1:1 phyllosilicate with a CEC of 5-10 cmol(+)/kg. It is a member of the kandite group of clay minerals

kastanozems (FAO)
soils with a thick dark brown or dark grey mollic horizon, overlying a cambic, calcic, or argillic horizon

krotovina
filled burrow of a rodent, usually in steppe soils and andosols

laterite

old term for iron-rich mottled gley material in tropical soils that hardens upon exposure and is used to make bricks. The current term is 'plinthite'. The term laterite has been widely used for strongly weathered tropical soils. Lateritic weathering is synonymous to ferrallitic weathering.

lateritic soils

old term for soils with plinthite, later erroneously used for Ferralsols or Oxisols.

latosols

an old term for Oxisols or Ferralsols

layer silicate

see phyllosilicate

lepidocrocite

γ-FeOOH, a common iron mineral in hydromorphic soils. Dimorphous with goethite.

leptosols (FAO)

shallow soils on consolidated materials, lacking diagnostic subsurface horizons

ligand

an organic group/molecule that can bind metal ions

lignin

and aromatic polymer; organic substance of woody cell walls. Lignin is a major precursor of soil organic matter.

lixisols (FAO)

soils with an argillic horizon, CEC <16 cmol/kg clay and base saturation >50%

luvisols (FAO)

soils with an argillic horizon, CEC >16 cmol/kg clay and base saturation >50%

mackinawite

FeS, a black sulphide mineral, formed under reducing conditions; precursor of pyrite

maghemite

γ- Fe_2O_3, a magnetic iron mineral of strongly weathered tropical soils on mafic rocks and of burnt soils

magnetite

$Fe(II)Fe(III)_2O_4$, a common magnetic mineral in intermediary to mafic igneous and volcanic rocks. It is resistant to weathering.

mafic rock

an igneous rock with more than 45%, but less than 66% SiO_2, usually dark-coloured

mean residence time (MRT)

the time that (a fraction of) organic matter remains in the soil. Also: the mean age of (a fraction of) humus in soils that are in equilibrium with their environment.

mesic temperature regime

soils with a mean annual temperature above 8°C and a difference between summer and winter temperature of more than 5°C at 50 cm depth

mica group of minerals

phyllosilicate minerals with 2:1 structure. The rock-forming minerals biotite and muscovite, and the clay minerals illite and vermiculite belong to this group.

mineral soil

(part of) a soil that predominantly consists of inorganic material

mineralization
transformation of an element from an organic to an inorganic state, e.g. oxidation of organic matter.

moisture ratio
volume of water divided by volume of solids in a given soil sample

mollic epipedon (USDA) or mollic A horizon (FAO)
a relatively thick dark coloured A horizon with at least 1% organic carbon and a base saturation of >50%; low in phosphate.

mollisols (USDA)
dark-coloured, base-rich mineral soils. All have a mollic epipedon overlying an argillic, natric or cambic horizon.

montmorillonite
a Mg-containing dioctahedral phyllosilicate clay mineral of the smectite group. It has a relatively high CEC.

natric horizon (USDA), natric B horizon (FAO)
a special kind of arg(ill)ic horizon which has prismatic or columnar structure and more than 15% saturation with exchangeable sodium, or with more exchangeable sodium + magnesium than calcium + exchangeable acidity. Natric horizons usually also have illuvial humus

necromass
the amount of dead organisms in the soil

nitosols (FAO)
soils with a deep argillic horizon that contains at least 35% clay

nontronite
a Fe(III)-rich 2:1 clay mineral of the smectite group.

ochric epipedon (USDA) or ochric A horizon (FAO)
a surface horizon that is too light in colour, too low in organic carbon, too thin, or too hard to qualify as mollic or umbric epipedon

opal
an amorphous hydrated silica mineral. It occurs in plants (phytoliths) and can also be formed by precipitation from solution. Opal-A is fully amorphous; opal-CT has some structural elements of cristobalite and tridymite (both SiO_2)

oxic *(not the adjective in soil classification)*
having molecular oxygen as part of the environment

oxic horizon (USDA), oxic B horizon (FAO)
a mineral subsurface horizon at least 30 cm thick with more than 15% clay, little or no primary alumino-silicates or 2:1 clay minerals, and low water dispersible clay. Typical properties are the presence of 1:1 clays, hydrated oxides of iron and aluminium, a low CEC (<16 cmol/kg at pH 7) and <10 cmol exchangeable cations per kg clay.

oxisols (USDA)
red, yellow, or grey mineral soils usually of high rainfall tropical and subtropical regions. They have an oxic or a kandic horizon within 2m of the surface, but do not have an argillic or spodic horizon overlying the oxic. Plinthite may occur within 30 cm of the soil surface.

paleosol
a soil that has formed under circumstance different from the present ones. Paleosols can occur at the surface and be subjected to present day soil formation (polygenetic soils), or be buried.

palygorskite
a chain-lattice silicate with fibrous morphology which may form in soils of (semi)arid regions. Also called attapulgite.

palynology
see pollen analysis

pedal soil
soil consisting of distinct structural elements such as blocky or prismatic aggregates

pediment
eroded surface, usually covered with material of (coalescent) alluvial fans

pedogenic
formed as a result of soil formation

pedotubule
biogenic pore, usually filled

pedoturbation
mixing of soil material by burrowing animals

pergelic temperature regime
soils with a mean annual temperature lower than 0°C

perhumid
said of continuously humid climates

perudic moisture regime
a moisture regime which implies that water moves through the soil in all months that it is not frozen; the moisture tension is usually lower than that at field capacity

petrocalcic horizon (USDA, FAO)
a continuous, indurated calcic horizon

petroferric material
an indurated crust, equivalent to petroplinthite

petrogypsic horizon (USDA, FAO)
a continuous, indurated gypsic horizon

petroplinthite
an plinthite layer that is indurated through exposure to the air; usually with lateral iron accumulation.

pF curve
a curve that depicts the relation between suction at which moisture is held, and the moisture content of the soil (the pF is the negative logarithm of the suction in cm water pressure)

phaeozems (FAO)
soils with a thick, black mollic horizon, but without a calcic or gypsic horizon

pH-dependent charge
that part of the total charge of the soil (both negative and positive) which is affected by changes in pH. Organic matter, allophane and sesquioxides are the major soil constituents with pH-dependent charge.

phenolic acid
an organic acid in which the acidic group is bound to a phenol ring (a benzene ring with one H substituted by OH)

phosphate retention
the capacity of (allophanic) soils to rapidly fixate phosphorus from a solution of specified concentration

phyllosilicate
mineral with a layered structure in which the SiO_4 tetrahedra are linked together in infinite 2-dimensional sheets. The tetrahedral sheet (T) is linked with a octahedral sheet (O) of either $Mg(OH)_2$ (brucite) or gibbsite ($Al(OH)_3$). Phyllosilicates are subdivided into T-O-T (2:1; smectites, micas), T-O-T-O (2:1:1 or 2:2; chlorites), and T-O (1:1; kandites) minerals

phytolith
secreted mineral matter in living plants, e.g. opal or calcium oxalate. Upon decay of the plant, the mineral is set free and may accumulate in superficial soil horizons

pisolitic
said of rounded concretions with layered structure

placic horizon (USDA), thin ironpan (FAO)
a thin (<0.5 cm), black or dark reddish layer cemented by iron, by iron and manganese, or by an iron-organic matter complex. It is a wavy layer more or less parallel to the soil surface. It is restrictive to penetration by water and plant roots. It forms at a contact between aerated (lower) and unaerated (upper) soil.

planar void
void between two adjacent, more or less planar, surfaces of soil aggregates such as blocks or prisms

planosols (FAO)
soils with an albic horizon abruptly overlying an argillic or natric horizon. The albic horizon has signs of wetness.

plinthite (USDA, FAO)
an Al and Fe-oxide-rich, humus-poor and highly weathered material consisting of a mixture of clay, quartz and other diluants. It occurs as dark red mottles, usually in platy, polygonal or reticular patterns. On exposure to repeated wetting and drying, plinthite changes irreversibly to an ironpan or to irregular aggregates. Hardened laterite, whether pisolitic or vesicular, is not plinthite.

plinthosols (FAO)
soils with more than 25% plinthite in a horizon that is more than 15 cm thick and occurs within 40 cm of the soil surface

podzols (FAO)
soils with a spodic B horizon

podzoluvisols (FAO)
soils with an argillic horizon with a broken upper boundary due to tonguing of the E horizon into the B horizon.

pollen analysis
analysis of the pollen content of deposits and soils, to reconstruct vegetation history or for dating.

polygenetic soil
a soil with evidence of essentially different phases of soil formation

pseudogley
see surface water gley

pyrite
FeS_2, cubic, a common mineral in anoxic coastal sediments

reduced
often used to indicate anoxic soil conditions

salic horizon (USDA)
a horizon of accumulation of salts (>2%) that are more soluble than gypsum in cold water

saprolite
the weathered part of a solid rock that has retained its original structure

secondary minerals
minerals that are formed upon weathering, e.g. clay minerals, hydrated oxides

sepiolite
$Mg_4(Si_2O_5)_3(OH)_2.6H_2O$, a mineral similar in structure to palygorskite; it forms in high-pH, Mg-rich and Al-poor pore waters

sesquioxide
an oxide with a metal:oxygen ratio of 2:3, e.g. Fe_2O_3 and Al_2O_3. Also used to describe iron, aluminium, and manganese (hydr)oxides in the soil

siderite
$FeCO_3$, a mineral that forms under reducing conditions, e.g., bogs

silcrete
a superficial cementation by silica (chalcedony or opal), occurring in arid regions

silica
silicon dioxide(SiO_2), either in hydrous, dissolved form (H_4SiO_4) or as mineral or amorphous precipitate

silicate
a mineral with an Si-O framework and other cations, often including Al. Literally: salt of silicic acid (H_4SiO_4) and a base

slickensides
polished or grooved surfaces on structural elements in soils dominated by swelling 2:1 clay minerals, resulting from sliding of soil mass

smectite
a group of swelling 2:1 clay minerals, with substitution mainly in the octahedral layer and high CEC (80-100 cmol(+)/kg). To the smectite group belong, e.g.: montmorillonite (Mg-Al), beidellite (Al), saponite (Mg), nontronite (Fe), hectorite (Mg,Li), and sauconite (Zn)

SOC
soil organic carbon

soil ripening
physical and chemical changes that are related to loss of water and compaction in young, water-rich sediments

solonchaks (FAO)
soils, excluding those from recent alluvial deposits, having a high salt content and lacking diagnostic horizons other than an ochric or histic epipedon, or a cambic, calcic, or gypsic horizon

solonetz (FAO)
soils with a natric horizon with signs of wetness and an abrupt texture transition at the top of the natric horizon

soluble salts
salts present in the soil that are more soluble than gypsum, e.g. NaCl, KCl, Na_2CO_3, Na_2SO_4. These salts cause a high osmotic pressure in water

SOM
soil organic matter

sombric horizon (USDA)
a dark-coloured subsurface horizon with illuvial, humus that is not associated with aluminium as in the spodic horizon or with sodium as in the natric horizon. It has a low cation exchange capacity and base saturation

spodic horizon (USDA), spodic B horizon (FAO)
a mineral subsurface horizon that has one or more of: a subhorizon thicker than 25 mm continuously cemented with a combination of organic matter with iron or aluminium or both, or uncemented but with significant accumulation of amorphous material (aluminium with organic matter, with or without iron) relative to crystalline aluminium and iron oxides in the clay fraction.

spodosols (USDA)
mineral soils with a spodic horizon. Equivalent to 'podzol'. A fragipan or an argillic horizon may occur below the spodic horizon, while some have a placic horizon in or above the spodic horizon or fragipan. Many undisturbed spodosols have an albic horizon

stagnic properties
properties related to the stagnation of water on top of an impervious soil horizon

stoichiometric
refers to the numbers of atoms of different elements in the smallest entity of a mineral or substance, or in a chemical reaction in which those minerals or substances take part

sulfuric horizon (FAO)
a horizon with low pH (<3.5) and jarosite mottles due to the oxidation of pyrite

surface water gley (pseudogley)
reduction and oxidation processes due to a temporary, perched water table. Pseudogley is frequently accompanied by weathering of clay (ferrolysis), as, e.g., in planosols and podzoluvisols

tepetate
a hardpan, common in climates with a pronounced dry season, cemented by recrystallized weathering products of volcanic material

tephra
general term for fragmented, unconsolidated volcanic products of all sizes

thermic temperature regime
soils with a mean annual temperature between 15 and 22°C, and a difference between summer and winter temperature of more than 5°C at 50 cm depth

thixotropy
a reversible gel-sol transformation. Property of a material that is solid (gel) when in rest and becomes viscous when agitated or pressed. This is a typical property of allophanic soils that seem firm and dry when in place and turn soft and wet when disturbed

TOC
total organic carbon

toposequence
a laterally adjacent sequence of soils of similar age and formed in similar parent material under one climate, having different characteristics due to variations in relief and drainage (see also catena). In practice, a toposequence has a limited vertical range (e.g. up to 500 m), because climate tends to change with altitude.

udic moisture regime
the udic moisture regime implies that in most years the soil is not dry in any part for as long as 90 consecutive days

ultisols (USDA)
mineral soils that have an argillic horizon or a kandic horizon with low base saturation because of strong leaching and great age

ultramafic (ultrabasic)
rocks containing less than 45% silica, virtually no quartz or feldspar and composed essentially of ferromagnesian minerals and metallic oxides, e.g. serpentinite, peridotite

ustic moisture regime
the ustic moisture regime is intermediate between the aridic and the udic regime. There is limited moisture, but the moisture is present at a time when the climate is otherwise suitable for plant growth

vermiculite
a group of non-swelling 2:1 phyllosilicates with dominant substitution in the tetrahedral layer. Vermiculites have a high CEC (140-160 cmol(+)/kg). Minerals with vermiculite-like behaviour (strong K fixation) may form upon weathering of illite

vertisols (USDA, FAO)
clay-rich mineral soils that have deep, wide cracks when dry and high bulk density between the cracks. In addition they may have slickensides, gilgai, or wedge-shaped peds. The clay fraction is dominated by smectite clays (high linear extensibility). The soils have high CEC and base status.

vivianite
$Fe_3(PO_4)_2(H_2O)_8$; a common phosphate mineral in eutrophic bogs

void ratio
the ratio of the volume of void space to the volume of solid particles in a given soil sample

water ratio
the ratio of the volume of water to the volume of solid particles in a given soil sample

weatherable minerals
those minerals that are easily hydrolysed under wet atmospheric conditions: micas, feldspars, amphiboles, pyroxenes, phosphates, zeolites, carbonates, olivine, volcanic glass, etc.

xeric moisture regime
the moisture regime of mediterranean climates, with moist, cool winters and dry, hot summers.

xerosols (FAO)
non saline, semi-desert soils with a weakly developed ochric epipedon and one or more of: a cambic, an argillic, a calcic, or a gypsic horizon

yermosols (FAO)
desert soils. They are like xerosols, but have a very weakly developed ochric epipedon

zero order reaction
a chemical process of which the speed is independent of concentration of the reactants, e.g. radioactive decay

zonal soils
soils that are in equilibrium with the prevailing climate and the natural vegetation in a climatic zone. Typical zonal soils are chernozems, kastanozems, ferralsols. See also *intrazonal* and *azonal* soils.

zonality
the concept that each climatic zone is characterised by the presence of a specific soil type. It is one of the classical concepts of Russian pedology.

INDEX